T0234153

From Algorithms to Hardware Architectures

Karim Abbas

From Algorithms
to Hardware Architectures

Using Digital Radios as a Design Example

 Springer

Karim Abbas
Faculty of Engineering
Cairo University
Giza, Egypt

ISBN 978-3-031-08695-3   ISBN 978-3-031-08693-9   (eBook)
https://doi.org/10.1007/978-3-031-08693-9

This Springer imprint is published by the registered company Springer Nature Switzerland AG
The registered company address is: Gewerbestrasse 11, 6330 Cham, Switzerland

To mom, for saying Karim meet books,
To dad, for showing me how to be me,
To Sehar, for giving me the passion to do it,
To Radwa, for the illusion of geek is cool,
To Sophia, for the dance interludes and music choices,
To Julia, for the philosophy of life,
To Lily, for milk stains and hope.

Preface

There are two types of people, those who count complexity in multiplications and those who count complexity in multipliers. The gap in the middle is very hard to bridge.

People who work on algorithms focus on getting them to work. Cryptographers design cyphers that are hard to break. People who work on machine vision want to recognize an object within a frame. If you are studying machine learning, you spend most of your time getting the python code to classify cat pictures as cat pictures and dog pictures as dog pictures. And if you are designing a channel decoder, you are focused on correcting as many bit errors as possible and reaching the limits of channel capacity.

This focus on function is necessary, because without it, we would not have good algorithms to implement. But at some point, a good idea becomes a product, and products must be efficient. They should run fast so that they become part of an enjoyable user experience. They should consume low power so that they run on a small battery-powered mobile device. And they should be cheap to make so that they sell and make money.

So, to turn a good idea into a good product, hardware designers get involved. Hardware implementations are much more efficient than anything running on software, but the road to a dedicated hardware platform for your algorithm is not easy. Getting any microchip to work is challenging. It is an iterative multi-step process that requires experience and dedication. Hardware designers focus on finding and breaking critical paths, detecting and fixing violations, distributing the clock network, drawing a layout that is acceptable to the fabrication facility, making sure the silicon actually does what the simulation says it does, reducing the area of the chip, minimizing power dissipation, and inserting test structures that allow the finished chip to be observed in detail.

There is something missing between systems designers who focus on the algorithm and hardware designers who focus on the chip. There is an area of knowledge in the middle which bridges the gap. There should be a common language where the algorithm designer can think ahead to how complexity translates in hardware and the hardware designer can leverage the algorithm to make efficient chips.

A systems designer will count complexity in terms of number of arithmetic operations. They might say something like this algorithm requires ten thousand multiplications, and they will use multiplications to compare their algorithm to other peoples'.

A hardware designer will not by default think in terms of multiplications. Instead, they think in terms of multipliers. A multiplier is a relatively complicated CMOS circuit that multiplies the contents of two registers to produce an output in a much larger register. The multiplier can produce outputs at a certain rate that will differ depending on the hardware platform on which it is realized.

There is a step in the middle where the multiplications are transformed into the number of required multipliers. At the end of the day, we need the final product to be good, so we should use as few multipliers as necessary. An algorithm designer can help this by reducing the number of multiplications (operations) necessary, or perhaps by realizing that some of the multiplications are special. The hardware designer can help by utilizing multipliers so that they are doing something useful most of the time. But the best deal is when you can leverage knowledge in both areas to improve the final product. If you can look ahead to the hardware implementation, maybe you can see what makes some multiplications "special" in the algorithm. By knowing something about the algorithm, you can pick the hardware architecture that best suits it.

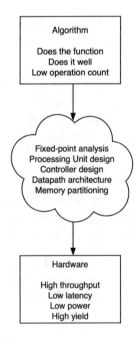

But this is only a small part of a vague but critical process that happens when complex algorithms are moved to hardware. Things in this foggy cloud include:

- *Fixed-point modeling*. Algorithms are prototyped using easy-to-use high-level programming languages on general purpose processors. These setups use floating-point registers with a large word length to allow accurate number representation. Hardware floating-point arithmetic is slow and inefficient and tends to defeat the purpose of moving to hardware. Thus, hardware arithmetic uses much smaller registers with a fixed position for the binary point. This allows arithmetic to run much faster. But the transformation between floating-point and fixed-point arithmetic can devastate the accuracy of the algorithm and is rife with design choices that require human intervention.
- *Determining the processing units (PUs)*. Algorithms describe a sequence of arithmetic and logical operations. Someone needs to break this down into a finite set of operations that are repeated to form the whole algorithm. These operations must be translated into parallel hardware processing units (PUs). Sometimes the PU is trivial, perhaps performing addition or multiplication. Sometimes it does something without a known hardware implementation, so you invest some time to find a trick to implement it efficiently.
- *Determining number of PUs*. Software prototyping is serial. Hardware is good because it is parallel. To move between the two, you budget how many PUs you will need. This is where multiplications are turned into multipliers. You must be aware of the throughput we need from the system, the clock rate of the PU (which means you know the implementation platform), and the complexity of the algorithm. This is where everything starts to come together.
- *Drawing the block diagram*. A block diagram uses PUs as building blocks and sketches the hardware as it directly maps the parallelized signal flow of the algorithm.
- *Determining architecture*. But a direct implementation as dictated by the block diagram is rarely if ever good enough. Most of the time, it is either too fast or too slow. A designer solves this by determining the final architecture. If the hardware is too slow, you can use pipelining or parallelism. If it is too fast (and yes this is just as bad), you can apply hardware reuse.
- *Memory design and partitioning*. The architecture you pick will determine your memory requirements. Memories are used to arrange data, delay it, or store it in intermediate steps. Someone must figure out how to allocate and partition these memories so that they can be accessed fast enough and so that their addresses are provided in every cycle.
- *Introducing flexibility and reconfigurability*. Good hardware, especially in complex systems, must be flexible and reconfigurable. We must be able to adapt or use it for different "modes." This requires fundamental decisions at both the PU and the controller level.
- *Mapping out control*. Complicated circuits contain memories that need to be addressed, signal flows that need to be routed, and flexible PUs that need to be configured. A controller must constantly observe the status of the circuit and

provide the necessary controls on a cycle-by-cycle basis. Controller design requires familiarity with both the hardware and the algorithm and can make or break the entire system.

This hazy set of steps is not the focus of hardware design books, and it is not in the scope of algorithm design books. And for good reason, both kinds of books have other very important things to do. This is why I wrote this book: to make this vague area in the middle more tangible.

But this book is equally about the design of digital radios. The implementation of radios is an important topic and could take multiple volumes to cover. But radios are an excellent example of the problem we described above: how to get something from a high-level description to hardware.

By radio we mean any platform that communicates wirelessly over the radio spectrum. This includes all Wi-Fi, cellular, and Bluetooth platforms, which should be enough to tell us why "radios" are important. Everything is either wirelessly connected or someone is doing research to get it connected. And by everything, I mean everything. Obviously, mobile phones, tablets, computers, video game consoles, vehicles, and televisions are all connected to a wireless network. But also, certain seemingly "dumb" electronics like home appliances are getting connected. With the rise of embedded electronics and wearables, even completely inert objects like clothing will connect to some sort of wireless network.

This means that there is a gold rush. And whenever there is a gold rush, there is scarcity. The scarce resource in wireless communications is bandwidth, and everyone is trying to capture more bandwidth to do one thing: get more data through. More data rate means more diverse applications, a better user experience, better quality of service, more exciting products that sell a lot, and more money.

Which is why radios are important. They are everywhere doing everything. So, they are trying to extract the last good bit of information out of every Hertz of bandwidth. To do this, radios play incredible tricks with the shape and nature of data. This makes radios interesting on the algorithmic level, and thus, particularly exciting to move to hardware.

Which is why we will be using them as an example. The "radio" we will be discussing in this book is particularly the digital part of the physical layer baseband. This part of the radio communicates on one side with higher layers that use general purpose processors. On the other side, it communicates with an analog section of the physical layer that deals with real signals transmitting over the air.

The complexity of radios leads to some interesting challenges that can be extended to other systems. For example, while discussing channel coding in Chap. 6, we find that the arithmetic is simple, but the challenge lies in partitioning the software-friendly algorithm into processing units that can be parallelized in hardware. When we discuss OFDM, we find that the processing unit is simple. The signal flow graph and the block diagram are also trivially related, and the algorithm is parallel by nature. But when we calculate throughput requirements, we will find a huge gap which forces us to implement hardware reuse, memory partitioning, and complex control. When discussing MIMO decoding on the other hand, there is no obvious hardware corresponding to the arithmetic of the algorithm. Thus, we take a step back and think about how to design a processing unit that efficiently does the not so obvious math we are required to do.

You can use this book in two ways. You can read it serially, in which case you will get a very good background in the implementation of digital radios, and along the way you will hear an interesting story about how a complex system is moved to hardware. Or you can read sections and chapters in isolation. Most chapters can stand alone and be read independently. Parts of the book can be broken down into three "types":

- Radios (R): This covers algorithms and models of Wi-Fi and cellular wireless systems. These sections assume very little background in communication theory. Thus, they are an opportunity for hardware engineers who want to target the implementation of radio physical layers. But more generally, these sections demonstrate to what degree a hardware engineer has to get involved in the algorithm they are implementing and how sometimes knowing the theory can substantially change the hardware.
- Implementation (I): These are sections that discuss how radio transceivers are implemented in digital hardware. While we will discuss the details of implementation, the aim is to provide general guidelines about concepts that can be extended to any algorithm.
- Generalization (G): These are sections that are more independent from digital radios. They cover how algorithms move to hardware in more explicitly general terms. These sections are more abstract in their approach and are thus easier to generalize.

The table below lists sections and chapters and the category: radio (R), implementation (I), or generalization (G). Think of a row in the table as a topic that can be read independently.

	Cat	Topic
Ch. 1	R	*Introduction to radio communication.* This chapter lays the foundation for radio communications. We cover the layer model for partitioning radios. We discuss the baseband physical layer as it relates to the layers directly above and below it. So, we discuss the radio frequency section and how its analog nature affects the baseband. We also discuss the medium access control layer as the source and drain of the physical layer. The MAC layer has a software nature while the RF section is very "physical." This leaves the baseband in a very interesting position
Ch. 2	R	*Wi-Fi modeling.* This chapter lays the theoretical and mathematical foundation required to understand the radio physical layer. We introduce integral concepts to Wi-Fi like packet-based communication. Most of the concepts in the chapter extend to any MIMO-OFDM radio. An interesting exercise at the end of the chapter is how to design a quality metric that characterizes a hardware system in a fair manner
3.1–3.6	G	*Fixed-point arithmetic.* How to move a high-level system-level simulation to a model that allows us to know the size of registers and arithmetic processing units. Of particular concern is how the size of processing units grows rapidly in algorithms that have a sequence of operations, particularly a sequence of multiplications

(continued)

3.7–3.9	G	*Foundations of hardware implementation.* We start by discussing different platforms that can be used to implement algorithms. The efficiency and complexity involved in implementing something on a GPU is on a different scale from that involved in ASIC design. We also get into a deeper discussion of what we mean by complexity, speed, and power in hardware. But ultimately this part of the book is about introducing the different hardware architectural strategies that will be expanded upon for the rest of the book
Ch. 4	R	*Wireless channel model.* The wireless receiver does most of what it does because of the channel. We develop a model that is deep enough to cover most channel effects but easy enough to understand that it can be practical
5.1–5.5 and 5.8	R	*Baseband modulation.* Baseband modulation is the process of squeezing more bits into the same bandwidth. When we see baseband signals on a constellation diagram, we visualize the channel effects in Chap. 4 in an intuitive graphical way
5.6 and 5.7	I	*Baseband modulation implementation.* How to slice observations at the receiver to decide the bits that were sent. We start to see preliminary issues that affect implementation, such as the trade-off between number of PUs and their complexity
6.1–6.7	R	*Channel coding.* Theoretical foundation for channel coding. We classify channel codes and consider how the insertion of redundant bits into a message can give it error detection and correction capability
6.8–6.13	I	*Implementation of Viterbi decoder.* The Viterbi algorithm exposes us to a very interesting case study in moving algorithms to hardware. The arithmetic involved in Viterbi is very simple. However, the challenge comes in partitioning the algorithm into different PUs and figuring out how to string these PUs together to form a whole. In other words, the signal flow graph of the algorithm does not directly relate to its block diagram
Ch. 7	R	*Orthogonal frequency division multiplexing.* To communicate with high bitrate, modern radios use wideband channels. Wide channels are usually frequency-selective, especially in urban environments. OFDM is one of the most efficient ways to deal with this frequency selectively without letting computational complexity explode
Ch. 8	G	*Algorithm to hardware.* In this whole chapter, we take the implementation of FFT as a case study for how an algorithm is moved to hardware. We discuss how a signal flow graph is transformed into a block diagram, and how fixed-point analysis can affect the design. But more importantly, we figure out how the block diagram is transformed into hardware. This opens the possibility of hardware reuse, pipelining, parallelism, and direct implementation. We also discuss different control strategies and how memories should be allocated alongside the datapath
Ch. 9	R	*MIMO decoding.* Most modern radios use multiple antennas either at the transmitter or the receiver or both. This chapter sets the ground rules for how communication can take place across these "MIMO channels." We see algorithms with complicated arithmetic that cannot be directly mapped to hardware the way the FFT was
Ch. 10	G	*Complex algorithms to hardware.* In some cases, the processing units of an architecture are simple, but the overall design is complicated. This is the case with the Viterbi decoder. In some cases, the PU is simple, and the connection is regular, but budgeting resources is complex. This is the case with FFT. In MIMO decoding, nothing is simple. The arithmetic we do is taxing and requires us to design custom processing units. The architecture might be regular but requires careful thinking. And hardware reuse with its associated control and memory challenges will always have to be used

(continued)

| Ch. 11 | I | *Estimation and synchronization.* A wireless receiver reverses what the channel did to the message that the transmitter sent. To do this, it must figure out what the channel is like. This chapter will force us to look deeper into the 802.11n physical layer standard, which exposes us to the state of the art in how systems engineers communicate their requirements to implementation engineers |
| Ch. 12 | R | *Advanced topics in wireless communications.* Most of the book uses Wi-Fi as an example. This chapter will illustrate how cellular systems differ from Wi-Fi. But more importantly, it demonstrates just how similar the two are and how easy it is to extend the lessons learnt from Wi-Fi to the implementation of similar systems |

Giza, Egypt   Karim Abbas

Contents

List of Abbreviations

ACK	Acknowledge Packet
ACS	Add-Compare-Select
ADC	Analog to Digital Converter
AGC	Automatic Gain Control
ASCII	American Standard Code for Information Interchange
ASIC	Application Specific Integrated Circuit
AWGN	Additive White Gaussian Noise
BER	Bit Error Rate
BMU	Branch Metric Unit
BPSK	Binary Phase Shift Keying
CMOS	Complementary Metal Oxide Semiconductor
CORDIC	COOrdinate Rotation DIgital Computer
CP	Cyclic Prefix
CRC	Cyclic Redundancy Check
CSMA-CA	Carrier-Sense Multiple Access with Collision Avoidance
CSMA-CD	Carrier-Sense Multiple Access with Collision Detection
CTS	Clear to Send
DC	Direct Current
DFT	Discrete Fourier Transform
DIF	Decimation in Frequency
DIT	Decimation in Time
DL	Downlink
DSP	Digital Signal Processing/Processor
DSSS	Direct Sequence Spread Spectrum
DTFT	Discrete Time Fourier Transform
EHF	Extremely High Frequency
eNodeB	Extended Node Base station
FEC	Forward Error Correction
FFT	Fast Fourier Transform
FH	Frequency Hopping
FPGA	Field Programmable Gate Array

FS	Fourier Series
FSM	Finite State Machine
FT	Fourier Transform
GI	Guard Interval
HT-LTF	High Throughput Long Training Field
HTML	HyperText Markup Language
HT-SIG	High Throughput Signal field
HT-STF	High Throughput Short Training Field
ICI	Intercarrier Interference
IDFT	Inverse Discrete Fourier Transform
IFFT	Inverse Fast Fourier Transform
ISI	Intersymbol Interference
ISM	Industrial, Scientific, and Medical [band]
L-LTF	Legacy Long Training Field
LNA	Low Noise Amplifier
L-SIG	Legacy Signal field
L-STF	Legacy Short Training Field
LTE	Long-Term Evolution
MAC	Medium Access Control
MCS	Modulation Coding Scheme
MIMO	Multiple Input Multiple Output
MIMO-BC	MIMO Broadcast
MIMO-coop	Cooperative MIMO
MIMO-MAC	MIMO Multiple Access
MISO	Multiple Input Single Output
ML	Maximum Likelihood
MMSE	Minimum Mean Square Error
MPDU	MAC Protocol Data Unit
MRC	Maximal Ratio Combining
MSDU	MAC Service Data Unit
MU-MIMO	Multiuser MIMO
OFDM	Orthogonal Frequency Division Multiplexing
OFDMA	Orthogonal Frequency Division Multiple Access
PA	Power Amplifier
PAPR	Peak to Average Power Ratio
PER	Packet Error Rate
PHY	Physical layer
PMU	Path Metric Unit
PRB	Physical Resource Block
PSD	Power Spectral Density
PSK	Phase Shift Keying
PU	Processing Unit
QAM	Quadrature Amplitude Modulation
QPSK	Quadrature Phase Shift Keying

QRD	QR Decomposition
RF	Radio Frequency
RTS	Request To Send Packet
SC-FDMA	Single Carrier Frequency Division Multiple Access
SD	Sphere Decoding
SHF	Super High Frequency
SIC	Successive Interference Cancellation
SIMO	Single Input Multiple Output
SISO	Single Input Single Output
SNR	Signal to Noise Ratio
STBC	Space Time Block Coding
SVD	Singular Value Decomposition
TCP	Transmission Control Protocol
THF	Tremendously High Frequency
UE	User Equipment
UHF	Ultra High Frequency
UL	Uplink
VGA	Variable Gain Amplifier
XML	Extensible Markup Language
ZF	Zero Forcing

Chapter 1
What Is a Radio?

1.1 Our Most Valuable Resource, the Radio Spectrum

In the year 1862, James C. Maxwell caused a paradigm shift. It is surprising that he did not invent anything new. What Maxwell did was to collect, organize, and synthesize a body of knowledge about electricity and magnetism that had long since been known. He described how electric and magnetic fields behave in intuitive integral and differential equations, and through the entanglement of the two, he made an astounding discovery.

An electric field is an effect in an area where a charged body will experience a force. A magnetic field is an effect in an area where a magnetic material will experience a force. Magnetic fields are more intuitive for most people because everyone has observed the phenomenon while playing with magnets.

Maxwell's first two laws are so simple that they feel like axioms (Fig. 1.1):

- If you take a small volume of space and enclose it, the electric field coming out of this volume will depend on the net charge contained within. The direction of the field will depend on the polarity of the charge. The magnitude of the electric field will depend on the amount of contained charge. The shape of the electric field will depend on the distribution of the charge within the volume
- Magnetic field lines do not begin or end; instead they either make complete loops or extend to infinity. This is a very esoteric way of saying something much simpler: magnetic charges do not exist; instead magnets exist as dipoles. In other words, magnets, unlike electric charge, cannot be unipolar. Every magnet must include a North and South pole, and the two must exist together. Thus, the net magnetic field lines coming out and going into a volume around a magnetic dipole must be net zero. But what if the volume you draw cuts a magnet in half? Would you be creating a unipole containing only the included pole? The answer is no; the cut magnet will always create an opposite pole at the location where it is cut, and the net field lines crossing the surface must be null.

K. Abbas, *From Algorithms to Hardware Architectures*,
https://doi.org/10.1007/978-3-031-08693-9_1

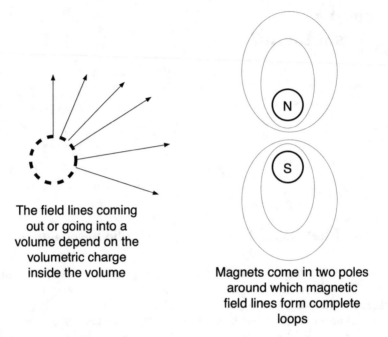

The field lines coming
out or going into a
volume depend on the
volumetric charge
inside the volume

Magnets come in two poles
around which magnetic
field lines form complete
loops

Fig. 1.1 Maxwell's first two laws

There is nothing groundbreaking so far. In fact, both laws were fully understood well before Maxwell. Even the third and fourth laws were well known. But it is in their implications that the genius lies. The third and fourth laws are:

- A time-varying magnetic field will generate an electric field.
- A time-varying electric field will generate a magnetic field.

There is obviously intertwining between these two laws. A time-varying electric field generates a time-varying magnetic field. This time-varying magnetic field generates a time-varying electric field, which in turn will regenerate a magnetic field. These two phenomena thus *could* start a feedback loop that *may* potentially maintain them indefinitely as they travel in space. It was this realization that led Maxwell to the understanding of electromagnetic waves.

Maxwell deduced that electromagnetic radiation was essentially paired electric and magnetic fields. The two fields are transverse and sustain each other. This allows them to travel forward in space in the form of radiation or waves.

Maxwell used the third and fourth laws to calculate the speed at which these "electromagnetic" waves travel in vacuum. During the same time period, the speed of light was independently measured. The two numbers turned out to be the same. This confirmed something people were already suspecting; visible light is electromagnetic radiation of a range of wavelengths to which the human retina is sensitive. This realization completed our classical understanding of the nature of electromagnetic radiation.

The electromagnetic spectrum covers forms of radiation that we consider radically different (Fig. 1.2.) At the lower end, we have radio and microwave frequencies. This is followed by infrared, then visible light, ultraviolet, x-rays, and finally gamma rays. Although the popular conception of each of these radiations is different, they all have the same fundamental nature in that they are electromagnetic waves. But because they differ in frequency and wavelength, they interact with the surrounding environment in very different ways. Different radiations are generated differently, are absorbed by the environment differently, and also diffract and reflect differently at interfaces.

To decide which of the "types" of EM waves in Fig. 1.2 we can use for wireless communication, we must determine what we need from radios. Our list of demands is:

- The waves must be easy to generate, radiate, and receive in antennas.
- They must not be harmful to humans or detrimental to the environment.
- They must be able to travel long distances without much loss of amplitude.

Waves can interact with dimensions comparable to their wavelength. At the higher end of the EM spectrum, x-rays and gamma rays have incredibly short wavelengths. So short in fact that they are comparable to human cells or even to DNA molecules. This allows them to mutate human chromosomes or cause cellular damage. Thus, gamma rays, x-rays, and parts of the UV band are out of the question.

Visible light is useful in extremely short-range wireless communication and is very useful for wired communication (fiber optics). However, for robust wireless communications over long distances, visible light will suffer from severe attenuation. It can also be very disruptive because it is detected by human eyes.

This leaves us with the radio spectrum (Fig. 1.3).

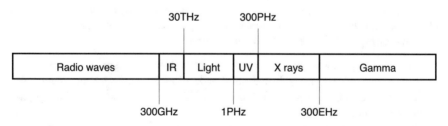

Fig. 1.2 The electromagnetic spectrum

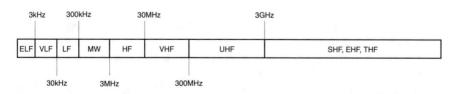

Fig. 1.3 The radio spectrum, not to scale. Lower sub-bands are much narrower than they look here

Table 1.1 Division of radio spectrum into bands according to the ITU

Band	Range	Applications
ELF, SLF, and ULF	Up to 3 kHz	Mainly submarine communication
VLF	3–30 kHz	Navigation
LF	30–300 kHz	RFID
MW	300 kHz–3 MHz	AM radio
HF	3–30 MHz	Shortwave radio, amateur radio, airline navigation
VHF	30–300 MHz	FM, terrestrial TV
UHF	300 MHz–3GHz	Wi-Fi, Bluetooth, television, microwaves, ZigBee, satellite radio, GPS, astronomy, cellular communication
SHF	3GHz–30GHz	Wi-Fi, cellular communication including 5G, digital video broadcasting, satellite
EHF	30GHz–300GHz	5G cellular communication, Wi-Fi (802.11ad)
THF	300GHz–3THz	

Bandwidth is everything. Larger bandwidth is the most straightforward way to increase throughput in digital and analog communication. Fortunately, the radio spectrum is very wide, potentially allowing wideband communication. It is hard to set a lower bound for the radio spectrum because it pretty much extends all the way down to zero hertz. At the higher end of the radio band, it merges hazily into infrared.

But the radio spectrum is crowded because everyone is always trying to use it. Every terminal will try to occupy as much bandwidth as it possibly can. To better manage this very wide and crowded "radio band," we divide it into multiple smaller sub-bands that are suitable for specific applications. There are many standards by which the radio frequency can be divided, and Table 1.1 shows one way, namely, the classification used by the International Telecommunication Union (ITU).

The radio spectrum offers a fundamental trade-off:

- Properties of path loss (Sect. 4.5) and absorption by atmospheric gases means that loss grows with frequency. Lower sub-bands allow longer-range communication.
- Lower band means lower available bandwidth. Higher sub-bands allow higher bitrate communication.
- Antenna size is directly proportional to wavelength. Higher sub-bands allow the use of smaller antennas.

Thus, very low-frequency bands are not interesting for commercial radio communications. The main reason is antenna size. Mobile wireless terminals in consumer electronics need extremely small, effective, and inconspicuous antennas to be commercially and practically viable. At the extreme low end of the radio spectrum for ELF, SLF, ULF, and VLF, the size of the antenna used means that the only practical applications are navigation on extremely large vehicles. This is particularly true in submarines, where the long range of these waves combined with propagation properties in water makes the band practical.

The MW, HF, and VHF bands were, and in some cases still are, used for broadcast in radio and television networks. But just a glance at the size of television antennas is a good indicator of why these bands are not preferred for mobile platforms.

Somewhere near the SHF band, there is a sweet spot where both longer-range communication is effective and small antennas are possible. This is the most sought-after area of the spectrum for Wi-Fi and mobile communication. The topmost four bands are of particular interest for Wi-Fi communication. We will discuss each in detail below.

Ultrahigh Frequency (UHF)
This is the first band of interest for Wi-Fi and cellular communications. The higher end of this band is known as the microwave range, while the lower end is known as decimeter wave. Lower ends of UHF communication are used for terrestrial television transmission. The higher end is extremely crowded. It is used by Bluetooth and is vulnerable to radiation from microwave ovens. But more importantly, this higher end, particularly around the 2.4-GHz point, is where one of the two ISM bands is found. This band is used by Wi-Fi, especially in legacy applications and in access points that need to cover large areas.

Super High Frequency (SHF)
This is the second of the sweet spots in the spectrum. This band combines advantages from lower frequency with those of higher frequency. From lower-frequency bands, SHF inherits the low absorption and long propagation range. Yet the frequency is high enough that antennas are of reasonable size. In fact, antennas in this range share a lot of properties with EHF antennas, allowing them to be directional and to allow frequency reuse, which is why this band is also crowded. All forms of broadcast and long-range communication will use this band. However, it is still free from some of the strong interference in the UHF band. The SHF band is one of the two bands commonly used by Wi-Fi. When given the option, it is generally safe to assume that the SHF band will offer better communication than UHF despite the lower range.

Extremely High Frequency (EHF)
Waves in this band suffer from significant atmospheric absorption. Path loss generally increases with frequency, but the EHF band is particularly devastated by resonant peaks from oxygen and water vapor. This significantly reduces the range of EHF waves to a few kilometers even with significant transmit power. At the higher end of the band, we cannot have any meaningful communications for more than a few meters.

However, the EHF band has some promising properties, including surprising advantages that arise from its short range. Because the wavelength is extremely small, very small antennas can be used. The antennas can be made more directional or large antenna arrays can be used to beamform. Combined with the low communication range, this allows frequency reuse. The lower end of this band is used for cellular communication in 5G networks (Sect. 12.6). In Wi-Fi, the lower end is suggested for use by the 802.11ad standard (Sect. 2.6).

Tremendously High Frequency (THF)
This band encroaches into the infrared regime. Waves in this band have terrible properties. They cannot penetrate most barriers and are absorbed by atmospheric gases. It is barely usable for any form of wireless communication. We do not even have reliable ways to generate, radiate, or receive such waves. There are very few real applications that use this band. However, there are exploratory attempts to standardize wireless communications in THF.

1.2 Getting People Out of Each Other's Way, the Layer Model

The layer model of communication systems is a very helpful tool because it divides levels of engineer involvement into discrete domains. In the preface, we discussed how engineers need to combine some knowledge of system design with hardware design to be able to come up with successful radios. However, if you look at a modern cell phone, it becomes clear why it is impossible for a single person to take care of the entire design.

A modern cell phone is not really a phone. It is a wirelessly connected general-purpose computer. It is running an operating system; the operating system allows applications written using a high-level language to use the processor. Most apps running on a cell phone will also communicate through the Internet. Thus, the app needs to access the spectrum, something that a lot of other mobile platforms lying around also do. Thus, there must be some "procedure" that all such mobile platforms follow to arbitrate who gets to access the wireless channel and when. If we get hold of the wireless channel, we need to understand how to push the binary data produced by the app through the channel in such a way that allows us to retrieve it meaningfully at the receiver.

Managing all of this can become very confusing for an app designer. The app designer uses a very high-level programming language suite to design their application. They care about things like screen gestures, graphics, responsiveness, user experience, and GUI design. And as well they should. If we ask app designers to also care about regulating access to the channel and figuring out how to push raw data through, they will give up, or at the very least they will make much worse apps.

Similarly, an engineer designing a radio is focused on pushing data through the ether and cannot be bothered with what application is currently using the channel. If they must care whether the application is a Battle Royale or a text chat app, they will not be able to do their core job: designing the radio. The layer model allows everyone to do their job in isolation while also being able to talk to others. Layer models share some basic characteristics:

- Layers are numbered from 1 to N, with 1 being the lowest layer.
- A "lower layer" is a layer closer to the physical transmission environment.
- A "higher layer" is one closer to the user experience.
- At the transmitter, data can only flow from a higher layer to a consecutively lower layer.
- At the receiver, data can only flow from a lower layer to a consecutively higher layer.
- Layer N is always at the application level and is called the application layer.
- Layer 1 is always in contact with the channel and is called the physical layer.
- Each layer is handled by an independent designer/engineer.
- An engineer at a specific layer can be completely unaware of what is going on in the other layers; this allows them to work independently and focus on the task at hand. There is one exception discussed in the bullet below.
- Data is passed between layers in the form of packets. Each packet at a specific layer-to-layer link must be formatted in a specific way. This is one of the few cases where a designer must think about what happens at other layers.

In very simple terms, a layer model allows designers at different levels to be isolated from each other. This frees the app developer to focus on developing the app and the radio designer to focus on designing the radio. However, to guarantee data can be transferred smoothly between layers, there is a certain format that data must be put into. This is the only case where a designer at a certain level must think of the other levels: they expect data from other layers to arrive in a certain format, and they hand data to other layers in a certain format.

The complete isolation of layers is not set in stone. While designers *can* focus on a single layer, they rarely do. For example, designers involved in the physical layer often also design the immediately higher layer. It is also true that allowing designers to think in terms of cross-layer design can sometimes open optimization choices that might otherwise be invisible. It is also unrealistic to think that each layer will have an assigned designer. Most designers can and will get involved in implementing a stack of several layers. In short, we can and should consider all layers to be isolated, but we must be aware that in practice this is an abstraction rather than a rule set in stone.

The prototypical and most extensive layer model is the Open Systems Interconnection model or OSI (Fig. 1.4). OSI divides the communication system into seven distinct layers in two categories. The two categories are the media layers, which deal with more tangible aspects of access to the channel; and the host layers, which deal with how data is preprocessed for transmission at the host. The media layers are, namely, physical, data link, and network. The host layers are, namely, transport, session, presentation, and application. We will discuss each layer in detail from top to bottom.

Layer 7, Application
This is the highest-level layer. It is the only layer that the user-side application can directly access. The application layer does not actually have anything to do with the application running on the processor unless the said application is trying to access the communication channel.

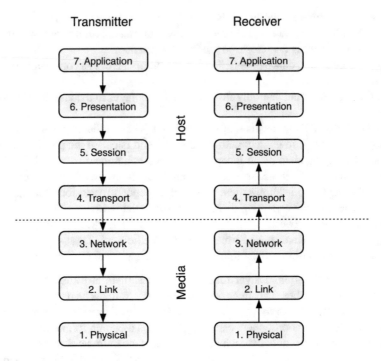

Fig. 1.4 The OSI layer model

The job of the application layer is to take in a request for communication from an application and determine basic information about the needed communication, such as the onboard resources, the high-level description of the source and destination, and the nature of the communication. The layer then takes the data to be communicated from the app and arranges it into its own protocol data unit (PDU). However, the application layer makes no effort to change the nature of the payload.

Layer 6, Presentation
This layer is critical. It has only one function: to take data from the application layer and make it fit for transmission. And by that we mean doing all preprocessing that should be aware of the nature of the data. Once data leaves the presentation layer, all lower layers will not and cannot care about the nature of data in the payload.

Compression is an operation that cannot be agnostic of the nature of data. The sparsity, patterns, and periodicity of data will affect the compression technique used. Audio, video, and text may be more efficiently compressed by different algorithms. Thus, compression must be done in the presentation layer.

Reformatting is also an integral component of the presentation layer. Consider, for example, a case where you need to reformat all data into ASCII, XML, or HTML. This will have to be done at the presentation layer because you are changing the nature of data.

Layer 5, Session

This layer will start and end any session. This involves establishing a connection between the computer and a remote computer, as well as ensuring that communication between the two takes place gracefully. There is particular concern with the way sessions are ended, with unexpected terminations being an issue. The session layer handles the need for retransmission and establishes checkpoints to ensure the two terminals are aware of where they stand. In the Internet communication paradigm, the session layer is associated with the TCP protocol.

Layer 4, Transport

The transport layer accepts data from higher layers and segments it into variable length segments. It then transmits these segments reliably between two terminals. This includes handling of acknowledgments and handshaking, as well as deciding when to send the next segment.

Layer 3, Network

The network layer charts a path between the two nodes of communication. This is particularly important in networks with many nodes. The network layer will not guarantee quality of service but will rely on higher layers to ensure this. Instead, the network layer will only find a short graph between the two nodes.

Layer 2, Link

This layer assumes a physical connection has been established between the two nodes. Thus, the link layer only concerns itself with the synchronization of transmission.

In more practical terms, the main function of the link layer in Wi-Fi is medium access control, or MAC. MAC is the process of determining when terminals get access to the shared transmission medium. The MAC layer is of particular importance to this book because it is the layer directly above the physical layer. Thus, it represents the input to everything we deal with at the transmitter and the drain for everything we produce at the receiver. We will discuss the MAC layer in Wi-Fi in more detail in Sect. 1.4 and for cellular systems in Sect. 12.2.

Layer 1, Physical

When we talk about baseband radios, we are mostly talking about the physical layer. This is the layer closest to the channel or transmission medium. The physical layer deals with a single question: how to make raw ones and zeros obtained from the MAC layer ready for transmission over the air. The physical layer will reshape the signal so that it has better spectral properties, more error correction ability, and more throughput. For most of this book, we will be dealing with a particular part of the physical layer: the physical baseband. We will discuss the other part of the physical layer (RF and mixed signal) in Sect. 1.3.

1.3   A Signal as It Really Is, Radio Frequency Signals

Figure 1.5 shows a representation of the physical layer of a radio. The physical layer is divided into two big blocks, the baseband and the radio frequency (RF). We will be covering the baseband in this book. However, a lot of what we *must* do in baseband is dictated by things that happen in the RF section. Thus, understanding what each part of the physical layer does and how they interact with each other is important. In this section we will give a brief introduction to the distinction between RF and baseband and what it is exactly that the RF section does.

The baseband section of the physical layer is a digital processor. At the transmitter, it is handed raw zeros and ones by a higher layer, usually the MAC layer. It hands over complex numbers to the RF section. To understand why it does so, and what these complex numbers mean, read Sect. 2.1.

The baseband radio prepares the signal for the channel. It introduces redundancies and modifies the signal to allow more data to be packed in the same bandwidth and to allow errors to be detected and corrected if they happen. So, to reiterate, the baseband does the following:

- Preparing the signal to travel over a frequency-selective channel
- Packing more bits into the same bandwidth to increase spectral efficiency (how many bits per second pass through each Hz)
- Performing channel coding, allowing the receiver to detect bit errors if they occur, and correct them if possible
- Performing detection and synchronization, by defining packet and symbol starting points, and managing offsets between the carrier frequencies at the transmitter and the receiver
- Allowing the receiver to estimate and invert channel state conditions, which is particularly important in wireless channels where such conditions are fluid

These functions are complicated, and were it not for the fact that we perform them in digital domain, they would have been impossible to handle. The main issue with baseband processing is how to translate the complicated algorithms required into digital hardware. This will be the topic for most of this book and will be used as an illustrative example for how complex systems in general are translated into hardware.

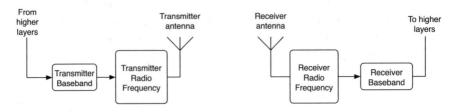

Fig. 1.5 RF and baseband in a radio

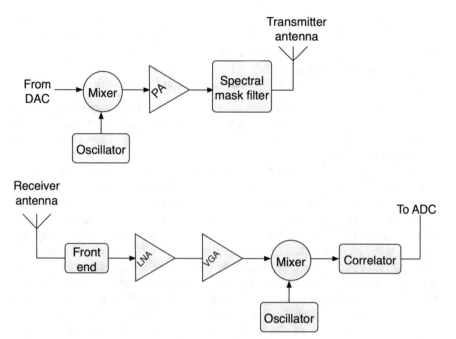

Fig. 1.6 Components of the RF section at the transmitter and the receiver

The radio frequency (RF) section takes the complex numbers from the baseband and sends them over the air. Signals sent over the air are real and analog. Signals in the baseband are digital. So, between the baseband and RF sections, there must be data converters. In the transmitter there is a digital-to-analog converter (DAC) that accepts data from the baseband and outputs corresponding analog waveforms to the RF. At the receiver there is an analog-to-digital converter (ADC) that accepts waveforms from the RF and provides the corresponding digital output to the baseband receiver. The design of these converters is a challenging field of its own.

DACs are relatively simple, but it is in ADCs where most of the challenge of mixed-signal design lies. The main metric of interest in an ADC is the number of bits. The ADC divides the range of its analog input into a number of discrete levels. This number of levels will define the size of the ADC digital output word. We will find out that this is a critical limiting factor of the performance of the overall radio.

Figure 1.6 shows the building blocks of the transmitter and receiver RF sections. The transmitter is tasked with taking the signal from the baseband up to the RF carrier frequency where it is radiated from the transmitter antenna. The receiver RF section is tasked with the opposite: taking the signal down from passband to baseband.

In the transmitter, there are three RF blocks (ignoring the spectral mask filter). There are many other components and subsystems in a realistic RF transmitter, but the ones shown in Fig. 1.6 are the most critical. We will discuss them in more detail below.

The Oscillator

Signal bandwidth for most practical standards is in the few dozens of megahertz. Band frequencies discussed in Sect. 1.1 were in the GHz range. These high frequencies are the carrier frequencies upon which the much narrower signal is mounted. This is done to allow the signal to be radiated from an antenna with reasonable size.

The crystal oscillator resonates and produces a sinusoid at the carrier frequency of interest. This high-frequency sinusoid is then used to raise the baseband signal to the carrier frequency. In Chap. 5, we will find out that we need two sinusoids at the same frequency offset by a phase of 90 degrees.

Two parameters are important for evaluating oscillators: phase noise and offset. Phase noise is a measure of spectral purity for an oscillator. A pure sinusoid has a spectrum with an impulse at a single frequency. All practical oscillators have a skirt around this central frequency that indicates the presence of other frequency components in the carrier. These additional components are thus "noise" and not the usual additive noise we deal with in wireless systems (Sect. 4.2). This noise is frequency selective, thus affecting the frequency of the produced tone from moment to moment in a stochastic fashion. The most common way to represent the noise in the skirt of Fig. 1.7 is as a variable phase component in the carrier that jitters from cycle to cycle.

An offset indicates that the frequency of the oscillator is never the nominal frequency as advertised by the vendor. Instead, there is a range of frequency tolerance around the nominal value called frequency offset. Offsets are usually stated in parts per million (ppm), which indicates how many Hz of offset is observed for each 1 MHz of the carrier. Frequency offsets might initially seem more innocuous than phase noise. After all, they are not stochastic. While offsets will vary from one oscillator to another, it is usually a constant for one oscillator. But in Chap. 7, we will see that frequency offset has a devastating effect on systems with multiple carriers.

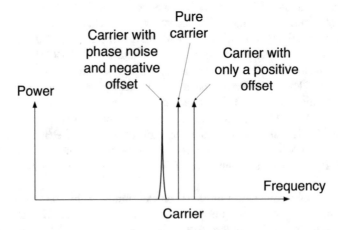

Fig. 1.7 Figure is not to scale. The middle tone is a perfect carrier. The tone to the right is the output of an oscillator showing frequency offset and no noise. The tone to the left shows a negative offset and a skirt representing phase noise

Figure 1.7 shows a tone at the expected frequency of the carrier. The tone to its right shows the output of an oscillator suffering from a positive offset. This figure is obviously not to scale because offsets are always too small to observe if the x-axis were drawn to scale. The tone to the left shows the output of an oscillator suffering from a negative offset as well as phase noise causing a skirt of tones to appear around the intended tone.

Mixer

This is an analog multiplier. It multiplies the carrier sinusoid by the analog baseband signal produced by the ADC. The mixer is a nonlinear circuit, and like all RF blocks, it is difficult to design. The carrier is a pure sinusoid at a much higher frequency than the baseband signal. Because it is a simple sinusoid, it carries no information. However, when the baseband signal is multiplied by the carrier, it is raised to the carrier's higher frequency. This allows the overall signal, now called the passband signal, to be radiated out from an antenna with a much smaller aperture. The carrier frequency for Wi-Fi is in the few GHz range and for modern cellular systems can go up to tens of GHz. The bandwidth of baseband signals ranges from tens of MHz to hundreds of MHz.

Power Amplifier

In terms of power consumption, this is perhaps the most important block in the entire transmitter. When the baseband signal is mixed with the carrier frequency, we call the result the passband signal (Sect. 2.1). This passband signal is then expected to travel a very long distance over the air between the transmitter and the receiver. In Chap. 4, we will see that travel over the air by necessity means a significant drop in signal power. This is not contingent on the channel being "tough" or "easy." Mere travel over the air entails a drop that is at least quadratic with the distance between the transmitter and the receiver. The only way to handle this is to increase the power of the transmitted signal, so that the attenuated received signal is still comfortably above the noise floor.

A power amplifier takes the small passband analog output of the mixer and amplifies it by a very large gain. Design of power amplifiers is extremely challenging. On the one hand, because their output signal has a very high amplitude, we must be careful with the linearity of the amplifier. On the other hand, maintaining good linearity significantly reduces the efficiency of the amplifier. We will consider this in more detail in Chap. 5, but for starters, know that the power burnt in the power amplifier could account for upward to half of the battery consumption of a mobile platform.

There is also another challenge in the design of power amplifiers: their gain must be variable, and user controlled. The level of the received signal will vary depending on the distance between the transmitter and the receiver. If the transmitter and/or the receiver is mobile, this distance will be variable. To guarantee a certain SNR at the receiver regardless of distance, we must guarantee that received signal power is the same. This can be done by allowing the level of the transmitted signal to be varied by manipulating PA gain, or by varying gains at the receiver, or both.

Spectral Mask

Every standard will impose restrictions on the power level of the transmitted signal outside the channel assigned to the user. This is necessary to allow multiple users to use multiple adjacent channels. The final transmitter filter makes sure the transmitted signal conforms with this standard by attenuating out-of-band power. All transmitters will have significant out-of-band power before filtering because sharp time-domain signals produce significant sidelobes in the frequency domain.

By inspection of Fig. 1.6, the receiver RF section seems to be significantly more complicated than the transmitter. This is true in RF, and it is true for the baseband physical layer as well as all higher OSI layers. The receiver will always contain more blocks than the transmitter, and receiver blocks with a corresponding transmitter block tend to be more complicated. This makes sense because:

- The receiver includes symmetrically opposite blocks to those in the transmitter. This allows us to reverse what happened at the transmitter, extracting raw bits to give to the MAC layer. These opposite blocks are either as complicated as the blocks in the transmitter, or more so.
- The receiver includes blocks without correspondence in the transmitter. These blocks handle impacts from the channel (Chap. 4).

The RF receiver blocks are as follows.

Front-End Filter

This is a relatively wideband filter that comes right after the receiver antenna. In Chap. 4 we will see that "noise" is a receiver phenomenon. It is also an extremely wideband phenomenon. The more bandwidth you allow into your RF chain, the more noise can interact with your signal. The front-end filter will allow the signal and a good-sized spectrum around it to pass, but it will attenuate out-of-band noise to the extent that it no longer matters in receiver calculations

Low-Noise Amplifier (LNA)

The LNA is pretty much the first significant block in the receiver RF chain. The front-end filter is so wideband that it has very little impact on the signal and is not particularly challenging to design. On the other hand, it is important that the LNA is the first block to process the signal and that the LNA has very good properties. The performance of the LNA could spell life or death for the whole receiver.

To understand why, we must understand the concept of noise factor. Any analog block, circuit, or subsystem will by necessity add noise to any input signal. A good analog block will add little noise; a bad analog block will add a lot of noise, but noise it must add. But how about an amplifier? Does it not magnify the amplitude of the signal? In that case does it even matter that it is adding noise? Perhaps it adds noise, but because it is amplifying the signal, it should also improve the signal to noise ratio (SNR).

Fig. 1.8 Amplifier adding noise ($n2$) to an already noisy signal

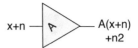

Let us consider the amplifier in Fig. 1.8. It receives a noisy input signal of the form:

$$x + n$$

This input signal x thus has an SNR of:

$$SNR = VAR(x)/Var(n)$$

The amplifier will amplify the input signal, and it will also amplify the input noise, for an output of:

$$Ax + An$$

But the amplifier is itself an analog block, so it will add its own noise after it has performed the amplification, for a total noisy output of:

$$Ax + An + n_2$$

Thus, the output SNR is worse than the input SNR and is equal to:

$$SNR = AVar(x)/\{AVar(n) + VAR(n_2)\}$$

The noise factor (NF) of an analog block is the degradation in SNR (measured in dB) seen through the block. NF must be nonzero and positive for any analog block above zero K. For example, for the amplifier above, the NF can be calculated as:

$$NF = 10\log_{10}(VAR(x)/Var(n)) - 10\log_{10}(AVar(x)/\{AVar(n) + VAR(n_2)\})$$

This still does not explain why we must put the LNA as early as possible in the chain. If the system is linear, then the order of blocks should not matter. Even if we consider noise, every block has its own noise, and all the noises are additive, so total noise should be the same regardless of the arrangement of blocks.

This is not true. Figure 1.9 shows two choices on the arrangement of blocks in an RF chain. In the first "choice," the LNA comes first; in the second it comes last. Let us assume for the sake of simplicity that the order of blocks has no impact on the function of the overall chain, and the only impact is on the chain noise factor. Without loss of generality, we can also assume the two blocks other than the LNA have unity gain.

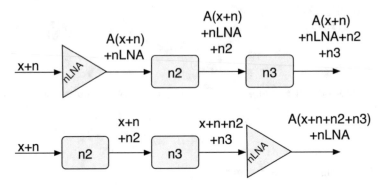

Fig. 1.9 LNA (shown as a triangle) at the beginning of a three-block chain (top), and LNA at the end of a three-block chain (bottom)

In the first arrangement, the output signal from the chain can be calculated if we assume the signal x enters the chain with associated noise n, leading to an overall chain output:

$$Ax + An + n_{\text{LNA}} + n_2 + n_3$$

With an output SNR of:

$$\text{SNR} = \frac{A\text{Var}(x)}{\{A\text{VAR}(n) + \text{VAR}(n_{\text{LNA}}) + \text{VAR}(n_2) + \text{VAR}(n_3)\}}$$

which can be restated as:

$$\text{SNR} = \frac{\text{Var}(x)}{\left\{\text{VAR}(n) + \frac{\text{VAR}(n_{\text{LNA}})}{A} + \frac{\text{VAR}(n_2)}{A} + \frac{\text{VAR}(n_3)}{A}\right\}}$$

In the second arrangement, the output signal from the chain is:

$$A(x + n + n_2 + n_3) + n_{\text{LNA}}$$

And output SNR is:

$$\text{SNR} = \frac{A\text{Var}(x)}{\{A\text{Var}(n) + A\text{VAR}(n_2) + A\text{VAR}(n_3) + \text{VAR}(n_{\text{LNA}})\}}$$

which can be restated as:

$$\text{SNR} = \frac{\text{Var}(x)}{\left\{ \text{Var}(n) + \text{VAR}(n_2) + \text{VAR}(n_3) + \frac{\text{VAR}(n_{\text{LNA}})}{A} \right\}}$$

The chain SNR clearly suffers if the LNA is placed late in the chain. The reason should be immediately obvious. If the LNA is early in the chain, it will amplify the input noise as much as it amplifies the signal. However, it will make the signal much stronger relative to the noise added by downstream blocks. If the LNA is placed late in the chain, it will amplify all the noise that came before it, not affording the signal any additional noise resistance. In other words, placing the LNA early does not grant the signal any advantage over antenna noise, but it does provide a lot of protection relative to all noise that comes downstream in the RF.

Variable Gain Amplifier (VGA)

The VGA is sometimes lumped in with the LNA, but it is useful to consider it a separate block. The VGA interacts with the ADC to provide automatic gain control (AGC). AGC is discussed in more detail in Sect. 7.3, but the idea is that the maximum amplitude of the received signal will vary based on distance between the transmitter and receiver. The VGA allows the gain of the amplifier to be tuned, providing more gain when the transmitter is far and less gain when it is near. AGC is critical to proper receiver operation.

Oscillator

As in the transmitter, the receiver has a carrier sinusoid to mix with. This is used in down-conversion to obtain the baseband signal from the passband signal. The carrier frequency of the receiver must be the same as that of the transmitter to allow the signal to be down-converted perfectly. However, minute differences between any two oscillators are inevitable. This frequency offset was discussed in detail under the transmitter oscillator. The offset is tiny relative to the carrier frequency, often measured in a few dozen ppm, but it will devastate multicarrier systems if left untreated.

Mixer

Like the transmitter, this multiplies the sinusoid generated by the oscillator by the passband signal with the objective of bringing the signal down to baseband. The receiver mixer will generate a high-frequency component as well as the baseband component, so correlation is required before the baseband signal can be isolated.

Correlator

This is not an RF circuit. It works on the down-converted signal in the analog domain. The correlator extracts the baseband signal by rejecting the high-frequency component of the down-conversion. Ideally, this should be performed by a matched filter. Realistically, a low-pass filter or an integration over a symbol duration does the trick. The output of the correlator is the input to the ADC. We will discuss this further in Chap. 2.

1.4   Avoiding a Fight Over the Channel, Medium Access Control (MAC)

The MAC or medium access control layer is a sub-layer within the link layer. In this book, we care a lot about the MAC layer because it is the layer directly above the physical layer. Thus, we must understand how to interface with the MAC layer, and we should at least know what it is trying to do.

The MAC does the following:

- It monitors handshaking and the receipt of acknowledgments. Thus, it is the first layer at which we can confirm or deny that the physical layer has done a good job.
- It establishes addressing, ensuring that the terminal is communicating with the correct access point and that it is receiving packets that are addressed to it.
- It controls access to the channel, ensuring that the terminal does not try to communicate over a busy channel.

The MAC layer is our first encounter with "intelligence" at the receiver side. This allows the physical layer to be relatively "dumb" in that it does not need to worry about channel access scheduling, addressing, validation of data, or even communicating over the correct link. This frees the physical layer to do the one thing it is required to do: push the most amount of useful data through the channel.

The MAC layer receives a packet of data from the upper layers. This is called the MAC service data unit or MSDU. This packet of data is then encapsulated, and a MAC-specific header is added to it to form a MAC protocol data unit or MPDU. In 802.11n and higher standards (Sect. 2.6), the MAC layer can gather multiple MSDUs into a single MPDU based on channel conditions.

If the MAC layer is dealing with an MSDU, we call the packet that results a data packet. The MAC layer also sends two other types of packets: control and management. Neither type of packets contains any user data. Control packets communicate controls for the flow of packets between the transmitter and the receiver, and thus include acknowledgment (ACK), request to send (RTS), and clear to send (CTS) packets. All these packets are used in handshaking. Management packets are used to establish connections between terminals and the access point.

A major issue with radios, especially those that use an unregulated spectrum such as Wi-Fi, is that they share a single physical channel. Thus, multiple terminals will try to use the wireless channel at the same time, and some form of multiplexing or multiple access scheme must be used. In Chap. 12, we will see time-division and frequency-division multiple access schemes used in cellular communication. But Wi-Fi is ad hoc in nature; we cannot use a centralized medium access scheme; instead terminals must rely on stochastic properties to assign access to the channel.

Early ethernet networks provided an answer for this through CSMA-CD (carrier-sense multiple access with collision detection), which proceeds as follows:

- Listen to the channel to determine if someone else is already transmitting, if the channel is clearly sent.

Fig. 1.10 In wireless systems a collision may occur even if it is not detected by the transmitter. We cannot do collision detection. Transmitters 1 and 2 can detect each other's transmissions. Transmitter 3 is invisible to transmitters 1 and 2 but will cause a collision at all receivers

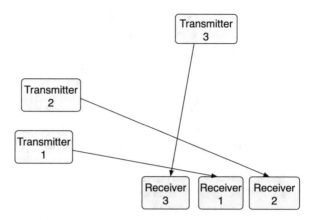

- While sending, continue to listen to the channel. If the channel continues to be clear till transmission ends, then we most probably succeeded.
- Because the transmitting terminal is listening, it will detect if any other terminal tries to transmit. This is a collision, and if one is detected:

  - The terminal stops transmitting.
  - It transmits a strong jamming signal that causes everyone else to stop transmitting as well.
  - Everyone quits and retries.

Listening for collisions in wireless networks is impossible. In Fig. 1.10, transmitters 1 and 2 are close to each other. If each tries to transmit at the same time, they will be able to detect each other, detecting the collision. But transmitter 3 is far from both other transmitters. If it communicates with receiver 3, then neither transmitter 1 or 2 will detect the signal. This is particularly true in modern standards where antennas are directional.

However, because all three receivers are close to each other, a collision will still happen at all receivers. Transmitters 1 and 2 are blind to this collision. So, nobody needs to listen, because even if they do, there will be plenty of collisions they will never detect. In simple terms, unlike in wired systems, the wireless transmitter will very often not see collisions that are visible at the wireless receiver.

The alternative to CSMA-CD in wireless systems is CSMA-CA where CA refers to collision avoidance. In collision avoidance nobody is ever listening while transmitting, and thus there will never be collision detection. Instead, we rely further on stochastic modeling to reduce the incidence of collisions:

- The transmitter listens to the channel, trying to detect any ongoing transmissions. If the channel is busy, the station continues listening.
- Once the transmitter detects a channel that is free for a minimum amount of time, it determines that the channel is clear; however, it will not transmit immediately.
- Once the channel is clear, the transmitter waits for a random period and then sends. Why wait? Because multiple stations might be listening simultaneously

waiting for the channel to clear. If they all start sending immediately, they will certainly collide. If each wait after the channel clears and the waiting periods are randomized, we can reduce the probability of collision.

- Once it starts transmitting, the transmitter will send the whole packet and will not stop in the middle no matter what. In fact, it will not even listen to the channel while transmitting. This is what sets it apart from CSMA-CD.
- Because the whole process is stochastic, there is no way to guarantee that a collision has not occurred. A collision could have occurred during transmission or due to the presence of blind nodes according to Fig. 1.10. The only way to ensure a collision has not occurred is to receive an acknowledgment from the receiver that it has received the packet correctly. The transmitter will thus wait for a certain amount of time to receive this acknowledgment; if the acknowledgment is received, the transmission was a success.
- If the acknowledgment signal times out, the transmitter will assume a collision has occurred. It will back off for a random period before sensing the channel again and repeating from the first step. This backoff reduces the chance that the exact same transmitters that collided will proceed to collide again because they are blind to each other

Wi-Fi MAC layers improve upon CSMA-CA by using RTS and CTS packets to establish a connection to the receiver. This prevents a situation where the transmitter is trying to send to a receiver that is not even visible (outage). This occurs when the transmitter and the receiver have moved into positions where there is no possible path for the signal. In terms that will become clear in Chap. 4, this occurs when there is deep fade or sustained shadowing. Such a situation will cause the transmitter to keep timing-out on acknowledgments, thus constantly assuming that collisions are occurring somewhere, where the problem is fundamentally with the quality of the channel between the transmitter and the receiver.

1.5   Making Ones and Zeros More Than Ones and Zeros, the Baseband Radio

The physical layer takes care of getting ones and zeros over the air from the transmitter to the receiver. This involves processing the bits so that they have properties that make them more suitable for traveling over a wireless channel. The physical layer is divided into two rough sections: the RF section and the baseband signal. The RF section was discussed in detail in Sect. 1.3.

The signal that travels over the air is called the passband signal. This is shown in Fig. 1.11 centered around the carrier frequency f_c. After down-conversion in the receiver and before up-conversion in the transmitter, the signal is in baseband. This is shown centered around DC (0 Hz) in Fig. 1.11. Assume that the passband signal has a bandwidth of 40 MHz; the baseband signal will have a bandwidth of 20 MHz because half of the band is consumed in negative frequencies.

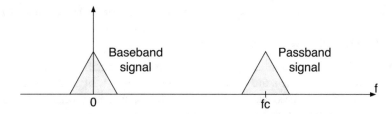

Fig. 1.11 Signal in passband and baseband

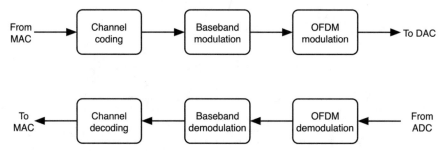

Fig. 1.12 Rough block diagram of the digital baseband

In fact, when we talk about "bandwidth," we must qualify whether we mean baseband or passband bandwidth. Unless otherwise stated, we will assume bandwidth always refers to passband bandwidth. One advantage of this convention is that every Hz of passband can accommodate 1 bps in BPSK. This will make our discussion of rates and capacity in Chap. 5 much easier.

So, what is the signal that we call "baseband" in Fig. 1.11? Is this the 1s and 0s that we receive in the MAC MPDU? That would be a terrible idea. Feeding the RF section a series of ones and zeros will result in a very inefficient transmission. There is something in the middle that transforms the ones and zeros from the MPDU into something that makes a little more sense. This "something" is the baseband radio, and it is the topic of most of this book.

Signals in the baseband are before the DAC at the transmitter and after the ADC at the receiver; thus, they are digital signals. This is very fortunate because it allows us to do a lot more processing than could ever be done in the analog domain. Figure 1.12 shows a rough view of the baseband radio. We will expand this view of the transmitter, and particularly the receiver for the rest of the book. But it is important to establish what the baseband radio, or for that matter any layer, is trying to do. There are only two things we care about:

- Getting as many bits across the channel in as little time as possible
- Getting this data across correctly

To do this, the baseband radio will do things that fall under one of the following categories:

- Adding error detection and correction abilities so that we do not get as many bit errors
- Pushing more bits within the same bandwidth

Wireless communication takes place over wireless channels. As we will see in Chap. 4, wireless channels are random and complicated. A significant part of the work of the wireless baseband receiver will thus be to detect what went wrong through the channel and try to reverse these effects.

Chapter 2
Basics of Baseband Radios

2.1 The Case for Complex Numbers, Signals in Baseband and Passband

In Chap. 1, we developed a simplified view of a typical radio. The physical signal travels through two domains: the baseband and the RF. In RF, the signal is carried on a high-frequency carrier. Raising the signal to higher frequency is necessary to allow it to be sent and received on antennas of reasonable size. However, most of the physical layer functions are performed in baseband. This includes modulation, detection, synchronization, equalization, and error correction. The reason is that the signal is digital in the baseband but analog in RF. The signal must be analog when it is about to be sent over the air; physical signals are all ultimately analog. However, the signal being digital in baseband allows us to perform a lot of complicated processing that would not be possible with analog electronics. In fact, digital radios are the main reason we can communicate the way we do today.

Assume that our baseband signal is x. In its simplest form, x is a binary sequence of zeros and ones. Assume $x = \{1, 0, 0, 1, 1, 0, 1, 0, 1\}$. The waveform of this baseband signal after it passes through a DAC is shown in Fig. 2.1. This is square waveform reminiscent of digital signals, but it is now a physical analog signal.

Thus, x is the vector of binary values that corresponds to the baseband signal in the digital domain, while $x(t)$ is the corresponding analog waveform of the digital signal when it is transformed through the ADC.

Note that x does *not* have to be a binary string. In Chap. 5 we will spend some time learning how and why the baseband signal can and should be transformed into something else before transmission. For example, there is no reason why the waveform in Fig. 2.1 could not be replaced by that in Fig. 2.2. In fact, the waveform in Fig. 2.2 has some desirable properties, not the least of which that it is zero mean.

The signal $x(t)$ shown in Fig. 2.2 corresponds to a baseband signal $x = \{0.5, -0.5, -0.5, 0.5, 0.5, -0.5, 0.5, -0.5, 0.5\}$. This shows that the baseband signal has now become a vector of decimal values instead of a binary vector.

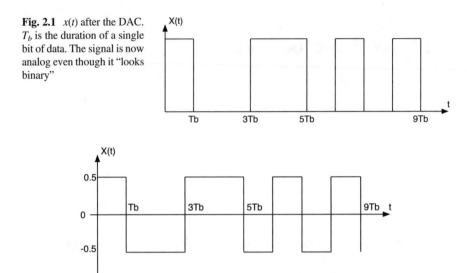

Fig. 2.1 $x(t)$ after the DAC. T_b is the duration of a single bit of data. The signal is now analog even though it "looks binary"

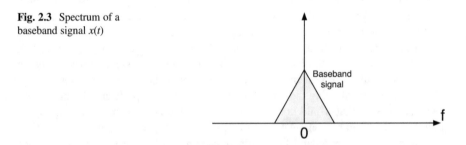

Fig. 2.2 The signal in Fig. 2.1 transformed to become zero mean. This can be done by subtracting 0.5 from the waveform in Fig. 2.1

Fig. 2.3 Spectrum of a baseband signal $x(t)$

Figure 2.3 is the spectrum of the analog version of the baseband signal $x(t)$ shown in Fig. 2.2. The spectrum is centered around zero frequency, as all baseband spectra should be. Note that the shape of the spectrum is just a conceptual representation; the actual spectrum will not be triangular.

How is this baseband signal transformed into an RF signal? We will see in Sect. 5.1 that this is done by multiplying the baseband signal by a sinusoid carrier of the required carrier frequency. Thus, at carrier frequency f_c, the passband signal is:

$$x_p(t) = x(t) \cos (2\pi f_c t)$$

Figure 2.4 is the spectrum of the RF signal $x_p(t)$. The "p" here stands for passband. Note that we are only showing the positive frequency component of the spectrum. Multiplying by the sinusoid moves the spectrum to a center frequency

Fig. 2.4 Spectrum of the passband signal $x_p(t)$. Only the positive frequency is shown

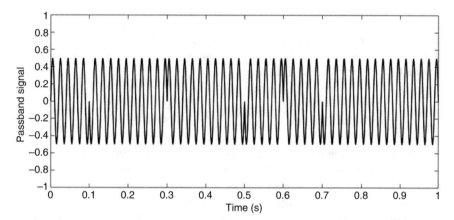

Fig. 2.5 Time-domain passband signal. The frequencies of both the carrier and the baseband signal are unrealistically low for the sake of illustration

around the carrier instead of DC. It also leads to doubling the bandwidth of the signal because the negative frequency components of the baseband now appear in the passband.

Figure 2.5 shows a time-domain representation of a passband signal. In general, each bit will include complete cycles of the carrier if the carrier frequency is a multiple of the inverse of bit duration. In other words, if $f_c = \frac{m}{T_b}$. In Chap. 5 we will consider this form of baseband modulation in more detail.

At the receiver, we should be able to lower the passband signal back to baseband before handing it over to the receiver baseband. This is done by the receiver RF, particularly, the mixer in the receiver RF. Thus, ignoring all channel effects, the receiver-side baseband signal is the low-frequency component of the output of the mixer.

The passband signal is:

$$x_p(t) = x(t)\cos\left(2\pi f_c t\right)$$

And the receiver mixer multiplies by the same sinusoid:

$$y(t) = x_p(t) \cos(2\pi f_c t) = x(t) \cos^2(2\pi f_c t)$$

This produces a DC-centered component, which is our desired baseband signal. It also produces a high-frequency component at double the carrier frequency:

$$y(t) = \frac{1}{2} \times x(t) \times \{1 + \cos(4\pi f_c t)\}$$

The high-frequency component can be removed by a low-pass filter or by integrating over a symbol duration. The reason we can get rid of the high-frequency component is that we integrate over an entire symbol duration and the system is designed so that a symbol duration contains whole cycles of the carrier. For the high-frequency component, this leads to integration over several entire periods of the sinusoid, reducing to null, specifically:

$$y(t) = \int_0^T \frac{1}{2} \times x(t) \times \{1 + \cos(4\pi f_c t)\} \, dt = \frac{x(t)}{2}$$

In Chap. 5, we will discover that multiplying by a cosine carrier or a sine carrier alone means we are missing an entire dimension we can use to transmit. This stems from the fact that sine and cosine functions are orthogonal to each other. Thus, when we try to correlate a cosine with a cosine, we end up with a value. When we try to correlate a cosine with a sine, we end up with null. The same is true for a sine. This is clear from the following four equations:

$$\int_0^T \sin(\omega t) \sin(\omega t) dt = T/2$$

$$\int_0^T \cos \omega t \cos \omega t dt = T/2$$

$$\int_0^T \cos(\omega t) \sin(\omega t) dt = 0$$

$$\int_0^T \sin(\omega t) \cos(\omega t) dt = 0$$

This occurs because the sine and cosine functions are orthogonal to each other. The two waves are offset by 90 degrees from each other. This means that we can send two independent signals x_i and x_q simultaneously in the same band if we mix one with a cosine and the other with a sine. This composite signal will effectively double our spectral efficiency. The composite signal $x(t)$ sent over the air is:

$$x(t) = x_i(t)\cos(\omega t) + x_q(t)\sin(\omega t)$$

To retrieve the original signal, we correlate with each of the cosine and the sine separately. This involves multiplying by each sinusoid and integrating over a whole symbol duration. Notice that in this section, a symbol duration is identical to bit duration. We will distinguish between the two in Chap. 5.

For the $x_i(t)$ component, called the in-phase component, correlation results in:

$$\int_0^T x(t)\cos\omega t\, dt = \int_0^T \big(x_i(t)\cos(\omega t) + x_q(t)\sin(\omega t)\big)\cos\omega t\, dt = x_i(t)/2$$

Similarly, for the sine component $x_q(t)$, called the quadrature phase component, we can retrieve the signal by:

$$\int_0^T x(t)\sin\omega t\, dt = \int_0^T \big(x_i(t)\cos(\omega t) + x_q(t)\sin(\omega t)\big)\sin\omega t\, dt = x_q(t)/2$$

In Chap. 5, we find that the magnitude of each of $x_i(t)$ and $x_q(t)$ may not be limited to -0.5 and 0.5. In fact, we can allow them to take any number of values, provided said values come from a finite set and can be represented using fixed-point numbers (Chap. 3). Thus, each of $x_i(t)$ and $x_q(t)$ can now be a sequence of decimal numbers. The sum of these two components is a sinusoid which can be characterized by its envelope (magnitude) and phase:

$$x(t) = x_i(t)\cos(\omega t) + x_q(t)\sin(\omega t)$$
$$= \sqrt{x_i^2(t) + x_q^2(t)}\ \sin\left(\omega t + a\tan\left(\frac{x_q(t)}{x_i(t)}\right)\right)$$

The envelope and phase will both vary from symbol to symbol, making sudden jumps. But how can we represent the baseband signal in such case? Representing it as two independent decimal number sets $x_i(t)$ and $x_q(t)$ is unwieldy and does not allow us to perform baseband digital processing.

Instead, the baseband signal is a set of complex numbers. The real part is the amplitudes of the in-phase components, while the imaginary part is the amplitudes of the quadrature component:

$$x_b(t) = x_i(t) + jx_q(t)$$

This makes sense since the phase difference between the real and imaginary parts of a complex number is 90 degrees, which is the same phase difference between sine and cosine. How do we express this in terms of baseband-passband transformation and back? Mixing at the transmitter involves multiplication by a complex phasor of the carrier:

$$x_{\mathrm{ph}}(t) = x_b^*(t)\, e^{j\omega t}$$

This creates a nonphysical phasor signal of the RF signal $x_{\mathrm{ph}}(t)$. It is important to notice that this signal has no real-world meaning and is just a mathematical representation. However, obtaining the real RF signal from this signal is as easy as taking the real part:

$$\begin{aligned} x_p(t) &= Re\left(x_{\mathrm{ph}}(t)\right) = Re\left(\left(x_i(t) - jx_q(t)\right) \times e^{j\omega t}\right) \\ &= x_i(t)\cos\ (\omega t) + x_q(t)\sin\ (\omega t) \end{aligned}$$

At the receiver, obtaining the baseband signal from the passband signal involves mixing again with the complex phasor of the carrier:

$$x_r(t) = \left(x_{\mathrm{ph}}(t)\, e^{-j\omega t}\right)^*$$

Much of the detail described in this section will only become clear once we discuss baseband modulation in Chap. 5, particularly quadrature-amplitude modulation. However, there are a few main takeaways that should already be clear:

- A baseband signal is represented as a vector of complex numbers.
- The real part of each complex number is the in-phase component of the signal and will end up being multiplied by cosine in the carrier.
- The imaginary part of the complex number is the quadrature component and will end up multiplied by sine in the carrier.
- The channel also has a representation in passband and a representation in baseband. In baseband, the channel coefficients are complex numbers the same way as the signal.
- Performing multiplications and additions in baseband can sometimes mimic physical phenomena happening in the passband.
- Each complex number will occupy a symbol duration in the physical signal. Again, understanding what a symbol is and how it is different from a bit is relegated to Chap. 5.

2.2 Noise and Signals, Random Variables, and Stochastic Processes, Mathematics

A random variable is a variable whose value cannot be known the same way a deterministic variable is known. Instead, we can only know the statistics and distribution of a random variable. To know a value for the random variable, we must observe the outcome of a random operation. Random variables are very important to wireless communications because most phenomena in the wireless system are stochastic in nature. For example, white noise is a Gaussian random process. Most forms of interference can be represented as random variables. Most channel effects are represented using random variables. Even signals sent by the user are random variables because the transmitted signal is not known at the receiver a priori.

The only way to fully characterize a random variable is to know its probability density function (pdf). The value of the pdf is the probability that the variable has a value in an infinitesimally small range. Thus, to find the probability that the random variable X lies in a range between $X1$ and $X2$, we integrate the probability density function in said range:

$$P(X1 < X < X2) = \int_{X1}^{X2} f_X(x)\, dx$$

And to guarantee normalization, the probability that the variable exists somewhere between minus infinity and infinity must be certain:

$$P(-\infty < X < \infty) = \int_{-\infty}^{\infty} f_X(x)\, dx = 1$$

The probability that the random variable takes a value less than or equal x is called the cumulative density function (CDF). It can be calculated from the pdf as:

$$F_X(x) = \int_{-\infty}^{x} f_X(x)\, dx$$

Next to the distribution, what we care about most are the statistics of the random variable. The most important statistic is the mean, which is also called the expectation. The mean is the weighted average of the random variable, giving us the average value we observe for the random variable:

$$E(X) = \overline{X} = \int_{-\infty}^{\infty} x f_X(x)\, dx$$

We also care about the moments of the random variable. Moments are the expected values of higher powers of X. The most important moment is the second moment, which is also called the variance, and is calculated as:

$$E(X^2) = \text{Var}\,(X) = \int_{-\infty}^{\infty} x^2 f_X(x)\, dx$$

We will discover that for a lot of signals, the second moment is also the power. When dealing with signals that contain a DC component, signal power can be obtained by calculating the second moment around the mean. This "centered" second moment is also equal to the variance if the signal is zero mean, which is the case for a lot of noise and signal phenomena. This "centered" second moment is calculated as:

$$E\left((x - \bar{x})^2\right) = E\left(x^2 - 2x\bar{x} + \bar{x}^2\right) = E\left(x^2\right) - \bar{x}^2$$

And its root is called the standard deviation, representing the spread of the signal around its mean:

$$\text{std dev} = \sqrt{E\left((X - \bar{X})^2\right)} = \sqrt{E\left(X^2\right) - \bar{X}^2}$$

One source of confusion when talking about random variables is the difference and relation between random variables and stochastic processes. A stochastic process is a group of random variables, usually indexed so that they form a function together. This is often a function in time, so that each instant in time is itself a random variable.

A prototypical example of stochastic processes is noise. Noise is a stochastic process indexed in time so that each time instant is itself a random variable. This is shown in Fig. 2.6. Here, we are generating the same noise process multiple times in parallel. Each graph is thus a noise signal. The y-axis is in arbitrary units; for noise this would normally be a scaled version of potential for example mV. The x-axis is often time. If we sample each of the curves at a specific time, each of them will have a different value. In fact, all these values at the specific time instant form a random variable with its own distribution.

To express this mathematically, consider each curve to be:

$$X(t)$$

This curve is not deterministic, nor is it a random variable. Each time we generate a curve, it will look completely different and will be random. It is not a random variable because it is not a single observation, rather an entire curve of observations.

However, if we take an observation at a specific time instant:

$$X(t_1)$$

And then repeat this observation many times by running the curve multiple times; then, this observation itself becomes a random variable. Thus, $X(t)$ is a random process, while $X(t_1)$ is a random variable. We can apply all the rules and observations we stated for random variables for the time instance of the random process.

If the probability density function of the distribution is the same for all time instances, the process is called strict-sense stationary. This does not mean that we will observe the same values for all time instances or even that we will reproduce the same values if we regenerate the process; it just means that the distribution that produces such values will be the same. In other words, the following two probability density functions are identical for any values of time:

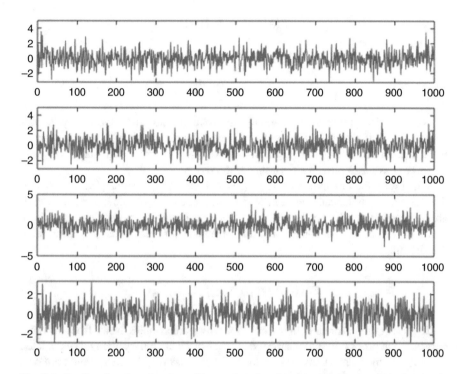

Fig. 2.6 Family of random functions. The *y*-axis is in arbitrary units. The *x*-axis is the sample number, which in continuous domain would be time in an appropriate unit

$$f_{X_1}(X(t_1)) = f_{X_2}(X(t_2))$$

Stationarity is a very important concept because it makes the mathematics a lot easier to handle. But it is also important because a lot of natural phenomena exhibit stationary properties. We are particularly interested in white noise, which is strictly stationary in most steady-state systems.

There are less strict definitions of stationarity that have to do with the statistics at each time instant. A first-order stationary process is a process where the mean is not time-varying, but the overall probability density function may not be constant with time. A second-order stationary process will have a variance and mean that are constant with time, but again not necessarily the probability density function. We can keep calculating higher-order moments to find higher-order stationarity. The ultimate form of stationarity is always strict-sense stationarity where the density functions are time-invariant. Naturally if the pdf is not time-varying, the moments will also be time-invariant.

As stated above, the random process we are most interested in is white noise. AWGN (Sect. 4.2) has a Gaussian distribution. If we assume the noise process is first-order stationary, then:

$$E(X(t_1)) = E(X(t_2))$$

If we assume the noise process is second-order stationary, then:

$$E\big(X^2(t_1)\big) = E\big(X^2(t_2)\big)$$

For Gaussian processes, it is enough to prove second-order stationarity to conclude strict stationarity. This is because a normal distribution is fully described by its mean and variance (Sect. 4.2). Thus, if we prove second-order stationarity for white noise, we have also proven strict-sense stationarity.

When we calculate moments above, what samples are we using in the calculation? A correct understanding of how random processes translate into random variables immediately answers this question. The samples used are samples drawn from different curves at the same time instance. Thus, to calculate the mean at a certain time:

$$E(X(t_1)) = \int_{-\infty}^{\infty} x(t_1) f_X(x(t_1)) \, dx$$

The integration is done with respect to x, which are samples drawn from different curves. If the process is stationary, or even first-order stationary, then this calculated mean is also the mean at any time instance. In other words, the mean would not be a function of t_1.

But how do we repeat a random function the way we did in Fig. 2.6 to generate a family of functions? In practice this is impossible. Let us consider a practical case where the random process is thermal noise in a resistor. If we start observing the noise in the specific resistor R1 at time $t = 0$ and end at time $t = 1s$, this gives us one curve. How do we get the other curves in Fig. 2.6?

We can measure another one second record of noise on R1, but this would not actually be a parallel curve as in Fig. 2.6. This would be a curve that starts at $t = 1s$ and ends at $t = 2s$. If the process is strictly stationary, this would be fine, but in all other cases the second curve would just be a continuation of the first curve rather than a "second" parallel curve.

We could also measure noise on a second resistor with the same value of resistance. The curve measured on resistor R2 can be recorded at the same times as that recorded on resistor R1. But this would still not be the situation in Fig. 2.6, because we would have two curves for two independent resistors, each of which represent an independent random process.

Thus, we cannot generate multiple simultaneous noise waveforms at the same antenna. The main problem with this is that it means there is no practical (experimental) way to calculate mean or variance of any observed random process. For the rest of this book, we will use algorithms that depend on knowledge of noise and signal power. This implicitly assumes the receiver can practically calculate moments of both. So, we are in trouble.

We have access to only a single noise waveform, which represents our only possible observation of the random process. What if we calculate the average of the random variable using the *time* integration of this waveform?

$$E(X) = \frac{1}{T} \times \int_0^T x(t) \, dt$$

where T is the window over which we are calculating the average. We can also attempt to calculate the second moment using the same approach:

$$E(X^2) = \frac{1}{T} \times \int_0^T x^2(t) \, dt$$

So, what did we just calculate? The mean and variance above are not always the mean and variance of random processes as discussed earlier. The operative word here is "always" because this means that these statistics are sometimes valid.

A necessary but not sufficient condition for the above moments to match the ensemble moments calculated earlier is that the process be second-order stationary. This is obvious because we are calculating a single value for mean and a single value for variance. This means that we are assuming that neither are a function of time, so we are implicitly assuming stationarity.

But more importantly, we are assuming that the moments obtained from a time average are the same as the moments obtained from an ensemble average, meaning that samples taken in time have the same moments as samples taken across different curves. A process that has this property is called ergodic.

White noise is both ergodic and stationary. This reduces the complexity of the mathematics involved and allows us to use time averages instead of ensemble averages. Both properties also apply for most channel effects (Chap. 4). We will also assume signals and interference are stationary and ergodic. This is much more of an approximation than for noise because signals often have time-varying properties.

Throughout this book we will often talk about independence, correlation, and covariance. These words are sometimes considered interchangeable, but they each mean something very specific. The most well-defined relationship between two random variables is their covariance. Covariance is a measure of the linear relationship between the two. High covariance means the two variables are likely to move together in the same direction. Thus, a high value of covariance means that knowing information about the behavior of one variable gives information about the value of the other. This does not mean certainty about either variable, because both are still stochastic. It just means some degree of mutual information. The covariance between two real random variables X and Y is defined as:

$$\text{Cov}(X, Y) = E\{(X - \overline{X})(Y - \overline{Y})\}$$

This can be simplified as:

$$\mathrm{Cov}\,(X,Y) = E(XY) - \overline{Y}E(X) - \overline{X}E(Y) + \overline{X}\,\overline{Y} = E(XY) - \overline{X}\,\overline{Y}$$

The word correlation is sometimes used interchangeably with covariance. Colloquially the two measure the same thing: the degree of dependence of the two random variables on each other. However, from the point of view of statistics, the two measure different things. The most used metric of correlation is the correlation coefficient, where the correlation coefficient between variables X and Y is defined as:

$$\rho_{X,Y} = \frac{\mathrm{Cov}\,(X,Y)}{\sigma_X \sigma_Y} = \frac{E(XY) - \overline{X}\,\overline{Y}}{\left\{ \sqrt{E(X^2) - \overline{X}^2} \sqrt{E(Y^2) - \overline{Y}^2} \right\}}$$

Covariance and correlation seem to be measuring the same thing; after all the only difference is the division by the product of the standard deviations of the two variables. However, this division is itself very important. The division leads to normalization of the covariance so that the correlation coefficient is now bound between $+1$ and -1. A correlation factor of $+1$ indicates the two variables are perfectly positively correlated so that an increase in one will certainly lead to a positive increase in another. A correlation factor of -1 indicates perfect negative correlation so that an increase in one leads to a certain decrease in the other.

So, what does a zero-correlation coefficient indicate? It indicates two random variables that are perfectly decorrelated. So, knowing information about one leads to knowing nothing about the other. But one should always be careful with terminology because two random variables can be decorrelated without being independent. Two random variables are independent when their joint probability function is the product of their individual probability density functions:

$$f_{X,Y}(x,y) = f_X(x)f_Y(y)$$

Knowing that the two variables are independent certainly indicates they are decorrelated. But the two variables can be decorrelated even if they come from the same process, which would make them dependent on each other. The clearest example of this is noise processes where each noise sample is completely decorrelated from all others (Sect. 4.2), but where they are all the product of the same random process and are not independent.

We will often refer to unnormalized covariance as correlation in this book. This is strictly not true but is not a catastrophic mistake. However, we will never call decorrelated random variables independent unless they are truly independent.

One special form of covariance we must consider is autocovariance. This is a covariance (or correlation if we are going to be colloquial about it) that affects a single random process. We correlate two samples at different times from the same process; thus:

$$R(X(t_1), X(t_2)) = \mathrm{Cov}(X(t_1), X(t_2))$$

If the process is strict-sense stationary, we do not care about the specific times $t1$ and $t2$; instead we care about the difference (interval) between the two time instances. This is because autocovariance is measuring how much relation there is between a sample and a sample that comes later. In case of stationarity, this degree of relation will only depend on the time difference not on the specific moments in time:

$$R(X(t_1), X(t_2)) = R(t_1 - t_2)$$

For real signals, this autocorrelation function will be even. This stems from the fact that covariance is a commutative operation, where the order of the variables X and Y does not matter. Thus, my relation to a sample that comes a time T later is the same as the relation of a sample that came T earlier to me.

For most of this section, we have dealt with continuous random variables. Indeed, most of our noise processes will be in continuous time, and when dealing with channel modeling, for example, we think about continuous time. However, the digital baseband processor is in discrete time. So, as soon as the signals and noise hit the ADC at the receiver, they become discrete-time random variables.

Dealing with discrete random variables requires a very small adjustment to the above equations. For the most part, we only replace integrations with summations. In fact, understanding discrete random variables is often easier than understanding continuous random variables. The simplest and first lessons in probability will deal with discrete experiments.

A discrete random variable has a probability function $P_X(x)$ instead of a probability density function. The probability function simply indicates the probability that one of the discrete possible values will be observed and is thus easier to understand than the probability density function. A cumulative distribution function can also be derived from the probability function as:

$$F(X) = \sum_{x=-\infty}^{X} P_X(x)$$

The mean can be calculated as:

$$\overline{X} = \sum_{x=-\infty}^{\infty} x P_X(x)$$

And the second moment is:

$$E(X^2) = \sum_{x=-\infty}^{\infty} x^2 P_X(x)$$

And in general, the mean of a function $g(x)$ of random variable X is:

$$E(g(x)) = \sum_{x=-\infty}^{\infty} g(x)P_X(x)$$

Covariance over a window of length L between two discrete signals X and Y can be calculated as:

$$\mathrm{Cov}(x, y) = \sum_{n=0}^{L-1} x(n)y^*(n)$$

Notice the conjugation on the second term. We have assumed the two signals are complex and zero mean. This works well with baseband signals and noise as we will see in upcoming chapters. Normalized autocorrelation between two samples in the same stationary process T seconds apart over a window of L samples is:

$$R(T) = \frac{1}{L} \times \sum_{n=0}^{L} x(n)x^*(n+T)$$

2.3   Receiver and Transmitter Architecture

Figure 2.7 shows a simplified view of the baseband physical layer transmitter and receiver in 802.11n. This is one of the IEEE standards used for Wi-Fi (Sect. 2.6). It has been mostly supplanted by 802.11ac. But 802.11n has a couple of advantages, which is why we will focus on it in this book. First, the 802.11n physical layer is easy to describe. But second, 802.11n is not oversimplified and includes all aspects representative of newer and older standards.

Figure 2.7 leaves a lot of details out. The figure is not too bad in its depiction of the transmitter, although we will expand on it in upcoming chapters. But the receiver is particularly sold short by the figure. To understand the real structure of the receiver, we must develop a good model for the channel. Only by Sect. 7.15 will we have a full image of the baseband receiver.

The 802.11n standard allows the use of either 20 MHz or 40 MHz bandwidth depending on regulatory restrictions in the geographical area of the user. At the transmitter, the antenna is sending out symbols at a rate of either 20MSps or 40MSps depending on bandwidth. The samples radiated out of the antenna contain data from the physical layer. This data is provided as an input to the physical layer from higher layers. In Sect. 1.4 we specifically described how the input to the physical layer is the MPDU as output by the MAC layer. The MPDU is pure binary data (zeros and ones). The output of the physical layer from the antenna is not binary; in fact it could be significantly different. What the baseband does to its input binary stream, and why, is the subject of this book.

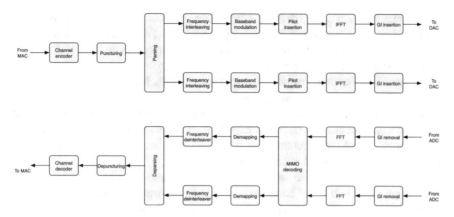

Fig. 2.7 Receiver and transmitter chains in 802.11n

The physical baseband does four things according to Fig. 2.7:

- Channel coding and decoding (Chap. 6). Channel coding adds redundancy to the message at the transmitter. This allows the receiver to perform error detection and correction on the message. Channel coding is essential for practical transmission through fading channels (Chap. 4).
- OFDM (Chap. 7). OFDM allows the system to transmit on very wideband channels. Frequency selectivity in the channel requires special preprocessing of the signal if prohibitive equalizers are to be avoided.
- Baseband modulation (Chap. 5). Baseband modulation squeezes multiple bits into a single symbol. This increases the spectral efficiency of the system, the rate that it can push in every Hz of bandwidth.
- Parsing, deparsing, and MIMO decoding (Chaps. 9 and 10). These are operations through which transmission is spread over multiple antennas instead of one antenna.

2.4 Why MIMO-OFDM?

In Sect. 2.6, we will note a progressive move toward the use of multiple antennas and wider bandwidth as we move to modern communication standards. This makes sense because new standards are released with only one aim in mind: pushing more good bits from the transmitter to the receiver (Sect. 1.5). This translates into a lot of good things at the application and commercial level. Higher throughput, for example, has changed the main method of video entertainment consumption into streaming. Local data storage is also being supplanted by cloud storage. In the future, high bitrate may also impact computing, with most user terminals being cheap weak frontends and most processing happening remotely. There might come a point in the future where a standard provides throughput that users perceive as "too much," but so far, if you provide more throughput, the applications that need it will come.

In Chap. 5, we will find that the fundamental limit on throughput is bandwidth. More bandwidth means proportionately more bitrate and at no cost provided you can overcome the channel selectivity. But bandwidth is limited, so we must find ways to squeeze more bits into the same bandwidth.

Higher-order baseband modulation does exactly this. It pushes more bits into the same bandwidth, thus increasing spectral efficiency. However, this will always come at the expense of higher BER. A denser constellation will always lead to more bit errors. We can use channel coding (Chap. 6) to correct such errors. So maybe we can keep using higher-order modulation combined with ever more capable channel coding to increase spectral efficiency indefinitely.

However, in Sect. 5.8 we see that there is a fundamental limit on the number of bits we can push in a Hz of bandwidth regardless of how great our error correction algorithm is. This is known as channel capacity. Thus, the ability of a combination of baseband modulation and forward error correction (channel coding) to increase bitrate is fundamentally limited.

So, we must find other ways to increase bitrate. One ingenious way to do this is to use multiple antennas. When multiple antennas are used at both the transmitter and the receiver, this is often called MIMO. True MIMO uses spatial multiplexing to increase channel capacity, thus allowing us to use the same bandwidth while transmitting a lot more data than a single antenna would have allowed (Sect. 9.5).

Channels that are good for MIMO have distinct spatial streams from each transmitter antenna to each receiver antenna (Sect. 9.5). This leads to multipath channels, which are by necessity frequency-selective (Sect. 4.4). These wideband frequency-selective channels are prohibitively complex to equalize. Thus, we use OFDM to relieve these equalization requirements (Sect. 7.1). Thus, we will usually find both MIMO and OFDM associated with cutting-edge Wi-Fi and cellular standards.

2.5 Packet-Based Communication

Figure 2.8 shows a very simplified view of the layer model discussed in Sect. 1.2. In this case, we only care about three layers: the application layer, the MAC layer, and the PHY layer. The PHY layer is the subject of most of this book. The application layer is interesting because it is the origin of the raw data being sent. The MAC layer is of interest because it is the layer directly above the PHY layer. It is the immediate source of binary data for the physical layer at the transmitter side, and the immediate destination for decoded data at the receiver side.

Based on the philosophy of the OSI layer model in Sect. 1.2, layers should be agnostic about functions, formatting, and inner workings of all other layers. Thus, at the transmitter the application layer prepares its data. It alone is aware of what this data means, and it alone can turn it into something useful. The application layer prepares a packet of data "payload" of a certain length and attaches an application layer header to this data. This header carries application layer-specific information

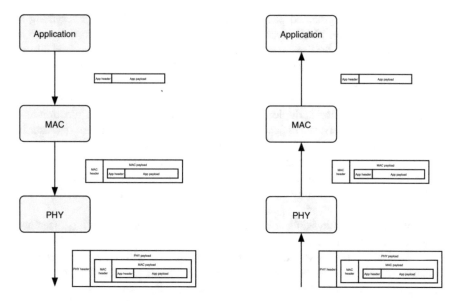

Fig. 2.8 Layer-based packetization

that might be useful for the application layer on the receiver side. The application layer then hands this packet down to the MAC layer.

The MAC layer receives the application layer packet blindly. In other words, it cannot and does not care to distinguish which parts are the application layer payload and which are the header. Instead, it deals with the entire packet as raw binary data. The MAC layer makes decisions about how this data can be sent. This will be based on channel conditions and the length of the packet. Once done, the MAC layer will attach its own header to its own payload. This header is useful only to the MAC layer on the receiver side since it tells it how the data is segmented and sent. The payload for the MAC packet is the entire packet from the application layer.

The MAC layer then hands over its packet to the PHY layer. The PHY layer handles how to effectively transmit raw binary data over the air. To the PHY layer, the entire MAC packet is the payload. The PHY layer adds its own header to this payload to form the final packet that will be sent over the air. The header in PHY will be discussed in detail, including specific training patterns and field formats in Sects. 11.2 and 11.3. However, as with the headers of other layers, the header of the PHY layer is only valuable to the receiver PHY layer. PHY headers perform two main functions: they give information to the receiver about choices made at the transmitter; and they help the transmitter figure out channel state information.

At the receiver side, the PHY packet is handled by the receiver PHY layer. The header is used to estimate and invert the channel. The receiver will then try to extract the original binary data from the PHY payload. If it succeeds, the PHY layer will hand this binary payload to the MAC layer. The MAC layer will recognize this as a MAC packet; it will look at the header and figure out if this is a packet of interest, if it is part of a larger segmented packet, and if it is a valid packet. If everything is according to plan, the MAC layer will hand the MAC payload up to the receiver application layer. The receiver application layer will divide the MAC payload into an application header and an application payload, finally being able to use the data to extract useful information.

It is important to point out that in almost all modern wireless communication systems, communication is duplex. This means that any platform can function as a transmitter and a receiver. This can sometimes be symmetric but is not necessarily so. By symmetry we mean that if platform A is transmitting to platform B, there would be no difference to the case where platform B is transmitting to platform A. By no difference we mean no difference in the packet structure, modulation, or transmitter/receiver structures. This is the case, for example, with Wi-Fi communication.

However, cellular systems are asymmetric. There is a substantial difference between the case where the base station is talking to the phone (downlink) and when the phone is talking to the base station (uplink). This will be discussed in detail in Chap. 12. It is also worth noting that the concept of half versus full duplex has no bearing on modern systems. All modern wireless systems act as both transmitters and receivers, but rarely would they do the two simultaneously.

The packet model discussed above leads to an interesting communication paradigm: packet-based communication. Packet-based communication is the method of communication used by all Wi-Fi systems. We will discuss how cellular communication diverts from packet-based communication in detail in Chap. 12.

In packet-based communication (Fig. 2.9), the transmitter will send data to the receiver in the form of time-limited packets. These packets correspond directly to the packets generated by the PHY layer, which is why we will spend time understanding the structure of PHY packets in Sect. 11.1.

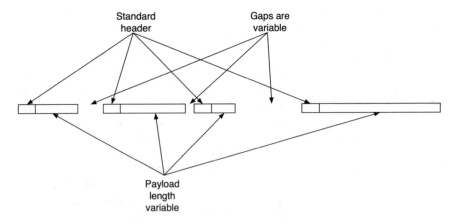

Fig. 2.9 Packet communication in a Wi-Fi system

The following applies to packet-based communication:

- Packets are of variable size. The transmitter is not bound to any length. The receiver cannot expect any length. Most standards specify a minimum length for a PHY packet, but there is no fixed-size limitation. The transmitter must inform the receiver of the length of the current packet in the PHY header; otherwise, the receiver will not know when to stop decoding.
- Times of arrivals and gaps between packets are variable, unregulated, and random. The transmitter will transmit when two things happen: it needs to transmit, and it finds the channel clear. The receiver does not know when packets will arrive. Most standards institute a minimum duration between two consecutive packets, but there is no maximum. This means the receiver must always be on the lookout for arriving packets.
- If the receiver detects an error in a packet, the MAC layer can ask the transmitter MAC layer to retransmit the packet. The MAC layer cannot ask for retransmission of part of the packet or a specific bit. Transmission is all or nothing.

Packet-based communication has a huge impact on many practical aspects of radio design. For example, in Sect. 2.7, we will discover that the most important performance metric that impacts our QoS is the packet error rate (PER) rather than BER; this is a direct corollary of the third bullet point above. In Sect. 11.8, we will spend some time understanding how the receiver detects that a packet is being sent over the air. This is a direct result of the second point above. In Sect. 4.6 we will look at Doppler spread, and in Chap. 9 we will learn how to deal with it in quasi-static channels. Both are direct results of the first and second points above.

2.6   The Quest for More Bits, Evolution of Wi-Fi

Wi-Fi is now ubiquitous. It represents our entry point to the Internet whenever we are in a "place." When moving between "places," we use a cellular network to get to the Internet. Thus, Wi-Fi is seen as the cheap, fast, stable, and reliable way to get to the Internet. But this is a very recent conception of Wi-Fi. As late as the early 2000s, wired Internet was seen as the go-to method for getting on the Internet. Wi-Fi was seen as a temporary, slow, and unreliable connection for when you want mobility.

There was a quick and well-defined migration in perceptions where Wi-Fi became the ole reliable and cellular networks became associated with mobility and incidental use. This is again changing as cellular communication systems become more available, economic, and reliable. This is especially true in developing economies where the deployment of high-quality cellular networks is more economic than laying down hundreds of kilometers of fiber-optic cables.

But it is important to understand where Wi-Fi started, how it developed, and why it became such a reliable method of connecting to the Internet. The main difference between Wi-Fi and wired ethernet connections is in the kind of channel each deals

Table 2.1 The evolution of Wi-Fi standards

Standard	Year	Band (GHz)	Bandwidth (MHz)	Maximum rate (Mbps)	MIMO maximum	Modulation
802.11	1997	2.4	22	2	No	DSSS/FH
802.11a	1999	5	20	54	No	OFDM
802.11b	1999	2.4	22	11	No	DSSS
802.11g	2003	2.4	20	54	No	OFDM
802.11n	2009	2.4/5	20/40	600	4x4	OFDM
802.11ad	2012	60	2160	6800	No	OFDM
802.11ac	2013	5	20–160	3460	8x8	OFDM
802.11af	2014	0.054-0.79	8	560	4x4	OFDM
802.11ax	2019	2.4/5	20–160	10500	8x8	OFDM

with. Wired connections deal with the channel imposed by a cable, which is bad. But it is nothing close to the unpredictability, challenge, and madness of a wireless channel (Chap. 4). Successive generations of Wi-Fi standards have overcome this fundamental challenge by sheer force. We kept pushing more bits through the channel and using deeper error correction. BER in wireless networks can be relatively high, but the overall performance of the connection is no longer observably different from wired connections in most situations.

But it all had to start somewhere. Wi-Fi is defined by the IEEE 802.11 standard. Indices of the standard indicate different releases of Wi-Fi (Table 2.1). But *the* original IEEE 802.11 was called just that, and it was first released in 1997. This standard was kind of strange. Its maximum bitrate was 2Mbps. It did not use OFDM, instead using DSSS or frequency hopping. This is substantially different from current Wi-Fi standards. However, it set the stage for what came later. It also instituted the use of forward error correction, packet structures, and the use of the ISM band for Wi-Fi, although it was only limited to the 2.4GHz band.

In 1999, the original 802.11 standard was "clarified." But simultaneously the 802.11a standard was released. 802.11a is very important because it uses OFDM. It laid the groundwork for *all* future OFDM Wi-Fi standards including the now popular 802.11n and 802.11ac standards. It uses QAM modulation and convolutional FEC, which are also features in all modern Wi-Fi.

However, there was a curious decision made for 802.11a. When using the ISM band, we have two choices: the 2.4GHz band and the 5GHz band. The 2.4GHz band has very strong interference because many standards use it. So, 802.11a uses the much quieter 5GHz band. However, because of the higher frequency, 802.11a waves are more absorbed by obstacles than corresponding 802.11 waves. This means that 802.11a radios suddenly had a much lower range. Therefore, Wi-Fi networks in the early 2000s could barely extend beyond the boundaries of a single room.

In 2000, 802.11b was released. This standard follows the same modulation technique of the original 802.11, namely, DSSS and FH. It uses the 2.4-GHz band and allows much higher data rates than the original 802.11 standard. Its maximum bitrate is around 11Mbps, which means it is inferior to 802.11a. However, working in the 2.4GHz band allows devices using the standard to have a much higher range. Despite the inferior modulation and throughput, adoption of 802.11b was fast, especially in office environments.

In 2003, the 802.11g standard was released. 802.11g is essentially 802.11a, except in the 2.4-GHz band. Thus OFDM, QAM, and FEC are used. This allows the standard to achieve throughput as high as 802.11a while preserving the range of 802.11b. Adoption was fast due to the urgent need for higher throughput than 802.11b.

The 802.11n standard was released in 2009. This standard includes all the basic elements of a Wi-Fi physical layer we have come to expect. It represents a proto-typical physical layer, and thus will be the basis of most of this book. 802.11n provides legacy support to 802.11a platforms and shares a lot of the legacy physical layer structure and packet format. The standard allows higher coding rates than 802.11a and provides optional higher-order QAM constellations. It also offers an optional 40 MHz bandwidth channel, doubling the spectrum of 802.11a. However, the major addition of the standard is allowing the use of multiple antennas (MIMO). 802.11n theoretically allowed throughput up to 600Mbps. Although adoption was fast, most platforms hardly approached the maximum allowable throughput, and most did not actually use true MIMO.

After 802.11n, Wi-Fi standards split into three paths:

- Standards that aim to develop on the ISM (2.4GHz/5GHz) OFDM paradigm.
- Standards that describe communication in the 60-GHz band. These are extremely high bitrate, extremely short-range standards.
- Standards that aim to use the decaying low-frequency spectrum of television broadcast to intelligently assign large bands to Wi-Fi.

In 2012, the 802.11ad standard was released. This is a very divergent standard describing possible implementations in the 60-GHz band. Adoption has been slow. Due to positioning at a very high frequency, the available spectrum is large and achievable bitrates can exceed 1Gbps. However, range is extremely short, and the signal cannot travel through any obstacles.

The 802.11 ac standard was released in 2013. It is the standard currently used by most Wi-Fi devices. It increases the possible bitrate to 1.3Gbps by improving all aspects of the physical layer of 802.11n. There are bandwidth options up to 160 MHz, up to 8 spatial streams in MIMO, and even denser QAM constellations. But the most fundamental change was the cross-layer addition of multiuser MIMO. In MU-MIMO, all platforms in the network intelligently act together to allow beamforming, reducing interference and improving access to the channel (Sect. 12.7).

802.11af was a standard released in 2014. This was the beginning of a series of standards that do something very interesting. It uses bands traditionally used by broadcast TV for VHF and UHF transmission. Because broadcast TV is no longer as ubiquitous as it once was, this means some spectrum is available at these frequencies. The long wavelength means that signals have incredible reach relative to the ISM band. The physical layer will typically resemble that of 802.11n/ac, but protocols are added to ensure medium access does not violate regulatory standards (Table 2.1).

Currently, the most important Wi-Fi standard is 802.11ax. This will be the successor to 802.11ac. It will support everything that 802.11ac supported, including MU-MIMO. However, there may be some fundamental changes to the way spectrum is allocated. For example, some proposals suggest an OFDMA approach in the downlink like that used for 4G LTE (Chap. 12).

2.7 Measuring Goodness, Quality Metrics: SNR, BER, Throughput, Energy, and Goodput

How do we decide if a radio is "good"? Words like good can be very troubling because they are subjective. Even if we find a well-defined objective metric to compare radios, metrics can be deceiving. In this section we will try to find a metric or a bunch of metrics from which we can understand how "good" a radio is.

Electronics are ultimately used in real-life platforms by real-life users. So, from the perspective of a user, what makes a cell phone good? There are only two things that I care about in a phone:

- How fast it gets the data across
- How often I need to recharge

We will shortly see that these two metrics are intrinsically related. But first we define what they mean and why we care about them.

The best way to characterize "how fast data gets across" is at the highest possible level, specifically at the level of quality of service observed by the user. For example, if I am streaming video, what is the highest resolution and framerate that I can stream without the need to buffer. When I attach a PDF to the email, how fast is it attached. When I click a link to find out why "reason number six will surprise" me, how long would it take for me to be disappointed. This is all that matters. How we translate these user experiences in terms of scientific jargon is secondary.

Battery life is also critical. If I am transmitting very fast but I need to recharge every couple of hours, then we are in trouble. The phone will not sell, and the radio is a failure, no matter how well it streams 8K video.

We need to characterize "speed" in terms we can measure. Our first instinct is to say that speed is throughput. And our first instinct is right. Throughput is the rate at which bits get across the channel from the transmitter to the receiver. Throughput for a radio is measured in Mbps, and speeds of up to Gbps are not uncommon in modern standards.

In Sect. 11.1, we will learn that in all communication systems, not all bits transmitted are data payload. In fact, a significant minority of bits will be dedicated to preambles and pilots used to allow the receiver to deal with the channel. But for

Fig. 2.10 A received "1" mistaken for a "0," in other words a bit error

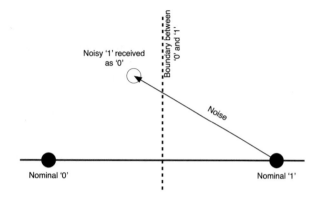

now, let us assume that every bit that manages to successfully get across is a data bit that is useful to the receiver at the application layer.

But a keyword above is "successfully." How do we know that the bit managed to get across successfully? In Chap. 4 we will study the effects of the channel on the signal, not the least of which is noise. Noise is added to the signal. Digital signals have an advantage over analog signals: noise margins. A logic "0" is not defined by a specific voltage level at either the transmitter or the receiver. Instead, a range of input voltages are recognized as logic "0," and a range of voltages are recognized as logic "1." This allows noise to be added to either a "0" or "1" signal while the signal is still correctly recognized at the receiver.

Figure 2.10 shows in-phase voltage levels on the *x*-axis. The two solid circles represent nominal voltage level for logic "1" and logic "0." If noise is added or subtracted from the solid circles on the way to the receiver, the received signal will move. There must be a boundary somewhere that separates the range of voltages we perceive as "0" from that we perceive as "1." With symmetric noise, this boundary is best set in the middle as shown on the dotted line in Fig. 2.10.

Thus, any received signal that lies left of the dotted line is interpreted as a "0" by the receiver. Any signal to the right is interpreted as a "1." Figure 2.10 shows a case where so much noise is added that it pushes a sent "1" left of the dotted line, shown as an empty circle in Fig. 2.10. This will be interpreted as a "0" by the receiver and a bit error occurs.

Thus, there is another aspect of communication that we care about. We not only care about how fast we are pushing bits into the channel but also how many of said bits arrive correctly. In Chap. 5, we will discover that there is a contrarian relation between the two. The faster we push bits through the channel, the more bits arrive incorrectly.

This quality of arrivals, the correctness or lack thereof the arriving bits, is characterized by bit error rate (BER). BER is defined as the ratio of incorrect bits at the receiver to total received bits. So, what level of BER is acceptable? BER is a function of the quality of service (QoS) required by the application and is not very easy to universally define. For example, real-time video streaming will require a lower BER than off-line audio transmission to give an acceptable experience to the user. However, there are few practical applications where BER above 1×10^{-6} is acceptable.

Knowing this, we can start improving our understanding of "speed." It cannot be raw throughput. The user does not really care about how many bits we are shoving into the channel. She only cares about the rate of useful (correct) bits she receives. These are the bits that will stream the video or recreate the email. Error bits must be retransmitted, which is a waste. So, combining BER and throughput, we define a new metric:

$$\text{Goodput} = (1 - \text{BER}) \times \text{Throughput}$$

Goodput is the rate of correct bit arrivals at the receiver. It will always be less than the throughput, but in a good radio the two should be close. There are two ways to increase goodput: increase throughput and decrease BER. Both can be achieved by doing one thing: increase the power of the transmitted signal. We will understand why this is the case for throughput in Chap. 5. But it should be immediately clear for BER. More signal power means that the levels for nominal "0" and "1" in Fig. 2.10 are more widely separated. This means more noise can be tolerated before an error occurs, which also means that for the same level of noise, errors will occur less often.

In fact, it is not the signal power that matters; it is the ratio between the signal power and the noise power. This ratio will define how often received signals cross the boundary in Fig. 2.10. This ratio is called the signal to noise ratio (SNR) and is usually measured in dB. We will discuss typical values for SNR in Sect. 5.8. However, we generally need the signal to be at least a couple of decades above the noise level for meaningful communication to occur.

Communication theorists can derive closed form relations between BER and the corresponding SNR for some modulation schemes. These relations almost always have a characteristic waterfall shape when BER is plot on the y-axis against SNR on the x-axis. A typical case is shown in Fig. 2.11. BER-SNR curves are always drawn with BER on a logarithmic scale and SNR in dB.

Any quality metric is made up of multiple variables. A good metric must involve a trade-off. A metric where all variables are in direct relation with the metric or all variables are in inverse relation with the metric must be missing something. In other words, a good metric must be improved by some of its constituent variables and worsened by some others.

Fig. 2.11 The shape of BER-SNR curves

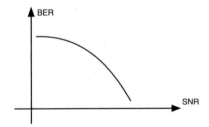

Goodput consists of two variables: BER and throughput. Both BER and throughput are ultimately dependent on SNR. If we increase SNR, we can increase throughput (Chap. 5), but we also improve BER. So, increasing SNR increases goodput doubly. So goodput on its own is not a good metric because it does not take account of the price paid for increasing SNR.

To improve this metric, we must somehow roll-in power consumption (at the transmitter) into the equation. Should we multiply goodput by power or divide by it? The answer depends on whether power and goodput are in an antagonistic or cooperative relation. In other words, does "improving" power also improve goodput or the opposite? Improving power means lowering it because higher power consumption is clearly undesirable. If we increase power, we improve goodput because we increase throughput and reduce BER. But increasing power is making power worse. Thus, power and goodput are antagonistic. When designing a quality metric, this generally means you should divide instead of multiply:

$$\frac{\text{Power}}{(1 - \text{BER}) \times \text{Throughput}} = \frac{\text{Power}}{\text{Goodput}}$$

To better interpret what this new metric means, we realize that $\frac{1}{\text{Goodput}}$ is the time it takes to receive one good bit:

$$\frac{\text{Power}}{\text{Goodput}} = \text{Power} \times T_{\text{Good bit}} = E_{\text{Good bit}}$$

This metric is the energy consumed for one good bit. If we use the transmitter power level, this will be the energy consumed by the transmitter to send one bit that is received correctly. If we use the receiver power level, it will be the energy consumed by the receiver. Typically, transmitter power is much higher than receiver power due to the need to power-amplify the signal (Sect. 1.3). However, recall that communication is typically two-way, so no one platform is always the transmitter or the receiver.

The energy per good bit is an excellent and important quality metric because it relates directly to battery life. Battery capacity is the amount of energy the battery holds for a full charge. If we define this as E_{Battery}, then dividing E_{Battery} by $E_{\text{Good bit}}$, we get the number of bits that can be transmitted/received before the battery runs out of charge:

$$\text{Bits transmitted} = \frac{E_{\text{Battery}}}{E_{\text{Good bit}}}$$

This can then be used to calculate the time it takes before the phone needs to be recharged. Assume, for example, that the number of bits calculated above turns out to be 10Gbit and that the user transmits 1Gbit every hour. This means the phone will need to be recharged every 10 hours. This calculation is very rough because it assumes the phone is only going to use power in radio communication and ignores all onboard processor requirements.

In Sect. 2.5, we discussed packet-based communication. One of the distinguishing properties of packet-based communication is that a single bit error in a packet means the entire packet must be discarded and resent again. This means that in packet-based systems, BER is not a good metric to measure QoS. Instead, we use a packet error rate, or PER. PER is the proportion of packets that arrive incorrectly at the receiver. An incorrect packet is a packet that contains one or more bit errors. It does not matter if the error is a single bit or all the bits in the packet. A single bit error is enough to throw the packet away.

PER levels tend to be much higher than BER levels. The reason is that multiple bit errors can occur in a single packet, but it does not happen the other way around. For example, consider the situation where we transmit 100 packets. Without loss of generality, assume all packets are equal length at 1000 bits. Now assume 2 bit errors occurred. The BER is $\frac{2}{100 \times 1000} = 0.00002$. The PER will depend on whether the two bits occurred in the same packet or in two different packets, but it will range between 0.01 and 0.02. Now assume 50 bits are erroneous; the BER is $\frac{50}{100 \times 1000} = 0.0005$. The PER will range between 0.01 and $\frac{50}{100} = 0.5$.

So, PER is generally much higher than BER for the same message. In general, it is also correct to assume that a higher PER can afford the same QoS than a much smaller BER. Notice that the way BER is measured can sometimes be confusing. In the above calculation, we calculate BER as if packet-based communication does not exist. But if we assume that all the bits in a bad packet are discarded (which is true) and then calculate BER, then this adjusted BER will match the PER.

Chapter 3
Multiplications to Multipliers: Performing Arithmetic in Hardware

3.1 Numbers on Computers and Numbers on Dedicated Hardware: Fixed-Point Registers

Binary numbers can be represented in different ways. Understanding number representation is the first step to understanding how we do arithmetic on these numbers, which in turn will lead to how arithmetic is performed in hardware.

For most of this book, our main concern will be to take an algorithm represented at some higher level of abstraction and transform it into a hardware implementation. The higher level of abstraction could range from a simulation in a programming language to an informal verbal description of what we want the circuit to do. But what we know about such high-level representations is that they have very little awareness of how hardware works, including what numbers look like.

First, let us consider a very simple question. If we have an N-bit register, what is the range of *unsigned integers* that the said register can store? The register can accommodate numbers from 0 to $2^N - 1$. Thus, 3-bit registers store integers from 0 to 7. A 6-bit register stores numbers from 0 to 63, and so on.

All the algorithms we will consider for radios deal with complex numbers. This follows from the description of baseband modeling in Sect. 2.1. Moreover, all numbers are signed. That is to say, the numbers will take positive and negative values. Thus, we must consider our options for negative number representation.

The most direct way to represent negative numbers is the sign-magnitude approach. In this approach, the sign of the number occupies the MSB, while the magnitude occupies the rest of the word. Assuming integer-only systems again, we can guess that the range of positive numbers that can be represented will be lower than in unsigned numbers.

For example, for a 4-bit unsigned register, the range of integers is 0–15. If the numbers are signed, the first bit is reserved for the sign, leaving three bits for the magnitude. This allows numbers from -7 to 7 to be represented. Thus, in the

© The Author(s), under exclusive license to Springer Nature Switzerland AG 2023
K. Abbas, *From Algorithms to Hardware Architectures*,
https://doi.org/10.1007/978-3-031-08693-9_3

unsigned regime, 16 numbers could be represented using four bits. In the sign-magnitude approach, only 15 numbers can be represented using 4 bits.

The one's complement approach is an alternative to signed number representation. Like the sign-magnitude approach, the first bit indicates whether the number is positive or negative. However, the magnitude of the number is not the same for positive and negative numbers; instead a complement is taken for all bit positions.

For example, in the 1's complement format, $+1$ in four bits is 0001. To obtain -1, first obtain $+1 = 0001$, and then invert all bits to obtain 1110. Extending this, we can obtain all possible numbers that four bits can represent:

```
0000 = 0
0001 = 1
0010 = 2
0011 = 3
0100 = 4
0101 = 5
0110 = 6
0111 = 7
1000 = -7
1001 = -6
1010 = -5
1011 = -4
1100 = -3
1101 = -2
1110 = -1
1111 = -0
```

It is clear why sign-magnitude and one's complement numbers have a lower range than unsigned numbers. This is because both signed number representations reserve two words to represent 0, one for $+0$ and one for -0. Since there is no difference between the two, both number systems are wasting their range.

The "best" signed number representation is two's complement. The big deal about two's complement is not only that it is a more efficient representation that makes no "waste" in representing numbers but also that it is friendly to arithmetic. In two's complement addition and subtraction are the same. You can use the same hardware to do the two, and you can mix positive and negative operands without any preprocessing or modification.

Let us consider how to represent negative and positive numbers in two's complement. Again, the first bit in the word is the sign. A word that begins with a "0" is positive; a word that begins with a "1" is negative. Positive number magnitude is the same as in unsigned numbers. The magnitude of negative numbers is a little more complicated and can be obtained in two ways:

- Write the positive number with a "0" sign bit. Invert all bit positions, and then add a "1" to the LSB.
- Write the positive number with a "0" sign bit. Examine all bits from LSB to MSB, leave all bits as is until you find the first significant bit from the LSB, keep it, and invert all upcoming bits up to and including the MSB.

Using either approach, we can find all the numbers that could possibly be represented by four bits:

```
0000 = 0
0001 = 1
0010 = 2
0011 = 3
0100 = 4
0101 = 5
0110 = 6
0111 = 7
1000 = 8  - 16 = -8
1001 = 9  - 16 = -7
1010 = 10 - 16 = -6
1011 = 11 - 16 = -5
1100 = 12 - 16 = -4
1101 = 13 - 16 = -3
1110 = 14 - 16 = -2
1111 = 15 - 16 = -1
```

The two's complement representation is so named because any negative number has a magnitude that is the complement of -2^N. It does not try to store 2 0's; thus it represents 16 numbers, although it is asymmetric, always representing one more negative number than positive.

Another important decision about binary number representation is whether to store numbers as fixed point or floating point. Floating-point is both a number representation and an associated set of arithmetic operations. It is the number storage method used by most general-purpose processors. It uses very wide busses and registers very efficiently and is unparalleled in the range of numbers that can be represented. Thus, it can simultaneously represent and operate on extremely large and extremely small numbers. This high dynamic range could be very helpful. For example, see Sect. 7.3 on why we spend a lot of effort on optimizing ADC dynamic range.

However, while floating-point number are extremely efficient, floating-point operations are not. Adding two floating-point numbers can be challenging. Thus, floating-point arithmetic units are slower and more power-hungry than their fixed-point counterparts. Additionally, floating-point processors only make sense with extremely wide busses. The kind of input wordlength we manage to extract from AGC and ADC (Sect. 7.3) never justifies floating-point representation.

A floating-point number consists of three parts as shown in Fig. 3.1. These are the multiplier, the base, and the exponent. For the sake of simplicity, let us assume unsigned integer numbers. If we have an 18-bit register, there are multiple ways

$$Multiplier \ X \ Base^{Exponent}$$

Fig. 3.1 Floating-point representation

which we can distribute the said 18 bits between the 3 components in Fig. 3.1. The simplest approach is to divide the available register equally between the three. Thus, each part in Fig. 3.1 is now six bits exactly. What range of numbers is this floating-point register capable of representing?

The smallest number is $0 \times 0^0 = 0$.
The largest number is $64 \times 64^{64} = 2.521728 \times 10^{117}$.

The range of numbers could obviously be huge depending on how the register is divided. To understand why this representation is called floating point, consider the case where each of the three registers forming the number is now signed. This allows the exponent to be negative, thus allowing us to represent purely fractional numbers. The position of the binary point is thus variable. It depends on the magnitude of the multiplier and base as well as the magnitude and sign of the exponent.

The bottom line is floating-point numbers can represent numbers with incredible and unparalleled precision. Numbers can be huge or minute. We can represent negative or positive numbers. And we can move the binary point's position at will by changing the exponent.

But floating-point numbers leave one question unanswered: how do we add two numbers? Addition is the simplest possible arithmetic operation. If we cannot do addition without a headache, we cannot do much else. Because they contain an exponent, floating-point adders are complicated and cumbersome, and the size of registers that would justify using them is not even possible in wireless receivers.

Now assume that we used all 18 bits in a single register. The range of integer numbers it could represent is:

The smallest number is $0 \times 0^0 = 0$.
The largest number is $2^{18} - 1 = 262143$.

This is a much smaller range than the floating-point range we discussed above. This method, where the register is not partitioned, is the fixed-point representation. All we mean when talking about fixed-point numbers is that the number is represented by a flat register where there is no exponent. However, the assumption above that the number is unsigned and purely integer does not need to hold. In fact, most fixed-point numbers are signed and most also contain integer and fractional parts. The fractional part is separated from the integer by a binary point.

The number of bits dedicated to the fraction and those dedicated to the integer is determined by the position of the binary point. When the position of the binary point is determined, it is fixed, and can no longer move at will, thus the *fixed-point* nomenclature. This causes fixed-point numbers to have a much narrower range than the corresponding floating-point number. However, operations are much simpler to perform on fixed-point registers.

In Sect. 2.1, we saw that the baseband representation of numbers means we deal with complex numbers. In Chaps. 4 and 5, we will discover that we must deal with fractional numbers. In Sect. 7.3, the wordlength of inputs to a receiver is determined and ultimately limited by the ADC and AGC. In Sects. 3.3 and 3.4, we will see that the area of an adder or a multiplier is a strong function of its wordlength. Combining all this information together, we must represent complex fractional numbers in fixed-point registers, and we must be cognizant at all points of how the wordlength of our registers grows, because this ultimately decides the size of the hardware.

Recall that as people who implement baseband radios, our input is an algorithm that comes from a systems engineer. Systems engineers have variable levels of exposure to hardware. But generally, we cannot assume they care about how the algorithm gets implemented, simply because they have a lot of other things to worry about. One of the first things we deal with while translating an algorithm into hardware is that algorithms always come in a floating-point form. Systems engineers generally use high-level simulation environments to allow quick prototyping, easy manipulation, and fast turnaround times. The said simulation environments always run on floating-point general-purpose processors.

Thus, we must find a way to quickly prototype algorithms in fixed point. This is called the fixed-point simulation, in which we translate every number and every operation in the original (system level) floating-point simulation into its fixed-point counterpart. This helps answer two related, complicated, and important questions: first, what wordlength should we use for registers at different points in the transceiver? And second, what price do we pay for this decision in terms of hardware area?

It is evident that the more bits we use in our registers, the closer we are to the floating-point algorithm. But more bits also mean bigger hardware. Bigger hardware costs more, is slower, and dissipates more power. But using smaller wordlength and smaller hardware incur a cost to performance. This cost is shown in Fig. 3.2. The BER curve for the fixed-point radio is shifted to the right relative to the floating-point radio. The difference in SNR between the two curves in Fig. 3.2 is the amount of extra power we must burn in the PA to perform as well as the original floating-point algorithm. In upcoming sections, we will develop an understanding for where this drop in performance comes from. We will also introduce a very simple way to perform fixed-point simulation, and finally we will discover why larger hardware is something we want to avoid.

Fig. 3.2 BER curve for floating and fixed point

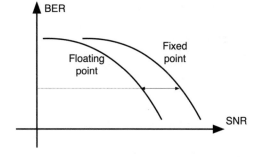

3.2   Modeling Fixed-Point Number Generation

To answer the questions in Sect. 3.1, we need to perform a fixed-point simulation. In a fixed-point simulation, we represent all numbers and operations as registers with a fixed number of bits. However, fixed-point simulation environments are difficult to come by. Instead, most high-level simulation environments run on general-purpose processors, which have floating point registers and processing units.

Thus, the question is: how do we represent fixed-point numbers and operations in a floating-point environment? To do this, we first make a fundamental observation: the position of the binary point in a fixed-point register does not matter if it is in the same position for all registers. This sounds very dramatic, but it is easy to see with examples.

Consider, for example, a four-bit register. First, assume all registers are pure integers. That is, there is no fractional part, and the binary point lies to the right of the entire word. For simplicity, and without loss of generality, assume the numbers are unsigned. The following four registers thus store:

```
1110 = 14
0001 = 1
0011 = 3
1001 = 9
0101 = 5
```

Now consider the case where the same registers with the same contents have a single fractional bit, and the remainder is integer. Thus, the binary point lies left of the LSB. This is equivalent to the integer numbers above being divided by 2:

```
111.0 = 14/2 = 7
000.1 = 1/2 = 0.5
001.1 = 3/2 = 1.5
100.1 = 9/2 = 4.5
010.1 = 5/2 = 2.5
```

Now consider the case where two bits form the fractional part, and two bits form the integer part:

```
11.10 = 14/4 = 3.5
00.01 = 1/4 = 0.25
00.11 = 3/4 = 0.75
10.01 = 9/4 = 2.25
01.01 = 5/4 = 1.25
```

Now, if only a single bit is integer:

```
1.110 = 14/8 = 1.75
0.001 = 1/8 = 0.125
```

```
0.011 = 3/8 = 0.375
1.001 = 9/8 = 1.125
0.101 = 5/8 = 0.625
```

And if the registers are purely fractional:

```
0.1110 = 14/16 = 0.875
0.0001 = 1/16 = 0.0625
0.0011 = 3/16 = 0.1875
0.1001 = 9/16 = 0.5625
0.0101 = 5/16 = 0.3125
```

Thus, the position of the binary point is inconsequential. If we move the binary point around, it is equivalent to a constant shift, which in decimal is expressed through a division or multiplication by a power of two. Again, the only thing that matters is that the position of the point is the same in all registers. This allows all registers to exist at the same "scale," which preserves the integrity of arithmetic operations.

Now, how do we generate fixed-point numbers in a floating-point environment? We will use a pseudocode that looks very similar to MATLAB script; however, it should be easy to adapt to any programming language. The "rand" command generates a random number drawn from a uniform distribution. Thus, the number a below belongs to the range $(0,1)$:

```
a = rand;
```

The number a is purely fractional, and it is floating point. This means that for all practical purposes, the fractional part of the number extends to an unlimited number of bits. Assume we want to convert this random number into an n-bit fixed-point register. Let us represent this fixed-point number as a pure integer. This is not a restriction and does not cause any loss of generality, because the position of the binary point is irrelevant and can be adjusted later through division by a power of 2.

To convert the random number above into a number with an integer part, we multiply it by $2^n - 1$:

```
a = a × (2^n - 1);
```

This converts the range of the number from $(0,1)$ to $(0, 2^n - 1)$. This number is unsigned because the original fractional number could not be negative. a is now a floating-point number with both an integer and a fractional part. To convert it into an n-bit fixed-point number, we truncate the fractional part:

```
afloating = a;
a = floor(afloating);
```

Thus, a is now a pure integer that is randomly drawn from the range 0 to $2^n - 1$. The entire range, and thus the number a, can be stored in an n-bit unsigned register. In general, an integer number lying in the range 0 to $2^m - 1$ can be stored in an m-bit register.

3.3   Fixed-Point Addition

Assume we generate two numbers using the approach above: a and b. Both numbers are stored in n-bit registers. We can add the two and store the result in a third register c:

```
c = a + b;
```

This operation is *floating point* even though both operands a and b are fixed point. This is because the simulation environment and processor used are floating point. But there is a special case where this operation is also considered fixed point. If a and b are n-bit and c is $n + 1$ bits, then floating-point and fixed-point additions are identical.

Figure 3.3 shows a ripple carry adder adding two operands each n-bit long. There are n-bits in the sum and an additional carry-out from the last full adder. Thus, the result must be stored in an $n + 1$ bit register if both operands are n-bit.

The result in Fig. 3.3 is the floating-point addition but is also the "noiseless" result of fixed-point addition. After all, the adder in Fig. 3.3 is fixed point. If the result from the adder in Fig. 3.3 is stored in a register that is "big enough," i.e., six bits, we say that it is noiseless.

But what sort of noise are we worried about? The problem in Fig. 3.3 is that the result is one bit larger than the inputs. We have three options to deal with this extra bit. The first is to accept that the output will be one bit longer and thus store it in a register that is one bit longer. Because the growth in wordlength for adders is limited, this is a very common approach. However, when we consider the dynamic range of multiplier outputs in Sect. 3.4, this approach will become impractical.

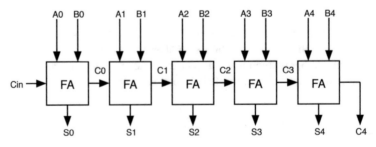

Fig. 3.3 n-bit ripple carry adder adding two n-bit operands. In the case shown, $n = 5$ and the result is 6 bits

The second option is to ignore the carryout from the adder. Thus, store only the n LSB bits and throw away the MSB. This allows us to store the results in a register only as large as the input registers. However, this exposes us to the danger of overflow. Overflow occurs when the final carryout (MSB of the result) is significant. Overflow is a failure of the adder. Consider, for example, a four-bit adder adding two unsigned four-bit numbers; we have two sets of possibilities. The first is that the carryout is zero, in which case the MSB truncated output is correct:

```
0000 + 0001 = 0001
0 + 1 = 1
0011 + 0100 = 0111
3 + 4 = 7
0111 + 0111 = 1110
7 + 7 = 14
1000 + 0010 = 1010
8 + 2 = 10
```

All the above operations produce correct results and produce them in only four bits. However, all the operations below produce incorrect results. We will show two results, the first is the "noiseless" result, and the second is the result after the MSB is truncated. Translating these equations into decimal readily indicates why overflow represents a total failure. The result in the four-bit register is completely unrelated to the correct result:

```
1000 + 1000 = 10,000= > 0000
8 + 8 = 16 = > 0
1110 + 1111 = 11,101 = > 1101
14 + 15 = 29 = > 11
0111 + 1111 = 10,110 = > 0110
7 + 15 = 22= > 6
0001 + 1111 = 10,000= > 0000
1 + 15 = 16= > 0
```

Overflow in unsigned numbers occurs when the MSB of the sum is true. This will generally occur when the operands have a large magnitude. If the size of the operand registers is small, overflow will occur frequently. If the operand registers are wide, then the very large values that cause overflow will be a less frequent occurrence. In floating-point adders, allowing outputs to overflow is thus a reasonable option. When a floating-point result overflows, we are dealing with absurdly large numbers or absurdly small magnitudes. In fixed-point systems registers are generally small, and thus overflow is frequent and allowing it is impractical.

The third way to deal with the extra bit is to truncate the LSB instead of the MSB. This creates a situation where the result is never a failure, but where sometimes the result is "a little off." Again, an example can help explain this:

```
1111 + 1111 = 11,110= > 1111
1111 + 1110 = 11,101= > 1110
```

In the first case above, the result should be 30; we store 15 instead. However, this does not mean that we have stored an erroneous result. By removing the LSB, we have effectively shifted the result to the right by a single bit, thus dividing by 2. So, to truly calculate the error due to truncation, we return the number again to the original scale. Thus, the error is $30 - 15 \times 2 = 0$. In the second case, the result is 29. We have stored 14 instead. If we calculate the error by returning the truncated number to its original scale, we obtain error $= 29 - 14 \times 2 = 29 - 18 = 1$.

Thus, truncation does not occasionally make the number wrong as in the case of overflow. Instead, it occasionally makes the result "noisy" by truncating a significant LSB. Through simple inspection, we deduce that even results will never be noisy when truncated, while odd results will always have an error of 1. This error or *noise* will be less significant for larger adders. It is a sort of noise introduced due to truncation. We call it quantization noise because it resembles the noise introduced by quantization in ADCs.

To summarize, adders increase the wordlength of the output by 1 relative to the operands. You have three options to deal with this:

- Store the result of *n*-bit adders in $n + 1$ bit registers. This is often a good solution for adders.
- Drop the MSB of the result and store the remaining *n*-bits. This can lead to overflow. This is a practical option only for floating-point adders.
- Drop the MSB of the result and store the remaining *n*-bits. This leads to quantization noise.

3.4 Fixed-Point Multiplication and Managing Wordlength Growth

Figure 3.4 shows a common parallel multiplier. This example is 4×4. Comparing this multiplier to the adder from the previous section, we can make important conclusions about their relative cost and performance; these are summarized in Table 3.1. The area of the multiplier, as measured in number of FA, is much higher than a comparable adder. The area also grows much faster in multipliers. The critical path of the multiplier is roughly double the length of an adder of comparable wordlength, which means it is only half as fast. Finally, the output of a multiplier has *double* the wordlength of the input wordlength if the input operands are of equal size. Thus, if operands are n-bit, the noiseless output is $2n$ bits. If the two operands have wordlengths m and n, the noiseless result is $n + m$ bits long.

The three strategies used to deal with wordlength increase from adders *cannot* be used with multipliers. If we preserve the noiseless output wordlength, we will have a disastrous growth in circuit sizes. When we want to reduce the wordlength of an output from an arithmetic block, it is not the size of the register we use to store the output that matters; registers are cheap anyway. The problem is that most algorithms do not involve a single mathematical operation but rather a series of operations that

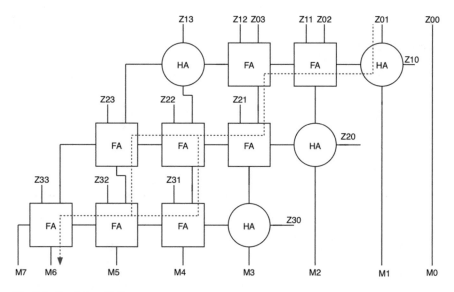

Fig. 3.4 Parallel multiplier

Table 3.1 Comparison of complexity of multipliers and adders. For the multiplier we consider half adders to be full adders

	Delay	Number of FAs	Output wordlength
Adder	$O(n)$	n	$n + 1$
Multiplier	$O(2n)$	n^2	$2n$

hand each other results. Thus, the output from the multiplier will be an operand for another multiplier. If we allow the multiplication result to remain at $2n$ bits, we must use a $2n$-bit multiplier for the next operation. This multiplier has four times the number of full adders as the first multiplier, and its output will be double the size of the first multiplier, requiring larger still multipliers down the line.

Thus, keeping outputs without trimming is not an option. We also have the option of throwing away extra MSB bits. This worked with adders, especially large adders because only a single bit needed to be thrown away. This limited the frequency of overflow. With multipliers, we are dealing with a completely different situation. Multiplication very often produces outputs with large magnitude; in fact outputs tend to bias toward the larger side. If we insist on keeping results the same size as operands, we end up with overflow occurring often. Thus, this is not an option either.

The only viable option we have is to drop bits from the LSB as in Fig. 3.5. This truncation will not make the result overflow. Instead, as we discussed with adders, it introduces quantization noise to the result.

Assume we generate two inputs a and b, both of which are stored in an n-bit unsigned integer register as in Sect. 3.2. If we multiply these two numbers and store

Fig. 3.5 Multiplier output truncation. The noiseless result is $2n$-bits. p bits are dropped from the LSBs, and only $2n-p$ bits are stored

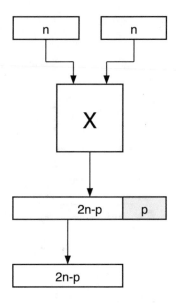

the untruncated result in register c, then register c must be $2n$ bits long. This is shown in the following line of pseudocode:

```
c = a × b;
```

Thus, c will be the result of multiplication if the multiplier were floating point, or if it were fixed point and the result is stored noiselessly in a $2n$-bit register. If we decide to truncate the result to m bits by throwing away p bits, then we know that $m + p = 2n$. To simulate this on a floating-point environment, we first divide the multiplication result by 2^p:

```
c = c/2^p;           .
```

This is equivalent to shifting the binary point p bits to the left. The result is still in $2n$ bits, except p of them are the fractional part, while m are the integer part. To truncate the result to only m bits, we get rid of the fractional part by flooring c:

```
c = floor(c);
```

This truncation will introduce quantization noise into the result. This sort of noise occurred only with odd results in addition in Sect. 3.3. However, with multipliers quantization error is higher and more variable. We need a way to calculate it. Noise is the difference between the noiseless result and the truncated result. However, notice that calculating the simple difference:

```
c_noiseless = a × b;
noise = c_noiseless-c;
```

will give an exaggerated and very wrong value for noise. This is because the two numbers now exist on different scales, and a simple subtraction will penalize the multiplier not only for the truncated bits but also for the difference in scale.

This is best illustrated by an example. Multiplying 14×15 gives a noiseless result of 210. The operands need four bits to be stored, and the result needs eight bits. Now assume we only have five bits available to store the result because we want subsequent operations to be five-bit operations. We truncate the result by performing *floor*$(210/2^3)$, where 3 is $8 - 5$. The result will thus be 26. The noise due to truncation is *not* $210-26$, not even close. To find the correct value of noise, we must have the two numbers at the same scale; in other words place their binary points in the same position. We have two options to do this, either dividing the noiseless number by 8 without flooring or multiplying the noisy result by 8. In the first case, the noise is $(210/8) - 26 = 0.25$. In the second case, the noise is $210 - 26 \times 8 = 2$.

So which value of noise is correct? Both! It depends on where you place the binary point. But to have a meaningful metric for the impact of quantization, we find a *ratio* rather than an absolute value for noise. This ratio is the ratio between the power of the signal, in this case the noiseless signal, and the power of the quantization noise. This can be calculated as:

```
QSNR = (c_noiseless)^2/(c_noiseless-c)^2;
```

This QSNR, or Quantization Signal to Noise Ratio, will be the same regardless of where you place the binary point. And this is the metric we care about. Notice that the original operands a and b were drawn from a uniform distribution. If we want to understand how the value of p affects QSNR, we draw many samples and take their average QSNR. This can be done through the pseudocode below:

```
operand_size = w;
noiseless_size = 2 × w;

for m = 1:noiseless_size-1
   signal_power = 0;
   noise_power = 0;
   for iterations = 1:max_iterations;
      a = floor(rand × 2^operand_size);
      b = floor(rand × 2^operand_size);
      c_noiseless = a × b;
      c = floor(c_noiseless/2^(noiseless_size-M);
      current_noise=c_noiseless-c × 2^(noiseless_size-M);
      current_noise_power=current_noise^2;
      current_signal_power=c_noiseless^2;
        signal_power = signal_power + current_signal_power;
      noise_power = noise_power + current_noise_power;
   end;
   qsnr(m)=10 × log10(signal_power/noise_power);
end;
```

Fig. 3.6 QSNR versus size
of result register

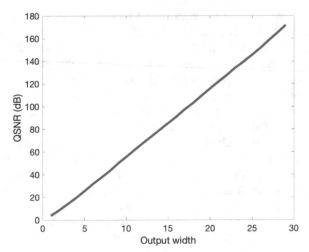

Figure 3.6 shows the result of the pseudocode above run for an input wordlength
(w) of 16 and 1000 samples. The x-axis sweeps the size of the output register used for
storage. The y-axis is QSNR in dB. The QSNR curve is linear and increases at a rate
of 6 dB/bit. This is expected from quantization theory and will be valid regardless of
the distribution of operands. There is no reason to make the result register size
smaller than that of the operands. We want to control the growth of results so that we
can control the growth of hardware. If the current multiplier is $n \times n$, there is no
reason to make subsequent multipliers smaller.

Notice also that you should stop calculating QSNR at an output size of $2 \times w - 1$.
This is one bit short of the noiseless size. The reason is, for any register at or above
the noiseless size, the QSNR will be infinite as the quantization noise drops to null.
Again, there is no reason to make the result any bigger than the noiseless size. Thus,
the range of outputs that makes sense is between w and $2 \times w$ bits inclusive.

So which value of QSNR is "good"? What m should we choose in the curve
above? There are two answers to this question. The first is that we should generally
choose values that introduce QSNR much higher than the channel SNR (Sect. 4.2)
observed by the system. This allows the noise due to quantization to sink far below
the AWGN level, thus making it negligible. A difference between QSNR and SNR
of 20 dB is acceptable, a difference of 30 dB is good, and a difference of 40 dB is
great.

But the above answer is misleading. It would be a good answer if we only had a
single multiplication operation in our whole system. The fact is our entire receiver is
fixed point. Dozens to thousands of fixed-point operations are performed to allow the
transceiver to work. There is no clear way to use QSNR as a quality metric in such
case. Instead, what we should do is simulate the entire system in floating point and
again in fixed point and compare the SNR-BER curves as in Fig. 3.2. Depending on
the loss of SNR we can afford, we decide the number of bits to use in each part of the

system. In other words, we must use experience, trial and error, and a very high-level quality metric to make these decisions.

From Table 3.1, we see that multipliers are a much "bigger deal" than adders. Multipliers have longer delay and occupy much higher area, thus dissipating significantly more power. When we translate an algorithm to hardware, we must be very careful when we observe a multiplier, while we can liberally use adders. Multipliers dominate the area of the circuit, determine its critical path, hog a lot of the digital power, and devastate dynamic behavior (the size of outputs).

Thus, it is very important when using a multiplier to make sure we really need a multiplier. This is relevant regarding the nature of operands. For example, consider the following three multiplications:

```
C1 = 12 × 3;
C2 = 4 × a;
C3 = a × b;
```

Only the third operation needs an actual multiplier to implement. The first is a constant; it only needs a register to store the integer 36. The second is a constant multiplier, where one variable operand is multiplied by the same constant coefficient all the time. This is not a parallel multiplier as shown in Fig. 3.4; it can be implemented using much simpler hardware depending on the nature of the coefficient:

```
D1 = 2   × a;
D2 = 4   × a;
D3 = 5   × a;
D4 = 7   × a;
D5 = 125 × a;
```

The above five operations can be implemented using adders and shifters. Some are simpler to implement than others. All use much less hardware than a full multiplier. For example, $2 \times a$ and $4 \times a$ can be implemented using shifters only, the first requiring a single bit shift, the second a 2-bit shift. The other operations require some combination of shifting and addition. For example:

$$5a = 4a + a$$

$$7a = 8a - a$$

$$125a = 128a - 2a - a$$

A deeper look at the parallel multiplier in Fig. 3.4 would have led to the above result. A parallel multiplier is doing long multiplication, which is also a bunch of shifts and adds. However, for the full multiplier, we assume that all summands are significant because we know neither operand ahead of the calculation. For a constant multiplier, we can make very important simplifications by nulling some summands. In short, when using a multiplier, do make sure you need a multiplier, because multipliers cost a lot.

3.5 Processing Complex Numbers

We will always deal with complex operands in the baseband physical layer. For a thorough discussion why, consult Sect. 2.1 and Chap. 5. Even noise, interference, and the channel coefficients are complex numbers. The real part of any operand will represent the coefficient eventually multiplied by a cosine RF carrier. The imaginary part will be multiplied by a sine carrier. Thus, we must reconsider how fixed-point numbers are generated, added, and multiplied when they are complex.

First, how is a complex addition performed? A complex addition is simply the addition of the real parts of the two operands into the real part of the result and the imaginary parts into the imaginary part of the result. Thus, a complex addition is two real additions:

$$a + b = (a_r + b_r) + j(a_i + b_i)$$

It follows that a complex *adder* is also two real adders. If the two operands are n-bit long per real and imaginary, then the result will be $n + 1$ bits long per real and imaginary, and the discussion on what to do with the extra bit in Sect. 3.3 is still valid. Thus, complex adders occupy double the area of a real adder, have the same critical path (and thus run at the same speed), and have the same dynamic behavior and quantization noise.

But how are complex numbers stored in hardware? They are stored in registers the same way real numbers are. You just need to store the real and imaginary parts separately. The hardware does not care if what you are storing is a complex number or two real numbers. You will have to take care of operations so that numbers are correctly interpreted as either. In Fig. 3.7, we give one alternative approach to storing complex numbers. Here a single 2n-bit register is used to store one complex number. The upper n-bits are the imaginary part; the lower n-bits are the real part. You can also use two separate n-bit registers to store the real and imaginary parts. The approach in Fig. 3.7 is conceptually stronger because it identifies the entire complex number as a single register. On the other hand, using two separate registers for the real and imaginary parts may make writing code easier. It does not make much of a difference which you use but use one strategy consistently.

Figure 3.8 shows how a complex multiplier is implemented using four real multipliers and two real adders. This can be easily deduced from the expansion of

Fig. 3.7 Complex register

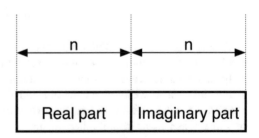

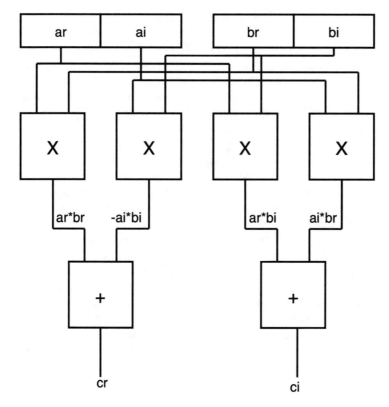

Fig. 3.8 Complex multiplier

complex multiplication shown below. Multipliers were seen in Sect. 3.4 to be a significant limiting factor to the power-delay performance in circuits. Figure 3.8 shows that complex multipliers are even bigger than complex adders, meaning the situation has gotten worse:

$$ab = (a_r + ja_i) \times (b_r + jb_i) = a_rb_r - a_ib_i + j(a_rb_i + a_ib_r)$$

The area of the multiplier in Fig. 3.8 is over 400% larger than a real multiplier with the same size of inputs. If the adders in Fig. 3.8 are noiseless, the critical path of the complex multiplier is also double that of the real multiplier.

To compare dynamic behavior, let us figure out the output wordlength of a *signed complex* multiplier. The complex multiplier is formed of four real signed multipliers and two real signed adders. Each of the real and imaginary parts of the operands is n-bits long. The n-bits can be thought of as $n-1$ bits amplitude and 1 bit sign. However, the discussion in Sect. 3.2 shows that there is an asymmetric behavior for two's complement numbers, with one more negative number than positive numbers.

Thinking in terms of an example, if each operand is four bits long, then each ranges from -8 to 7. This means the product of a real multiplier will range from -56

to 64. This requires eight bits to store. Seven bits are required to store the magnitude of 64, and an additional bit is required to store the sign bit. This is a very convoluted way to reach the same conclusion we reached for unsigned numbers: a real signed multiplier requires double the wordlength at the output to store the result noiselessly.

Now if the signed multipliers are used to build a complex signed multiplier, we use two real multipliers and then add up their results in one adder. Typically, this would require an additional bit to guard against the overflow of the adder. However, in this case we must consider the nature of the multiplier outputs.

Each of the real multipliers in Fig. 3.8 produces an output in $2n$ bits. Back to the 4-bit example, the range of possible outputs is from -56 to 64. The output is stored in eight bits. The range of numbers that can be represented in two's complement in 8 bits is from -128 to 127. The multiplier outputs when added will range from -112 to 128. To store these numbers, we *will* need an additional bit; thus 9 bits are required at the overall output of the complex multiplier.

To sum up the above discussion, when each of the operands of a complex multiplier is n-bits, the output of the complex multiplier is $2n + 1$ bits. This accounts for the signed nature of the numbers when stored in two's complement. The $2n + 1$ bits are divided as follows: $2n$-bits are required to store the outputs of the real multipliers; the additional remaining bit is required to guard against adder overflow.

There is an alternative implementation to complex multipliers that will be very useful in most of our applications. We can reach this alternative by adding and subtracting additional terms to both the real and imaginary parts of the product. We will then be able to take common factors to reduce the expression of the product:

$$ab = a_r b_r - a_i b_i + j(a_r b_i + a_i b_r)$$
$$= a_r b_r + a_r b_i - a_r b_i - a_i b_i + j(a_r b_i + a_r b_r - a_r b_r + a_i b_r)$$
$$= a_r(b_r + b_i) - b_i(a_r + a_i) + j\{a_r(b_r + b_i) + b_r(a_i - a_r)\}$$

The above product is significantly more complicated than a straight implementation of the multiplier. In fact, if we count the number of real operations, we need four real multiplications and six real additions. This is much worse than the original four real multiplications and two real additions.

However, notice that the term $a_r(b_r + b_i)$ is repeated in both the real and imaginary parts of the product. This term thus needs to be calculated only once, meaning we can save one multiplier and one adder. This reduces the hardware cost of the above complex multiplier to three real multiplications and five real additions. How does this compare with the original four real multipliers and two real adders? We have seen that multipliers are much larger than adders of the same input length (Sects. 3.3 and 3.4). Thus, it is generally preferable to use three additional adders to save a single multiplier.

This alternative implementation is tempered by the following caveats:

- Multipliers are significantly larger than adders only for large values of n. Thus, if the wordlength of the complex multiplier is small, the original implementation might be better.
- The critical path of the alternative implementation is longer than the original. For a straight complex multiplier, the critical path is an n-bit multiplier followed by a $2n$-bit adder. This is $4n$ FA delays. For the alternative implementation, we need an n-bit adder to prepare the initial sums, followed by an n + 1-bit multiplier and finally a $2(n + 2)$ bit adder. This is $5n + 6$ FA delays, which is not a negligible hit to speed.
- The area of the original implementation is $4n^2 + 4n$ full adders. The area of the alternative is $3n + 3(n + 1)^2 + 2(2n + 2)$ full adders. This simplifies to $3n^2 + 13n + 4$ FA. Thus, for a small wordlength, as discussed in the first point, the original implementation is both smaller and faster. Equating the above areas, we find that for any $n < 10$, a straight implementation works out best.

3.6 Processing Vectors and Matrices

In Chap. 9 we will find that the introduction of multiple antennas changes all arithmetic from scalar to matrix vector. This means we need to extend all the fixed-point arithmetic we did so far to matrices and vectors. This extension is much easier than it sounds. Vectors are stored element by element in related registers or memory locations. Each location will contain a real and imaginary part forming the complex element. Matrices are stored the same way. Interpreting the vector, scalar, or matrix nature of data is more about how it is processed than about how it is stored.

Consider vector and matrix addition operations:

$$\begin{pmatrix} a11 & a12 \\ a21 & a22 \end{pmatrix} + \begin{pmatrix} b11 & b12 \\ b21 & b22 \end{pmatrix} = \begin{pmatrix} a11 + b11 & a12 + b12 \\ a21 + b21 & a22 + b22 \end{pmatrix}$$

There is nothing special about the 2×2 matrix size. Addition is also element by element for row and column vectors and for matrices of any size. Matrix addition is in fact identical to scalar addition because it is done element by element.

Multiplying matrices is a little less direct. When we multiply two matrices, every element in the product is an inner product of a row vector from the first operand and a column vector from the second operand. This means that if we figure out dot product for vectors, we can extend its implementation to matrix multiplication of any size:

$$(a_0 \; a_1 \; a_2) \cdot \begin{pmatrix} b_0 \\ b_1 \\ b_2 \end{pmatrix} = a_0 b_0 + a_1 b_1 + a_2 b_2$$

Consider the inner product above, and for now think of all numbers as purely real. The product is a scalar that results from the summation of three terms. Each term in the result is a product of two elements from the original vectors. If the elements of the original vectors are n-bit each, the terms in the product are each $2n$-bit long. If there are m elements in the vectors, we are doing $m - 1$ additions; thus, we need $\log_2(m + 1)$ additional guard bits. In other words, the size of the result of the inner product is $\log_2(m + 1) + 2n$.

Now if the vectors consist of signed complex numbers, as would be the case in a MIMO communication system, each term in the result will be $2n + 1$ long. This will apply separately for the real and imaginary parts. We will need the same amount of guard bits for each of the real and imaginary rails. Thus, the result will thus need $\log_2(m + 1) + 2n + 1$ bits per rail.

3.7 Implementation Platforms

When you implement a circuit, its architecture and performance will be fundamentally affected by the platform. The platform is the type of hardware used to implement the circuit. Figure 3.9 shows the trade-off in "performance" and flexibility/cost of the different implementation platforms. Performance could be any metric that combines power and delay, such as power-delay product or the energy consumed to produce a single output. Cost and flexibility are contradictory and usually correlated. Cost could be cost in dollars or in man-hours. Flexibility measures how easy the

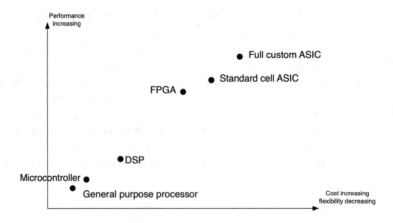

Fig. 3.9 Trade-offs in implementation

implementation is to get off the ground and how easy it is to modify or debug. This could also be measured in man-hours.

Implementation platforms come in all shapes and sizes and provide a continuum on the graph shown above. Points marked on the graph are the macro categories into which we group platforms. We will discuss each individually.

General purpose processor This is a processor as found on a computer or cell phone. General-purpose processors are extremely easy to develop for. When you develop an algorithm to run on a general-purpose processor, you do so in a high-level language that uses APIs to access the processor. There is a clear separation of the designer from the hardware. Development time is quick, and debugging is easy and supported by user-friendly tools. Fixes, updates, patches, and modifications are easy to develop and deploy. On the other hand, general-purpose processors are the "worst" at running the algorithm in terms of power and delay. A general-purpose processor needs to be everything for everyone, so it ends up doing everything less than optimally.

A GPU works by fetching instructions from an instruction memory, decoding such instructions, and implementing them on a generic ALU. It uses a large and cumbersome operating system. It must do this to provide the flexibility to run a wide variety of programs and to allow designers to be separate from the hardware. This all imposes the largest overhead among all the implementation platforms. For radios, the GPU can be used to apply higher layers of communication. The physical layer is most definitely not suited for this platform.

Microcontroller A microcontroller is a low-end no-frills microprocessor. It is used when there is no expectation of a complicated operating system or the need to develop complex software and applications. This reduces the overhead of the processor, thus raising its performance. On the other hand, the designer must be more aware of the hardware of the processor, and more so the hardware surrounding it. This complicates design relative to a general-purpose processor. What constitutes a microcontroller and when it is good to use it is a matter of semantics and debate. Microcontrollers find common use in embedded systems and as a state machine and regulator for simple applications where you would not expect electronics to be present. However, they should not be used in radios. Higher layers are best served by a GPU, while lower layers need more efficiency than the microcontroller can provide.

Digital signal processor The line between general-purpose processors, microcontrollers, and digital signal processors can be blurry. DSPs are distinguished from microcontrollers by being more powerful. They contain little overhead, do not run even a real-time operating system, and have multiple parallel ALUs with dedicated hardware for fixed-point or floating-point arithmetic.

DSPs can still be developed using high-level programming languages, but making full use of the hardware requires full knowledge of the structure of the ALU, and sometimes requires writing some assembly language. This makes DSPs significantly more efficient than a general-purpose processor, and significantly more complicated to develop for. DSPs enjoyed some historical popularity due to their low cost and how easy it is to use them. However, outside of prototyping, they have consistently been squeezed out on both sides of Fig. 3.9 by improvements in general-purpose processors and FPGAs.

Array structures Array structures are regular arrays of programmable elements. They could be onetime programmable or multiple-time reprogrammable. Array structures historically provided a middle ground between ASICs and processors. However, except for FPGAs, array structures are only a historical curiosity, again being squeezed out by improving costs and performance at both ends of the spectrum.

FPGA Short for field-programmable gate arrays. These are regular arrays of identical reprogrammable elements. The elements are programmed by a file provided by the user. The word "programming" in this case means configuring the hardware of the cells so that it most fits the hardware that the user wants. FPGAs are increasing in popularity both as prototyping and implementation platforms. Although they would not replace ASICs for highly dedicated tasks, it is conceivable that at least some of the jobs required by the lower layers can be delegated to FPGAs.

FPGAs started out being complicated to develop for, almost as difficult as ASICs. They were also inefficient, almost as much as DSPs. However, many factors have helped them surge in value and popularity. Technology scaling has led to constant improvement in the performance of FPGAs. Intelligent choices about cell design and the use of dedicated special purpose cells for memory and arithmetic have also helped improve power performance.

The thing that is most attractive about FPGAs is the development cycle. FPGA development parallels that of ASICs. However, because FPGAs have a more rigid structure, less things can go wrong. And because they are reprogrammable, if things do go wrong, you can easily fix them. This means that the development cycle for FPGAs is more agile than ASIC, but using good design practices, you also end up with design files that can be easily migrated to ASIC.

Standard cell ASIC Digital application-specific integrated circuits are usually implemented using the standard cell approach. The ASIC design flow is used to map the design described at a higher level to standard cells drawn from a vendor library. These cells are then arranged in neat rows, producing a layout from which fabrication photomasks can be derived. Standard cell ASICs provide near best performance, but require a lot of time, expertise, and money to develop properly. An ASIC of any kind requires mass production to make economic sense.

Full-custom ASIC In a full-custom ASIC, layouts can be hand-drawn, or tools can be overridden to produce a circuit that is very specific to what the designer wants. This approach is extremely rare for very large digital circuits. It is more common in analog circuits and when large corporations are trying to optimize a certain critical section of a processor. The lower layers of a commercial radio will combine standard cell and full-custom ASIC philosophies. Nothing can challenge the efficiency of a full-custom ASIC. Every part of the chip is there for a purpose, and no single thing is there as overhead. However, it is very hard to get a full-custom ASIC to work.

3.8 Hardware Architectures

If we want to transform the following pseudocode into hardware, how many multipliers should we use?

```
A = B × C;
D = E ×G;
F = K × L;
W1 = Z × X;
X1 = A1 × B1;
```

The immediate answer is five, but five what? The code obviously has five *multiplications*, but this does not mean that we will need five *multipliers*. Making this leap in thinking can be challenging because it requires knowledge about the hardware used, the algorithm, and quality of service requirements. But once we do make the leap, we will be a lot closer to deciding our architecture. We will consider four architectural options, and will expand upon them in Chaps. 8 and 10:

- Direct implementation
- Parallelism
- Pipelining
- Time-sharing (hardware reuse)

The first and easiest architecture is the direct implementation. In this case, we replace each step in the algorithm with a processing unit (PU). Thus, the pseudocode above leads to the implementation in Fig. 3.10. Notice that the pseudocode has five multiplications, and the direct implementation has five multipliers.

The rate at which we generate outputs in Fig. 3.10 will be the speed of the available multipliers, which will differ from platform to platform (Sect. 3.7). Thus, if

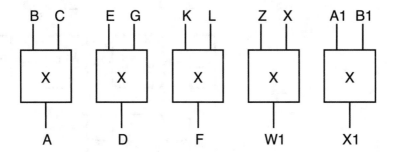

Fig. 3.10 Direct implementation of the multiplication network

we are in an ASIC CMOS technology where multipliers run at a rate of 100 MHz, the circuit in Fig. 3.10 will produce outputs at a rate of 100MSps.

If the throughput we require is 100MSps, then all is well and the direct implementation in Fig. 3.10 is perfect. However, this is almost never the case; the multiplier will either run faster or slower than needed. Both cases pose their own challenges and impose their own cost:

- Multipliers that run slower than needed are obviously a problem. We do not obtain throughput at the rate we require. This problem is solved either by parallelism, pipelining, or both. The solution will depend on the severity of the problem and the nature of the path that causes it.
- When a multiplier runs faster than needed, this is also a huge problem. In fact, this is the case we are more likely to encounter. This means the one-to-one implementation in Fig. 3.10 uses more resources than needed. This means more power is being burnt and more area is occupied than needed. One way to alleviate this problem is to back off the operating frequency of the circuit, allowing us to lower the power supply and substantially lower power dissipation (Sect. 8.14). But the more common solution is to time-share or time-multiplex the processing units among operations (Sects. 8.8 and 8.13). This is a very common strategy that leads to the focus of design moving from the processing units to managing memory bandwidth (Sect. 8.9) and complex controls (Sects. 8.11 and 8.12).

To delve deeper into these architectures, we must define what we mean by delay and power. Delay on the circuit level could either mean the propagation delay through a combinational logic block or the delay through a path consisting of multiple combinational blocks.

Most digital circuits consist of synchronous pipelines as in Fig. 3.11. Circuits that use multiple clocks require special attention and are beyond the scope of this book, but we will consider them briefly in Chap. 10. Figure 3.11 consists of CLBs, or combinational logic blocks contained between register pairs. Being synchronous, all registers in the pipeline use the same clock. The minimum amount of time between

Fig. 3.11 Paths, critical path, and operating frequency

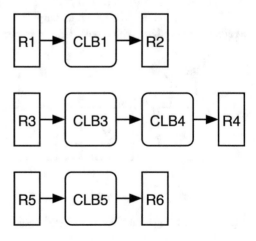

two active edges of the clock is the period of the clock. The period must allow enough time for data to propagate through the combinational logic blocks. Because we have only a single clock but multiple combinational blocks, the clock period will be:

$$T \geq \max \left\{ t_{\text{combinational logic}} \right\}$$

The above assumes there is no overhead to delay in the registers, which is a wrong but acceptable assumption. Any combinational delay between two registers is called a path. The path with the maximum delay is called the critical path, and it is very important because it dictates the performance of the overall circuit, for any circuit is only as strong as its weakest link.

So, when we talk about speed, what do we mean? The first and most obvious measure of speed is the reciprocal of the clock period. This is the clock frequency. But a more important measure of speed is throughput. Throughput is the rate at which meaningful outputs are produced from the circuit measured in MSps. In Figs. 3.10 and 3.11, throughput and clock frequency are identical, but when parallelism or time-sharing are used, this will not be the case.

Another measure of speed is latency. Latency is the number of cycles, or time in seconds that we take to produce a certain output in response to certain inputs. Or in other words, it is the number of cycles that the pipeline is not producing meaningful results as it is filling up. So, which is a more representative metric of speed, throughput, or latency? The answer depends on the application. In most applications throughput matters a lot more, with latency being a mere curiosity that is noticed only when the circuit is first turned on.

In baseband radios, particularly those that deal with packet-based communication, latency might matter more than in a typical application. Every packet, the receiver needs to start fresh, and we need to absorb the latency of the circuit again. Moreover, certain blocks in the transceiver chain have inevitable latency. This is latency that does not relate to the implementation but rather to the algorithm itself. For example, the interleaver (Sect. 6.13) and FFT/IFFT (Sect. 7.11) will always have considerable latency.

Latency can add up through the system, but why should we care? If we assume that we have continuous communication, we would not. But we have packet-based communication. When the packet ends, the standard expects the transmitter and receiver to finish certain things at certain times. For example, there is no maximum time between packets, but there is a minimum. In 802.11n, as soon as a packet ends, there is a possibility that another packet will follow 2 microseconds later. The receiver must have finished processing by then to be able to receive the new packet. If the summation of block latencies exceeds this budget, we run the risk of missing a packet. Also, the MAC layer requires us to send an acknowledgment packet within a certain amount of time; otherwise, the transmitter will consider us to have timed-out and will retransmit. Although this grace period for acknowledgment is much larger than typical latency, recall that the acknowledge packets must access the channel, which is going to take some time using CSMA-CA.

So, should we care more about throughput or latency? The answer again depends on the application, but we should be careful not to totally ignore either. In packet-based systems, we *must* take latency into consideration. In fact, the most effective measure of speed is how long we take to finish a certain function. Consider, for example, a receiver which has a throughput of 200Mbps operating at a frequency of 200 MHz. It has a latency of 512 cycles. How long does it take for the receiver to process a packet of 1000 bits?

The receiver operates at a frequency of 200 MHz, producing 200 Mbps; this means it is not producing multiple bits per cycle and is not taking multiple cycles to produce a bit. Instead, we are in a very simple situation in which the receiver produces a single bit per cycle. Next, translate the latency to time:

$$\text{Latency} = \frac{512}{200} \mu\text{sec} = 2.56 \ \mu\text{sec}$$

Now, we find the time it takes to produce the 1000 bits once the pipeline is full:

$$\text{Time} = \frac{1000}{200} \mu\text{sec} = 5 \ \mu\text{sec}$$

Thus, processing the packet will take 7.56 microseconds, of which 2.56 are consumed in an initial latency while the pipeline fills. This is the best measure of "delay" or "speed" we have because it combines the impact of both throughput and latency.

Next, let us discuss power and why it matters and how we can control it. Power dissipation combined with "speed" gives us an idea of how long a battery charge will last. For example, assume that a cell phone has a battery with a capacity of 2000 mA. hr providing a supply voltage of 1 V to the phone processor. Assume all the following:

- The phone receives packets of 1000 bits on average.
- Delay is the same as the discussion above both for latency and throughput.
- The processor consumes a power of 1 W *when it is running*.

The question is: how long the battery will last? Although we cannot answer how long in terms of time, we can answer how long in terms of bits. To receive 1000 bits, we take 7.56 microseconds. The battery provides 2000 mA.hr. at 1 V. This is 2000 mW.hr. because the potential is unity. Processing a single packet of data will consume 1×7.56 W. μsec of energy. Converting this energy into mW.hr will give a very small number. To deal with a more reasonable number, we calculate the energy needed to process 1 million bits. This will be:

$$7.56 \times \frac{1000 \times 1000}{10^6 \times 60 \times 60} = 0.0021 \text{ mW.hr}$$

Thus, the battery will run out after we process $2000/0.0021$Mbits $= 952$Gbits. While this might sound like a lot of data, it is not if the processor is being constantly used. For example, to stream a movie at 1080p, you will need to stream between 1 and 2 GB of data. This is upward to 20 Gbits. Streaming at 4 K is going to require even more data.

Power dissipation in a CMOS microchip is mostly consumed in active dynamic power. This is especially true in very small channel length transistors where leakage is controlled by FinFET structures. In this case power dissipation has the expression:

$$P = \alpha C f V_{DD}^2$$

where:

$\alpha =$ activity factor
$C =$ capacitance switched in transitions
$f =$ operating frequency
$V_{DD} =$ supply voltage

The activity factor is the proportion of cycles that the capacitance is actually switched. If the activity factor is 1, then we are assuming that the capacitance charges and discharges every cycle of the clock. The above equation tells us that the most effective way to reduce power dissipation is to reduce supply voltage.

It is also worth knowing that operating frequency is directly proportional to supply voltage:

$$f \propto V_{DD}$$

This is truer in long channel transistors than it is in more contemporary MOSFETs, but there is still a favorable relation between the two. Back to Fig. 3.10, assume that the direct implementation produced a higher throughput than needed. One solution to this problem would be to lower the operating frequency. This would allow us to proportionately reduce supply voltage, which results in a dramatic drop in power dissipation. Power will drop because frequency drops, but more importantly because of the drop in supply.

Now consider the code segment we looked at near the beginning of this section. There are five multiplications. Each of them is independent from the other, meaning none of the lines are waiting for the output from a previous line. When we compare the throughput of the direct implementation in Fig. 3.10 and the throughput required, there are two possibilities: the multipliers are faster than needed, and the multipliers are slower than needed.

Assume, for example, that we need each of the outputs at a rate of 100Msps and that each multiplier can work at a rate of 20Msps. This is an unrealistically slow

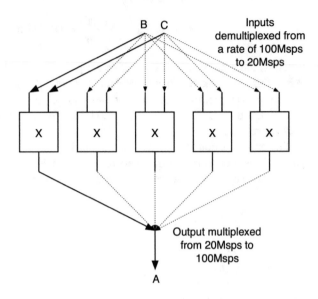

Fig. 3.12 Parallel implementation of the multiplication for output A

speed for a multiplier; typical values are over an order of magnitude higher; however, these numbers are used here only for illustration. In this case, the multipliers are much slower than needed. If we use a one-to-one implementation as shown in Fig. 3.10, we obtain a throughput that is only one fifth of the throughput we need. In such a case, we should consider parallelism.

Figure 3.12 shows how parallelism can be used to resolve this discrepancy. There are five multipliers in parallel, each working at a rate of 20 MHz. We produce five outputs at a time; this allows us to produce output at an effective rate of 100Msps. Parallelism is a straightforward proposition, and the trade-off is clear. We use more area and get proportionately more throughput. More area also means proportionately more power.

Because power and speed in parallelism are increasing at the same rate, power-delay product remains constant. This makes sense because parallelism is not a smart design choice, and it should not improve efficiency. It is also not a stupid design choice, so it should not harm efficiency.

Parallel architectures have one major challenge: providing inputs at the required rate. In Fig. 3.12, we are assuming that inputs can be provided at a rate of 100 MHz. This is not always the case. In fact, in radios, this is rarely ever the case. Inputs are obtained from the antenna, and they are provided only as fast as they arrive. So, in a 20-MHz system, we can only provide inputs at a rate of 20 MHz, in which case it would be impossible to even use parallelism.

The other architecture available to cover a throughput deficit is pipelining. This is better illustrated using the circuit in Fig. 3.11. Assume that the clock rate of this circuit is 100 MHz. The critical path is the middle path formed by CLB2 and CLB3. Assume also that each CLB has a delay of 5 microseconds. This means that the path formed by CLB2 and CLB3 has a delay of 10 microseconds, leading to a frequency of 100 MHz.

Now assume that we need to operate Fig. 3.11 at a rate of 200 MHz. This would require paths with delay 5 microseconds. This is fine for the paths formed by CLB1 and by CLB4. But we must break the path formed by CLB2 and CLB3 by inserting a pipeline register between the two. This creates two new paths out of the old critical path and increases frequency and throughput for the whole circuit.

Pipelining leads to a small increase in area but might improve throughput substantially. It leads to proportionately more power dissipation, an increase in latency (as measured in cycles), and more complicated control.

Now assume the opposite is true about rates: multipliers are faster than needed. For example, assume multipliers produce outputs at a rate of 100 MHz, while the output is needed at a rate of 20MSps. This is a common case in practical designs. Processing units are often much faster than we need them to be. If we use the one-to-one implementation, we end up with circuits that work so fast that we cannot even provide them inputs at the rate they require.

As mentioned earlier, one thing we can do about this is to reduce the operating frequency down to the level we need. Thus, only operate the multipliers at 20 MHz even though they can operate at 100 MHz. This means we have reduced their operation to a fifth of its potential. In Sect. 8.14 we will find a direct relation between operating frequency and power supply. We also saw a quadratic relation between active power and supply. Thus, reducing operating frequency from 100 to 20 MHz allows us to reduce the supply to a fifth of its original value. This leads to a decrease in power dissipation of ($1/(5 \times 5 \times 5)$).

This comes with its own caveat. Reducing power supply reduces the noise margins accordingly. On a more practical note, chips use dedicated metal planes to distribute power supply to different parts of the chip. The designer does not have full freedom to define any value for power supply on a continuum. It must be picked from a couple of supply levels that are available on-chip.

The more common approach when processing units are faster than needed is to time-share or reuse them (Sects. 8.8 and 8.13). In time-sharing, the processing unit is clocked at its native frequency and is used multiple times to produce an effective throughput that satisfies the requirements of the original circuit.

Figure 3.13 shows a single multiplier time-shared to perform the five multiplications described in the pseudocode at the beginning of this section. The multiplier is operated at 100 MHz; thus, it can accept a new input every 10 μsec. We must produce a full set of five outputs every 50 μsec. The single multiplier will finish producing the set of five outputs in five cycles, which is also 50 μsec. So, in effect, we are implementing the pseudocode at a rate of 20 MHz using only a single multiplier.

Time-sharing is efficient relative to other hardware management approaches, but it still has a cost. The cost lies in the complexity of managing data inputs and outputs to the processing unit. This is something that must be manually managed and optimized by the designer and cannot be automated. In Fig. 3.13, we are using multiplexers to manage inputs to the multiplier. The multiplexer selection lines are

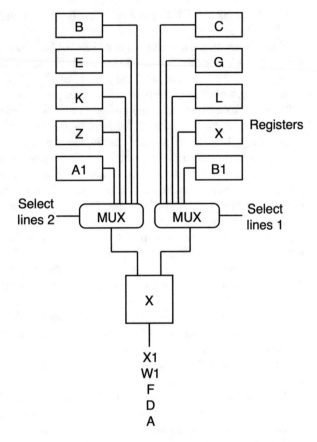

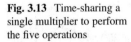

Fig. 3.13 Time-sharing a single multiplier to perform the five operations

generated by a state machine that the designer has hand drawn. In our example, the controller will be very simple: every cycle the select bus is incremented. At the output we must also manage where data goes. If data goes to registers, we enable the registers on a cycle-by-cycle basis.

The approach above using multiplexers and registers is not scalable. But we can see that the register-multiplexer combo is just a memory, and this is exactly what we must replace them with in problems of realistic size. Fig. 3.14 shows a conceptual view of the arrangement in the general case.

The input of the processing unit is connected to the output of a memory. Every cycle, we read from a memory location corresponding to the correct input. The read addresses are generated from an address generator that is associated with the circuit state machine. The output is written to a memory (not shown in the figure), and again the write addresses are obtained from the state machine managing flow in the whole design.

```
A1 = B × C;
A2 = A1 × D;
```

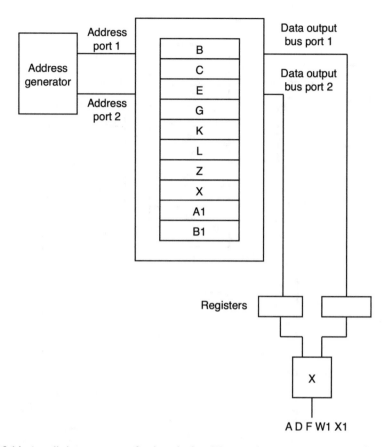

Fig. 3.14 A realistic arrangement for time-sharing. We are using a two-port memory for convenience, but we could just as easily use a one port memory

```
A3 = A2 × E;
A4 = A3 × F;
A5 = A4 × G;
```

Now consider the pseudocode above. This differs from what we have been dealing with so far in that the operations are interdependent. To be able to do the second operation, we must finish the first operation. How can we use memories and a multiplier to implement this? We are still assuming a five to one relation between processing unit speed and the required throughput.

The arrangement is shown in Fig. 3.15. Here, we are using the same memory to read and write. This is a requirement because the outputs in some cycles are the inputs to the next cycle. This is something that we will find to be true in most algorithms. This raises an important limitation and challenge of time-sharing: memory bandwidth.

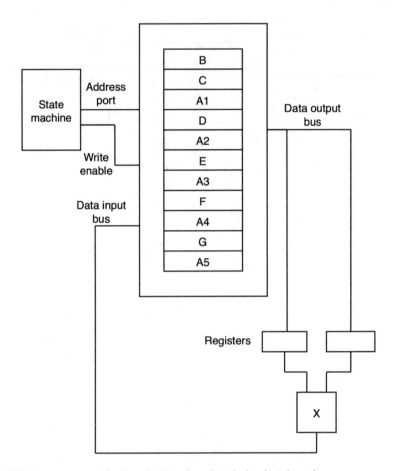

Fig. 3.15 An arrangement for time-sharing where there is data interdependency

In Fig. 3.15, the inputs need to be read from memory at a rate of 100 MHz, but the outputs also need to be written to it at the same rate. If the memory is single port, this means we need to clock it at three times the clock speed of the PU. If the memory is two-port, we need to clock at twice the rate, specifically 200 MHz. This means that for every cycle of the multiplier, the memory will be clocked twice; on one clock we will read two inputs from the two ports. On the second cycle, we will write the output using a single port.

But what if we cannot clock the memory at such a rate? We might object that this sounds impossible; a rate of 200 MHz is very easy for most RAMs. However, we are considering a very simple case in which we only have a single multiplier. This happened because we had five operations and the multiplier was five times faster than the operations. In most time-sharing implementations, we need multiple processing units.

In such cases, we need to clock the memory many multiples of times per processing unit cycle. This is necessary to provide all the inputs to the processing units and to write their outputs to the shared memory. This is the crux of the problem of time-sharing. PUs need to intercommunicate, and they can only do so through a shared memory. This puts an untenable stress on memory bandwidth.

Visiting the above pseudocode again, assume we use two-port memories, and we can only clock them at 100 MHz. We can only clock the memory as fast as the PU. Thus, we can only have two memory accesses per cycle (one per port). This will not do, because in every cycle, we need two reads and one write. But if we divide the total memory into two memories, we can solve this problem.

This is called memory partitioning, where we divide memory into banks, each with its own read and write ports. This alleviates contention on memory bandwidth because we can clock multiple memories in parallel. But memory partitioning requires familiarity with the application to decide how to assign data to the partitions.

For example, for the above code assume, we divide the memory into two partitions and decide to put A1 and D in the same bank. This would be a bad decision because in a certain cycle, we will need to write A1 to that specific bank and then read A1 and D from the same bank. This means three memory accesses per a single PU cycle. This will not work out because our maximum is two memory accesses per PU.

Thus, only when we properly distribute the data will we be able to resolve the memory bandwidth problem. The problem becomes even more challenging when dealing with a very large number of processing units all working together to perform a complicated algorithm with multiple steps. We will revisit time-sharing and memory partitioning again in Chap. 8 with a realistic design example.

3.9 Cost and Performance of Arithmetic

n-bit adders are often implemented using an architecture called the ripple carry adder. This is shown in Fig. 3.16. The critical path of the ripple carry adder passes from the LSBs of the operands all the way to the last sum bit. The delay of the adder

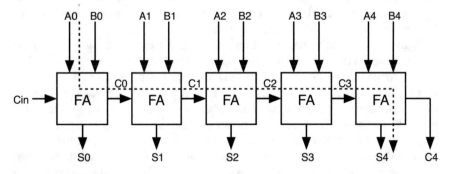

Fig. 3.16 Ripple carry adder

increases linearly with the number of bits. Thus, a 32-bit adder operates at roughly half the frequency of a 16-bit adder. There are adders whose delay increases slower than the ripple carry adder, but this is beyond the scope of this book.

Multipliers, at least full multipliers (Sect. 3.4), are implemented using the array multiplier architecture shown in Fig. 3.4. The array multiplier reflects the binary multiplication operation. AND gates generate partial products which are arranged in summands. Half adders and full adders then reduce the summands into a product. The delay of a multiplier increases roughly at double the pace of increase of input wordlength. Its area increases at a quadratic pace. So, multipliers are generally slower than adders of comparable length. Therefore, multipliers are generally part of the critical path of any processor or digital platform, and the speed of the integrated circuit is determined by the speed of the multipliers used.

We have talked about additions and multiplications as the only mathematical operations in digital circuits. This is a bold assertion, but it is not far from reality. When we consider an algorithm written in floating point by a systems designer, it will usually contain a liberal collection of operations supported by high-level programming. This includes an essentially endless list of possibilities, including but not limited to exponentiation, trigonometric functions, radicals, division, matrix decompositions, and combinations thereof.

There is no native way to implement such transcendental functions in hardware the same way we implement addition and multiplication. For both operations we simply map long addition and long multiplication directly into hardware using full adders, half adders, and simple logic. For other mathematical operations, we need to be a little more creative when moving to hardware.

In Chap. 10 we will use CORDIC to implement trigonometric and radical functions using only adders and shifters. This is an efficient approach for a large variety of functions. However, using lookup tables (LUTs) is a more general, albeit less effective approach, which can implement any function.

Figure 3.17 shows the simplest way to use the LUT approach in implementing functions. The LUT is a memory, which in this case need only be a ROM although it could also be a RAM, which is loaded with the correct values at the beginning of operation. The memory contains entries corresponding to $\log_{10}$ of the address. Thus, the location of address 3 contains $\log_{10}(3)$. The location of address 6 contains the value $\log_{10}(6)$, and so forth. This example is very limiting because it suggests that the independent variable x could only be an integer which increases by 1, corresponding to the address bus of the memory.

Figure 3.18 shows that this limitation is artificial. If there is a decoder between the LUT and the function input, then the function input could take any values and be mapped to the integer addresses. In the example shown, θ takes values $0 : \pi/20 : \pi/2$. This corresponds to 11 locations in the LUT. A translation block stands in between the memory and the input θ. It takes care of translating so that $\theta = 0$ goes to $x = 0$, $\theta = \pi/20$ goes to $x = 1$, $\theta = 2\pi/20$ goes to $x = 2$, and so forth.

LUT-based functions can be wasteful and memory consuming, or they can be efficient depending on the implementation. One major factor affecting this is the resolution we demand from the function. For example, in Fig. 3.18 we are using a

Fig. 3.17 Using a LUT to implement $\log_{10}(x)$ for $x = 1:10$

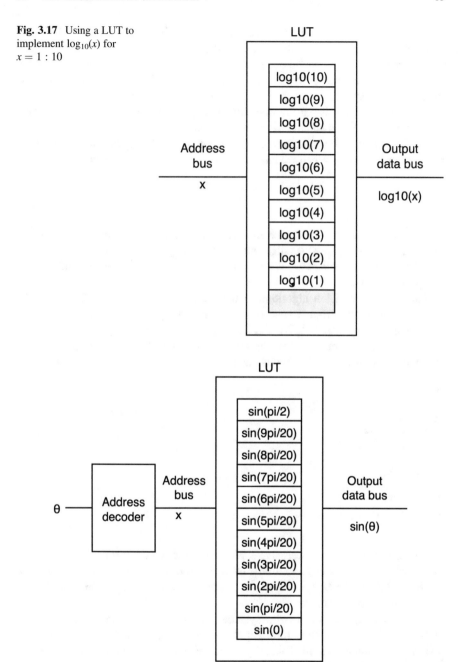

Fig. 3.18   LUT used to store the values of $\sin(\theta)$ where θ lies between 0 and $\frac{\pi}{2}$

resolution of $\pi/20$. Using a higher resolution will require a larger memory. Utilizing symmetries in functions can also increase efficiency substantially. For example, for trigonometric functions we do not need to store the entire range between 0 and 2π. Symmetries in the function mean we only need to store between 0 and $\pi/2$. We can then utilize the symmetries to derive the function in other ranges.

Now to division, it is our favorite child among the complicated functions. There are many algorithms available to implement long division. Most of the algorithms are well-described and characterized in software, but some of them can also be easily adapted for hardware. When we talk about "division," we mean the following form:

$$\frac{A}{D} = Q + \frac{R}{D}$$

Specifically, we are dividing a number A which is n-bits long by a divisor D. D will always be assumed smaller than A. Q is the quotient, the integer number of times that A is larger than D. We will also produce R, the remainder which is an integer less than D.

The simplest division algorithm is the successive subtraction algorithm. The divisor is subtracted from A successively. When the result becomes negative, we know that we have overshot. We add the divisor back to the register. The number of times we have subtracted is the quotient. The contents of the register are the remainder. The pseudocode for this algorithm is:

```
Q = 0;
R = A;
Repeat forever
  R = R - D;
  Q = Q + 1;
  If R < 0
    Break;
  End if;
End repeat;
Q = Q - 1;
R = R + D;
```

The problem with the successive subtraction method is that it has a nonconstant throughput. We may take any number of cycles to produce an output based on how much A is larger than D. This algorithm is more suited for software, and particularly floating-point processors.

When dealing with fixed-point systems, many other division algorithms exist with a constant and reliable throughput. One example of such algorithms is the restoring division algorithm. This is by no means unique or optimal, but it is easy to understand. The pseudocode for this algorithm is:

```
R = A;
D = D << N;
For i = N - 1:1:0
```

```
R=2R - D;
If R > 0
  Q(i) = 1;
Else
  Q(i) = 0;
  R = R + D;
End if;
End for;
```

The algorithm first loads A into the initial value of the remainder. It then makes D twice as wide as A by shifting it. We loop through all the bit positions of A. At each step we shift R to the left and try to subtract the divisor. If the result is negative, that bit of the divisor is null, and we must restore the value of remainder. If the value is positive, the bit of the quotient is one. This algorithm is very similar to successive division; the main difference is that the number of steps is fixed at the number of bits of the operand A.

Chapter 4
The Wireless Channel

4.1 Partitioning Wireless Channel Effects

Figure 4.1 shows a simplified view of a wireless communication system. The transmitter is radiating a signal out, and the receiver is trying to capture the said signal. In between, there is a wireless channel that changes and distorts the signal. Parts of the channel act as if they are a linear time-invariant system, parts will be dynamic, and parts are additive. Because the channel will substantially distort the signal, the receiver needs to reverse these distortions, which is why the receiver is more complicated than the transmitter.

If the receiver is to invert the effects of the channel, it must *know* what the channel is. And if it is to know what the channel is, then the transmitter must send data that the receiver knows in advance. This way the receiver will examine this a priori known data and deduce the distortions that the channel introduced. Therefore, we need preambles in the header and pilots within data. We will discuss how channel state information can be obtained at the receiver in detail in Chap. 11.

For the receiver to estimate channel state information (CSI), it must have a model of the channel. In other words, we should all agree on a mathematical equation relating the transmitted signal, received signal, and channel characteristics. It is neither necessary nor useful to lump all channel impacts in a single stroke. Channel impacts are physically independent, and thus they are easier to understand and model separately.

We will divide channel effects into the following:

- Additive white Gaussian noise (AWGN)
- Path loss
- Short-term fading effects
- Shadowing
- Doppler shift

K. Abbas, *From Algorithms to Hardware Architectures*,
https://doi.org/10.1007/978-3-031-08693-9_4

Fig. 4.1 Simplified view of
a wireless communication
system

We will discuss each of these effects in detail. But first we must develop a conceptual understanding of what each means:

- Path loss is probably the most intuitive channel effect. It is the natural loss of signal power the farther the signal travels.
- Fading is what distinguishes a crowded city from an open field. It is a local effect due to the presence of reflectors in the environment.
- AWGN is a fundamental physical phenomenon that adds random zero-mean noise to a system due to thermal motion of atomic and subatomic particles in the receiver RF.
- Shadowing is a long-term variation due to changes in the environment. It is the long-term partner of fading. Imagine passing through a long tunnel or behind a large mountain, the impact you observe is called shadowing.
- Doppler shift is yet another fundamental physical phenomenon where relative motion between a transmitter and a receiver will lead to a shift in the observed signal frequency. Doppler also models the motion of reflectors in the system; thus, it is a representation of the time variability of the channel.

All these effects are stochastic and impact the signal negatively. But this is where their similarities stop. Each effect comes from a different place, impacts the signal differently, and must be addressed by the receiver in different ways.

There are two effects that we will discuss in this chapter but ignore for the rest of the book. These are, namely, Rician fading and shadowing. Neither is particularly important for Wi-Fi. But it is worth noting that shadowing can be devastating in cellular systems, while Rician fading can destroy systems with multiple antennas.

4.2   Additive White Gaussian Noise (AWGN), a Fact of Nature

AWGN is the most fundamental and natural channel effect. It is present in all electronic systems whether or not they communicate wirelessly. Interestingly, AWGN is often misunderstood and is hard to deal with because at the fundamental level, you cannot get rid of it.

To understand AWGN it is best to consider each of the words that describe it. AWGN stands for additive white Gaussian noise. Additive indicates that the noise is *added* to the signal rather than multiplied by it. Thus, the model that describes AWGN is:

$$y = x + n$$

where x is the transmitted signal and y is the received signal. n is a scalar value representing the current noise sample. This equation is valid for both passband and baseband. In the baseband, which is where we will mostly use it, all the variables in the equation are complex numbers. x will come from a dictionary of possible symbols according to Chap. 5, while n can take a continuum of values.

The nature of n is described by the remaining letters of the acronym AWGN. It is a stochastic process whose distribution is Gaussian and whose power spectral density is white. AWGN is a fundamental and unavoidable effect that occurs whenever particles exist at any temperature above zero Kelvin. In such conditions the particle will acquire thermal energy equal to:

$$E_{\text{th}} = \frac{2}{3} kT$$

where T is the temperature in Kelvin and k is Boltzmann constant. This thermal energy is transformed into kinetic energy of an equal amount. For atomic and particularly subatomic particles, the mass is low enough that this amount of kinetic energy transforms into a measurable and meaningful thermal velocity v_{th}:

$$v_{\text{th}} = \sqrt{\frac{4kT}{3m}}$$

This velocity is very high for electrons and holes in a semiconductor due to their small effective mass. One would expect such a high velocity to lead to a large current. However, thermal *speed* might be high, but the direction of motion is a zero-mean random variable. Thus, the net motion of any particle under thermal velocity is null. The motion does, however, lead to a kind of noise called thermal noise, which is the physical manifestation of AWGN.

The W in AWGN stands for white. This is a description of the spectrum of the noise. Specifically, it is a statement on its PSD (power spectral density). The PSD of white noise is flat, at least in the band of interest. In fact, the power density of thermal noise is a constant for a given temperature at kT eV. At room temperature, this is roughly 25 meV. The whiteness of noise indicates a lot about its nature. For example, it indicates the noise does not discriminate based on frequency. It impacts all frequencies equally. It also indicates that the total amount of thermal noise in the system is directly proportional to its bandwidth.

But whiteness of noise can be viewed more intuitively in time domain. When PSD is sent back to time domain, we get the autocorrelation function. A flat (white) PSD corresponds to an impulse autocorrelation function at time 0. This is shown in Fig. 4.2.

The autocorrelation of AWGN exposes the challenge of this form of noise and hints at the most effective way to address it. An impulse autocorrelation means that a sample of AWGN noise has full correlation (indicated by the impulse in Fig. 4.2)

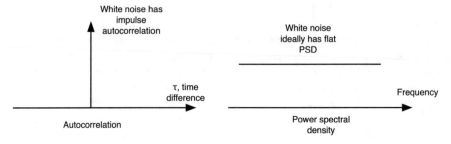

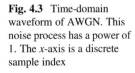

Fig. 4.2 Autocorrelation and PSD of AWGN

Fig. 4.3 Time-domain waveform of AWGN. This noise process has a power of 1. The *x*-axis is a discrete sample index

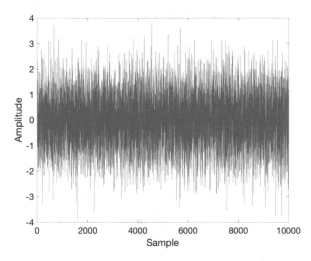

with itself. This is natural, expected, and true for any signal of any nature. After all, a sample gives us all the information about itself, and is thus fully correlated with itself.

What is unique and interesting about AWGN is that autocorrelation will drop to zero for any positive or negative time shift. This means that a sample of AWGN is fully decorrelated from all other samples of the same noise process no matter how close that other sample is to it in time.

The implication here is tremendous. If the current thermal noise sample is large, this means nothing for upcoming and preceding samples (Fig. 4.3). Knowing the previous sample tells us nothing about the current sample and vice versa. A sample could be small and be followed by a large sample, a small sample, one with a similar phase, one with a different phase, or any combination, with all possibilities equally likely.

This tells us that white noise is hard to deal with. After all, correlation is information. Knowing some correlation allows us to "prep" our receiver to better handle noise. Indeed, colored noise, noise whose PSD is not flat and whose autocorrelation is not an impulse, can be minimized with intelligent decoding techniques (see, e.g., SIC in Sect. 9.6).

However, white noise is not as bad as it sounds. In fact, for the same total power, white noise will have a lower average power for all frequencies than colored noise. Colored noise of equal total power could destroy a certain sub-spectrum where it has a PSD peak.

In fact, we will find that the best way to deal with white noise is to understand it and address it on its own terms. You can try to use features of AWGN to cancel it out, but you will fail, because AWGN has no features. However, the fact that AWGN is zero mean means that we could do something very simple to reduce it: take the average.

x, the sent signal, is not zero mean. If we assume that we sample x multiple times, then the average of all these x's is also x. On the other hand, the average of white noise tends to null the more samples we add. This leads to our fundamental theory in dealing with noise:

$$y = x + n$$

Taking averages:

$$E(y) = E(x) + E(n)$$

The average is taken over N samples:

$$\sum_N y = \sum_N x + \sum_N n$$

And if N tends to infinity:

$$\lim_{N \to \infty} \sum_N y = \lim_{N \to \infty} \sum_N x + \lim_{N \to \infty} \sum_N n$$

which for AWGN is:

$$\lim_{N \to \infty} \sum_N n = 0$$

And assuming we repeated each instant of the sent signal, this gives a received signal:

$$\lim_{N \to \infty} \sum_N y = \lim_{N \to \infty} \sum_N x$$

Fig. 4.4 Normal distributions. All distributions are zero mean. Their variances are 0.5, 1, 1.5, and 2 from the highest peak to the lowest peak

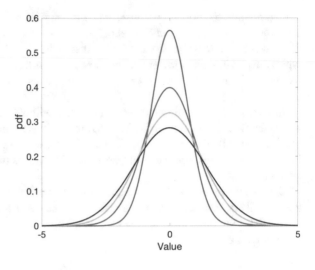

$$Ny = Nx$$

$$y = x$$

This is a self-evident but very important result. In short, it is not useful trying to outsmart AWGN; we should instead brute force smooth it out. But who said that AWGN is zero mean? This is where the G in AWGN comes in. The fact that the noise is Gaussian is an indication of the distribution of its amplitudes. Specifically, the pdf of the amplitude is:

$$f_N(n) = \frac{1}{\sqrt{2\pi\sigma^2}} e^{-\frac{(n-\mu)^2}{2\sigma^2}}$$

All normal distributions are fully characterized by mean and variance (or standard deviation) (Fig. 4.4). However, AWGN is always zero mean, which means it is only characterized by variance. Variance for AWGN is a direct measure of noise power kT. Thus, the pdf is:

$$f_N(n) = \frac{1}{\sqrt{2\pi\sigma^2}} e^{-\frac{n^2}{2\sigma^2}}$$

It is important to understand the relation, or rather lack thereof, between the whiteness and the normal distribution of AWGN. The normal distribution indicates differing likelihoods for certain magnitudes for a sample of noise taken randomly. Strictly speaking, if a large enough number of simultaneous identical noise processes are sampled, their histogram will form a normal distribution with a variance equal to their power. If the noise process is ergodic, which is always the case for AWGN, then the same applies if we take a large enough number of samples in time.

However, the increased likelihood of seeing certain amplitudes as opposed to others does not indicate any information about the following sample in time. In short, knowing that the current sample is large or small does not indicate anything about the magnitude of the following noise sample or any noise sample. But taken together as a population, the noise samples fit a Gaussian distribution distinguished only by the variance.

It is interesting to stop for a moment and consider why thermal noise has a normal distribution. In fact, what is so "normal" about it? Thermal noise occurs due to random Brownian motion of individual particles in a conductor. The zero-mean nature of thermal motion means that there is no current flow; however, the vibrations will lead to power dissipation. We do not, at this point, have any information about the nature of the individual motion of particles. But what is interesting is that we do not care.

The central limit theorem is a very important result from statistics. It states that for independent identically distributed random variables, the distribution of their summation tends to a normal distribution the larger the number, regardless of the individual distributions. The most striking fact here is that the result of the summation is always normal *regardless* of the individual distributions.

So, for our subatomic particles moving around in a conductor, we only care that they all have the same distribution and that their motions are independent. The summation of the effect of their motion *will* be Gaussian. Therefore, the Gaussian distribution is also called the normal distribution, because it is the "natural" distribution that a lot of natural phenomena exhibit, for example, heights in a human population, size of litter in cats, number of leaves on a tree, etc. The reason this distribution is very common and "normal" is that it is *always* the result with a large population given the distribution is identical and the individuals are independent.

AWGN is entirely defined either by the noise power (variance) or the SNR. The SNR is usually a better measure of noise because it combines the noise with the signal to give us a single metric that we care about. Signal to noise ratio, or SNR is defined as:

$$\text{SNR} = 10 \log_{10} \left(\frac{P_{\text{signal}}}{P_{\text{noise}}} \right)$$

SNR is measured in dB, and it is important to develop some intuition about what range of SNR makes sense when talking about radios. A negative SNR means the signal is below the noise floor, in which case the signal will drown in the noise and the BER will go through the roof (to be more accurate, it will tend to 0.5).

An SNR of zero means the signal and noise are at the same power level. An SNR of 10 means the signal power is ten times higher than the noise power. An SNR of 20 means the signal power is 100 times above the noise power. So, which of these levels is "good"? The answer depends on the quality of service required by the application and the structure of the radio. For most radios, SNR below 20 dB will not give a useful QoS. However, some radios *must* operate in SNR between 10 and

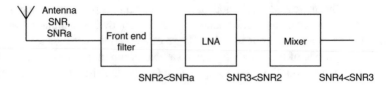

Fig. 4.5 The deterioration of noise factor in the receiver RF

20 dB. In such case, these radios will use very sparse constellations (Chap. 5) and deep coding (Chap. 6). On the other hand, many radios need SNR above 30 dB or even 40 dB to be useful. This is particularly true for radios that use a very large number of antennas (Chap. 9).

An important source of confusion is *which* signal power to use in the SNR calculation. Is this the transmitter power or the receiver power? It should be very clear before moving forward that we *always* talk about received signal power when calculating SNR. This may not immediately make sense, but it is very logical if we understand *where* AWGN occurs.

White noise is a receiver phenomenon. Specifically, it is the noise added to the signal as it moves through the RF section of the receiver. As shown in Fig. 4.5, as the signal moves through the RF section, it is moving as an analog signal through analog blocks. Each block will magnify or attenuate both the signal and the noise identically before adding its own noise. Thus, SNR deteriorates through the whole RF chain. But what is the *input* SNR to the RF chain? This would be the SNR at the receiver antenna.

Thus, thermal noise is a phenomenon of the receiver RF chain, particularly the receiver antenna. SNR is calculated using the received signal power rather than the transmitted signal power. Noise is *not* constantly being added and subtracted from the signal as it travels through the air. What we care about is ultimately the total noise power that gets added to the signal at the input of the baseband section. *This* is what defines our SNR. Deterioration in signal power as it travels from the transmitter to the receiver is accounted for under the umbrella of path loss and fading.

4.3   Single-Path Rayleigh Fading, Reflectors in the Environment

Fading is an umbrella term used to describe distortions that affect the signal due to reflectors and blockers in the environment. Fading does not consider loss of signal power due to radiation; this would be path loss. Rayleigh fading, which is the topic of this section, does not consider the presence of large blockers like mountains or large buildings; this is covered under shadowing in Sect. 4.8. Rayleigh fading only considers the presence of small but plentiful reflectors that do not completely block all paths between the transmitter and the receiver.

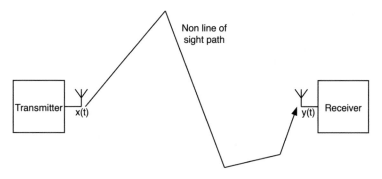

Fig. 4.6 A signal traveling along a single non-LOS path from transmitter to receiver

In Fig. 4.6, we see a signal traveling from a transmitter to a receiver. It is moving along a single ray. The ray is non-line of sight; it is not a straight line traveling from the transmitter to the receiver. Instead, there are multiple reflections along the way. According to Sect. 4.2, AWGN should be *added* to the received signal. But along the way, the sent signal is multiplied by a factor to account for what happened to it as it bounced along the way from the transmitter to the receiver. The factor is shown in the following equation as h:

$$y = hx + n$$

The above merits a few notes and a couple of questions:

- The factor h is called the channel coefficient, the channel factor, or just the channel.
- The channel factor does *not* account for path loss, which is a gross attenuation in amplitude that should be calculated from the distance the signal travels.
- Why does h multiply x? In other words, why is it not adding to it?
- We do not have any clues on the nature of h. Is it stochastic or deterministic? If it is stochastic, what is its distribution?
- The above is a baseband equation. All elements of the equation are complex scalars. Thus, when we write y, we know that we implicitly mean $y(n)$, or more appropriately $y(nT)$. The same applies for x, n (noise), and even h.

Let us make a first ditch effort at figuring out the nature of h. Examine Fig. 4.7; it shows an arbitrary time-domain signal moving through the channel in Fig. 4.6. Because the signal $x(t)$ has traveled through the channel, when it reaches the receiver as $y(t)$, its amplitude will change. The phase of the signal will also be different because we are considering complex numbers. However, the *shape and duration* of y are identical to those of x, because there is no reason they would not be.

What is the impulse response of the channel corresponding to Figs. 4.6 and 4.7? Impulse response is the response of the channel to an impulse input. Thus, it is y, when x is a delta-Dirac function, or an impulse in discrete domain. In other words, h is defined as:

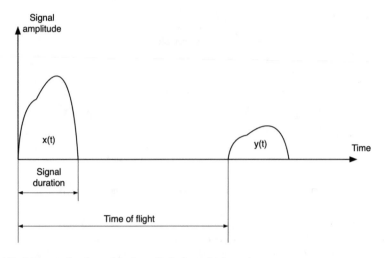

Fig. 4.7 Arbitrary signal moving through single-path channel

$$h(n) = y(n)|_{x(n)=\delta(n)}$$

We can go through a long derivation to prove what h should be, but the answer is intuitively obvious. An impulse input to the channel will also arrive as an impulse at the receiver, albeit with a different amplitude and phase. Thus:

$$h(n) = Ae^{j\phi}\delta(n)$$

We still need to know the nature of the amplitude and phase distortion coefficients A and ϕ, but at least we can agree that this impulse response describes the system in Fig. 4.6.

To understand the nature, or rather the distribution, of A and ϕ, we need to take a step back and think about the quality of the ray model in Fig. 4.6. Is this model describing anything close to what is happening in reality? The reality is that EM waves travel in wavefronts, not as straight-line rays. The ray model is a useful representation of what happens to signals in channels under certain conditions. But we must qualify the ray model and understand its limitations before we use it.

When a wavefront hits a reflector, it does not reflect in one coherent wavefront that reaches the receiver. Instead, it reflects as multiple versions. Each version will travel to the receiver differently, arriving at a different attenuation, phase, and time. They might add up constructively at the receiver, or they might act against each other. Depending on the delay each sees, they may even cause intra-symbol interference.

Figure 4.8 shows the channel impulse response in the presence of a single reflector. This is still the channel in Fig. 4.6, but instead of using a single ray, we recognize that there will be multiple reflections due to the multiple wavefront arrivals at the receiver. The impulse response will thus be a train of impulses, each

Fig. 4.8 Narrowband
single-reflector channel
impulse response

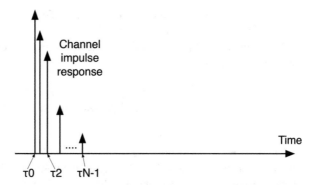

representing a single arrival at the receiver. The impulses will decay and die off, and
we will no longer observe significant signal arrivals after a time τ_{N-1}. This is
because the signal is traveling in only a single path, so we must eventually see the
signal dying off. Also note that we are not necessarily waiting for the arriving
impulses to completely stop; we are just waiting for them to fall below the noise
floor.

Thus, the impulse response, assuming N significant (above noise) arrivals at the
receiver is:

$$h(t) = \sum_{i=0}^{N-1} a_i e^{j\omega_c(t-\tau_i)} \delta(t - \tau_i)$$

A single impulse sent at the transmitter will arrive as N impulses with variable
amplitudes, phases, and delays. The channel is an LTI system, with h the time-
domain system impulse response. x is the input to the system and y is the output. The
correct system model would thus involve a convolution with the impulse response:

$$y = h * x$$

In this case because h is a train of impulses, the convolution is simple to calculate.
y will be multiple copies of x, each delayed to the corresponding impulse and scaled
to its amplitude and phase. In other words, there is no need to perform convolution,
and y can be calculated as:

$$y(t) = \sum_{i=0}^{N-1} a_i e^{j\omega_c(t-\tau_i)} x(t - \tau_i)$$

Notice that the carrier frequency is clear and glaring in the expression of the
impulse response. This means that contrary to our convention x and y above are not
baseband signals but rather complex passband signals. They are not the real

passband signals that travel over the air either. They are complex passband signals that are multiplied by the carrier phasor.

To obtain the real passband signal, the signal that travels over the air, we obtain the real part of y above:

$$y(t) = \text{Re}\left\{ \sum_{i=0}^{N-1} a_i e^{j\omega_c(t-\tau_i)} x(t - \tau_i) \right\}$$

But what is the corresponding baseband signal? As we already discussed earlier, the baseband signal is a series of complex numbers, which when multiplied by the carrier phasor gives the complex passband signal. In other words, if we can extract the carrier phasor from the summation above, we will be left with the baseband signal:

$$y(t) = \text{Re}\left\{ e^{j\omega_c t} \sum_{i=0}^{N-1} a_i e^{-j\omega_c \tau_i} x(t - \tau_i) \right\}$$

And the baseband received signal is:

$$r(n) = \sum_{i=0}^{N-1} a_i e^{-j\omega_c \tau_i} x(nT - \tau_i)$$

At no point above did we make any assumption about the nature of x. It could have any pulse shape, and it can be continuous or discrete. In the above equation, r is the received baseband signal suffering from a single reflector when the fact that we are dealing with wavefronts is taken into consideration.

If signal duration is much longer than the time periods separating wavefront arrivals, then $T \gg \tau_i$ for all i. This allows us to perform a critical approximation called the narrowband approximation:

$$r(n) = \sum_{i=0}^{N-1} a_i e^{-j\omega_c \tau_i} x(nT - \tau_i) \approx \sum_{i=0}^{N-1} a_i e^{-j\omega_c \tau_i} x(nT)$$

$$r(n) = x(nT) \sum_{i=0}^{N-1} a_i e^{-j\omega_c \tau_i}$$

This is called the narrowband approximation because it assumes the signal duration is much longer than channel discrimination. Later in the chapter, we will find out that in frequency domain, this means that signal bandwidth is much narrower than bands in which the channel response has variations. We will rename r as y to circle back to the model we introduced near the beginning of the section:

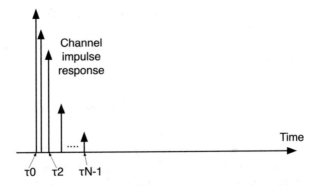

Fig. 4.8 Narrowband single-reflector channel impulse response

representing a single arrival at the receiver. The impulses will decay and die off, and we will no longer observe significant signal arrivals after a time τ_{N-1}. This is because the signal is traveling in only a single path, so we must eventually see the signal dying off. Also note that we are not necessarily waiting for the arriving impulses to completely stop; we are just waiting for them to fall below the noise floor.

Thus, the impulse response, assuming N significant (above noise) arrivals at the receiver is:

$$h(t) = \sum_{i=0}^{N-1} a_i e^{j\omega_c(t-\tau_i)} \delta(t - \tau_i)$$

A single impulse sent at the transmitter will arrive as N impulses with variable amplitudes, phases, and delays. The channel is an LTI system, with h the time-domain system impulse response. x is the input to the system and y is the output. The correct system model would thus involve a convolution with the impulse response:

$$y = h * x$$

In this case because h is a train of impulses, the convolution is simple to calculate. y will be multiple copies of x, each delayed to the corresponding impulse and scaled to its amplitude and phase. In other words, there is no need to perform convolution, and y can be calculated as:

$$y(t) = \sum_{i=0}^{N-1} a_i e^{j\omega_c(t-\tau_i)} x(t - \tau_i)$$

Notice that the carrier frequency is clear and glaring in the expression of the impulse response. This means that contrary to our convention x and y above are not baseband signals but rather complex passband signals. They are not the real passband signals that travel over the air either. They are complex passband signals that are multiplied by the carrier phasor.

To obtain the real passband signal, the signal that travels over the air, we obtain the real part of y above:

$$y(t) = \text{Re} \left\{ \sum_{i=0}^{N-1} a_i e^{j\omega_c(t-\tau_i)} x(t - \tau_i) \right\}$$

But what is the corresponding baseband signal? As we already discussed earlier, the baseband signal is a series of complex numbers, which when multiplied by the carrier phasor gives the complex passband signal. In other words, if we can extract the carrier phasor from the summation above, we will be left with the baseband signal:

$$y(t) = \text{Re} \left\{ e^{j\omega_c t} \sum_{i=0}^{N-1} a_i e^{-j\omega_c \tau_i} x(t - \tau_i) \right\}$$

And the baseband received signal is:

$$r(n) = \sum_{i=0}^{N-1} a_i e^{-j\omega_c \tau_i} x(nT - \tau_i)$$

At no point above did we make any assumption about the nature of x. It could have any pulse shape, and it can be continuous or discrete. In the above equation, r is the received baseband signal suffering from a single reflector when the fact that we are dealing with wavefronts is taken into consideration.

If signal duration is much longer than the time periods separating wavefront arrivals, then $T \gg \tau_i$ for all i. This allows us to perform a critical approximation called the narrowband approximation:

$$r(n) = \sum_{i=0}^{N-1} a_i e^{-j\omega_c \tau_i} x(nT - \tau_i) \approx \sum_{i=0}^{N-1} a_i e^{-j\omega_c \tau_i} x(nT)$$

$$r(n) = x(nT) \sum_{i=0}^{N-1} a_i e^{-j\omega_c \tau_i}$$

This is called the narrowband approximation because it assumes the signal duration is much longer than channel discrimination. Later in the chapter, we will find out that in frequency domain, this means that signal bandwidth is much narrower than bands in which the channel response has variations. We will rename r as y to circle back to the model we introduced near the beginning of the section:

$$y(n) = x(nT) \sum_{i=0}^{N-1} a_i e^{-j\omega_c \tau_i} = x(nT)h$$

We can answer some of the questions we asked earlier about h. First, why are we multiplying h by x? In fact, we are not. The channel h is convolved with the transmitted signal. However, because we are dealing with a single path, there is a single impulse in the channel and convolution reduces to multiplication.

But haven't we been discussing that the channel is not really a single impulse? It never is, but as shown above, under the narrowband assumption, the multiple impulses at the receiver are so close to each other that their differential times at the receiver are negligible relative to the symbol time.

Another question is that intuitively, from Fig. 4.6, the single impulse should arrive after a delay at the receiver. The delay is equal to the time of flight the signal takes from the transmitter to the receiver. However, as we will discover from packet-based communication, the receiver's "zero time" is defined as the time it starts detecting an arriving signal. Thus, we can ignore the time of flight.

There is still one unanswered question, and it is a big one. What is the nature of h? From the derivation above, h is the summation of multiple complex numbers. This makes it a complex number:

$$h = \sum_{i=0}^{N-1} a_i e^{-j\omega_c \tau_i} = \sum_{i=0}^{N-1} a_i \cos(\omega_c \tau_i) - \sum_{i=0}^{N-1} a_i \sin(\omega_c \tau_i)$$

$$h = \Psi + j\Phi$$

where:

$$\Psi = \sum_{i=0}^{N-1} a_i \cos(\omega_c \tau_i)$$

$$\Phi = -\sum_{i=0}^{N-1} a_i \sin(\omega_c \tau_i)$$

To understand the distribution of h, we must figure out the distributions of Φ and Ψ. Each is the summation of many random variables, making the whole ordeal cumbersome beyond any practical limit.

However, you might have noticed something here we used in Sect. 4.2. The multiple arrivals of impulses at the receiver are all random variables. For each path the number of individual arrivals N is large, and each arrival is independent of the others. This is all very suggestive of the central limit theory. However, we must be careful here, for the central limit theory is only valid for real signals. Thus, we can only apply it on Φ and Ψ individually. Each of these variables consists of the superposition of many independent, identically distributed variables. Thus, each of

Φ and Ψ is a normal distribution. There is no reason why either distribution would have a different variance from the other; thus, they both have the same power. Note that the beauty of the central limit theory is that we did not stop to contemplate the distribution of a_i or the phase.

Thus, h is the summation of two Gaussian zero-mean variables with identical variances. h is complex, so it is impossible to define a distribution for its whole. Instead, we will characterize either its real and imaginary parts or its amplitude and phase. We have already deduced that its real and imaginary parts are Gaussian.

The characterization of amplitude and phase is more useful because it is compatible with the multiplicative model for the channel, with the magnitude of the channel multiplied by the magnitude of the sent signal, and the phase added to the phase of the signal. Thus, we separate the channel h into two useful components A and ϕ:

$$h = \Psi + j\Phi = Ae^{j\phi}$$

$$A = \sqrt{\Psi^2 + \Phi^2}$$

$$\phi = \tan^{-1}\left(\frac{\Phi}{\Psi}\right)$$

When the two components of a complex random variable are Gaussian, the phase has a uniform distribution. Thus, ϕ is uniformly distributed from 0 to 2π.

The magnitude A has a distribution called the Rayleigh distribution. The pdf of the Rayleigh distribution is:

$$f_X(x) = \frac{x}{\sigma^2}e^{-\frac{x^2}{2\sigma^2}}, x > 0$$

And its CDF is:

$$F_X(x) = 1 - e^{-\frac{x^2}{2\sigma^2}}, x > 0$$

Its mean is:

$$\mu_X = \sigma\sqrt{\frac{\pi}{2}}$$

And its variance is:

$$\mathrm{Var}(X) = \left(2 - \frac{\pi}{2}\right)\sigma^2$$

where σ is the standard deviation of the formative Gaussian distributions. Note that the Rayleigh distribution is nonzero mean even though its constituent Gaussian distributions are zero mean. First, how is this possible, and second, does this make

Fig. 4.9 Rayleigh
distributions. The standard
distributions of the
constituent normal
distributions are 1, 1.5, and
2 from top to bottom

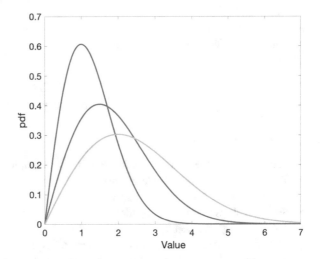

Fig. 4.9 Rayleigh distributions. The standard distributions of the constituent normal distributions are 1, 1.5, and 2 from top to bottom

sense in terms of fading? It is perfectly possible, even though the formative distributions are zero mean, their summation may not, and in fact should not be zero mean. If the two variables are independent, they add up and subtract in ways that give their complex summation a positive amplitude. Second, it completely makes sense that the arriving signal should not be multiplied by a zero-mean variable. If this were the case, the received signal would be completely dead, and we would never be able to communicate.

Figure 4.9 shows multiple Rayleigh distributions. There is an intrinsic relation between the mean and variance of a Rayleigh random variable. When one rises, the other must too, because both are controlled by the variance of the formative Gaussians. A path is thus solely defined by the variance of the formative Gaussians. This determines the mean magnitude of the multiplicative channel factor h. A "strong" path is one with a larger mean; a weak path is one with a smaller mean.

But Fig. 4.9 is often tricky. One's instinct would be to define the distributions with higher and stronger peaks as strong paths. In fact, these are paths with very low mean and are the weakest paths. Stronger paths lie to the right of the diagram. It is our instinct to ignore them because they have lower peaks. However, it is important to recall that the y-axis in Fig. 4.9 is the magnitude of the probability, not the magnitude of the signal, and that for Rayleigh distributions as the mean rises, the variance also grows, bringing the peak probability lower.

The phenomenon we have described here is called Rayleigh fading. The signal fades as it travels through the channel. The effect is represented by a multiplicative factor whose magnitude is Rayleigh distributed.

The baseband channel model so far is:

$$y = hx + n$$

Fig. 4.10 Frequency-
domain representation of a
single-path, flat-faded
channel

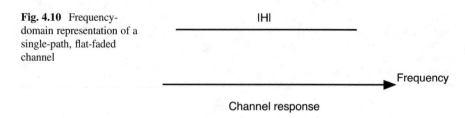

Note the following:

- The factor h is the channel coefficient and is a complex number.
- The magnitude of h is Rayleigh distributed.
- The phase of h is uniformly distributed.
- So far, h shows no time dependence.
- AWGN is represented by the addition of n.
- The above equation is valid in both time and frequency domains. Because we are multiplying the system response with the input, it is strictly a frequency-domain equation, but as we discussed above, convolution in time domain reduces to multiplication if we have a single path.
- Note that noise is added *after* taking fading into consideration. This suits the definition of SNR we made earlier. Noise and signal powers are measured *at the receiver* after the signal has passed through the channel.

Figure 4.10 shows the frequency-domain representation of the channel H. This is simply the Fourier transform of the impulse response. Because the impulse response is a single impulse, the frequency response is flat. This kind of fading is often called flat fading because the frequency response is flat. All frequencies are magnified or attenuated by the same factor. Recall that modeling a system in frequency domain is always done by multiplying the input by the system frequency response; thus, the frequency domain system model is:

$$Y = HX$$

A single-path channel is always synonymous with a flat-fading channel. Both mean the same thing. Dealing with fading at the receiver requires us to find out the value of h and then divide y by this value to obtain the best possible estimate for x, in other words:

$$\widehat{x} = y/h$$

We will discuss this in more detail in Chaps. 5 and 9. But it should be clear that for a flat-faded channel, we only need to estimate a single complex number h to characterize the channel. In the next section, we will contrast this with the situation where there are multiple paths.

4.4   Crowded Environments, Multipath (Frequency-Selective) Channels

Figure 4.11 shows a wireless channel that allows communication between the transmitter and the receiver through multiple paths. As with the single-path approach, the ray model shown in the figure is an oversimplification. The reality is that each of the paths consists of multiple wave arrivals. But under the narrowband assumption, each path will ultimately be seen as a single impulse by the signal.

Figure 4.12 shows the impulse response of the channel depicted in Fig. 4.11. The drawn amplitudes of the impulses are only representative because they are all random variables. There are three distinct paths in Fig. 4.11. This is reflected as three impulses in Fig. 4.12. Each impulse has an arrival time, amplitude, and phase.

As discussed in Sect. 4.3, the amplitudes of each of the impulses will have a Rayleigh distribution. A Rayleigh distribution is fully described by the variance of its constituent Gaussians. The higher the Gaussian variance, the higher the mean of the Rayleigh. A path with a higher Rayleigh mean is a stronger path that carries more signal energy to the receiver. Figure 4.12 suggests that path 1 is the strongest, while path 3 is the weakest. This makes some sense since path 1 travels the least distance and reflects the least number of times. However, this will not always be the case. Sometimes a shorter path will get there by reflecting off highly absorbent bodies,

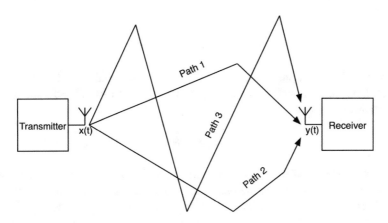

Fig. 4.11 Multipath channel

Fig. 4.12 Impulse response of a three-path channel

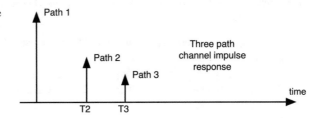

thus arriving early but with a low amplitude. Note that signal strength also does not dictate instantaneous relative strength. There might be time instances in which path 3 will be the strongest, but these are less probable than instances when path 1 is the strongest.

The arrival times in Fig. 4.12 correspond to the times of flight of the three paths in Fig. 4.11. The shorter path will always arrive earlier; the quality of the reflectors has no impact on this. It is useful to think of the arrival time of the first path as zero time. As we will see in Chap. 11, synchronization means that receivers consider their zero time to be the first instance when they recognize there is a signal over the air.

To understand what happens to a signal as it suffers from the channel in Fig. 4.12, we go back to the original definition of system response, where the input is convolved with the channel impulse response. We can no longer multiply the signal and the impulse response in time domain because the channel impulse response is not a single impulse:

$$y = h * x$$

In the case in Fig. 4.12, the channel impulse response consists of three impulses, with random phases and amplitudes. Each of the amplitudes is a Rayleigh distribution; each phase is a uniform distribution. Each path is also characterized by an arrival time. The arrival times can be considered deterministic in the short term but will vary in the long term (see Doppler shift in Sect. 4.6). Thus, the impulse response is:

$$h = A_1 e^{j\phi_1} \delta(t) + A_2 e^{j\phi_2} \delta(t - T_2) + A_3 e^{j\phi_3} \delta(t - T_3)$$

Fortunately, the impulse response is still a series of impulses and convolution is simple. Each impulse will create a copy of the signal at its time and scale it by its magnitude and phase:

$$y = h * x = A_1 e^{j\phi_1} x(t) + A_2 e^{j\phi_2} x(t - T_2) + A_3 e^{j\phi_3} x(t - T_3)$$

The three copies of the signal may add constructively, destructively, or both at different times depending on the phases and the arrival times. Figure 4.13 shows a specific example. The same sent signal from Sect. 4.3 is used. The top of Fig. 4.13 shows scaled and delayed versions of this sent signal at the positions of the three path impulses. The bottom of the figure shows the addition of the three copies, representing y. We are assuming in-phase arrivals. There are three conclusions we can draw from Fig. 4.13:

- As in single-path fading, the magnitude and phase of the received signal are different from those of the sent signal.
- Unlike single-path fading, the shape of the received signal is completely different from that of the sent signal. This immediately indicates that the multipath channel is much "worse."

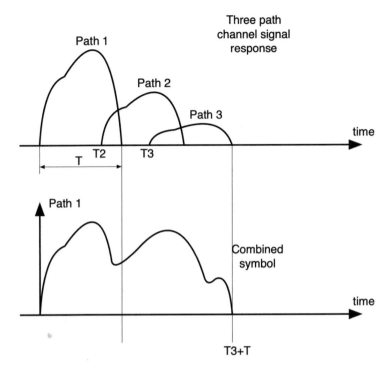

Fig. 4.13 Multipath channel causing delay spread in a signal. Top, three signal arrivals corresponding to three paths. Bottom, combined symbol as seen by the receiver

- The sent signal ends at time T. The received single-path signal in Sect. 4.3 also ended at time T. The received signal in Fig. 4.13 ends at time $T+T3$. Thus, the received signal bleeds for some time after it was supposed to end. The duration of the bleed depends on the arrival times and magnitudes of the multiple paths. This phenomenon is called delay spread.

Due to delay spread, multipath channels are dangerous not only for their impact on a particular signal but because of what they do to other signals. Figure 4.14 shows three signals $y1$, $y2$, and $y3$ received in sequence. The sent signals $x1$, $x2$, and $x3$ are T seconds long each. But after passing through the channel, they each suffer from delay spread. This leads to $y1$, $y2$, and $y3$ extending beyond T. As shown in Fig. 4.14, this causes each received signal to bleed into the following, interfering with it. This is a phenomenon called intersymbol interference (ISI) and can devastate BER if not addressed properly.

Figure 4.15 shows *a* frequency response for a multipath channel. Note that the shape of the channel response will vary significantly based on the number, phase, and arrival times of the multiple paths. However, we do know one thing about the channel response: it will not be flat. Thus, the channel in Fig. 4.15 is known as a frequency-selective channel. Multipath channels and frequency-selective channels are one and the same.

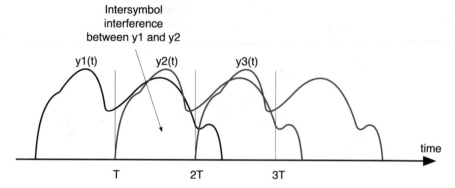

Fig. 4.14 Intersymbol interference due to multipath

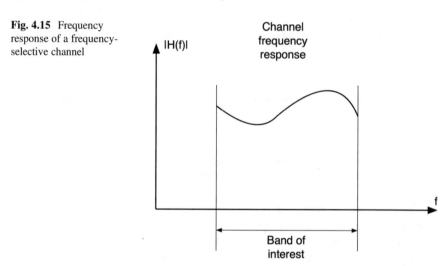

Fig. 4.15 Frequency response of a frequency-selective channel

Figure 4.16 shows an input signal X going through a frequency-selective channel H in frequency domain. The output signal Y does not only change its amplitude and phase but also in shape. This is a more fundamental impact than that in Fig. 4.10, where the flat channel preserved the shape of the signal spectrum.

Getting rid of the frequency-selective channel in frequency domain is as direct as in the single-path case: we must divide by H. However, as Fig. 4.17 shows, this is much easier said than done. In this case we are not dividing by a single number, we are dividing by a function of frequency. Since we must eventually do this in hardware, we should be skeptical about our ability to do so.

So how about trying to solve the problem in time domain instead? This is a more productive approach. If we do know the channel impulse response h, then we can fully characterize all the impulses. We can then create delayed versions of the signal and subtract them from the total signal to obtain a single copy.

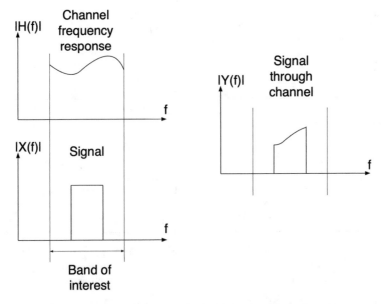

Fig. 4.16 Signal going through a frequency-selective channel

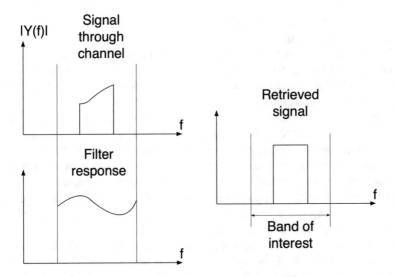

Fig. 4.17 Solving frequency selectivity in frequency domain

Thus, if we know that the received signal is a result of three paths, and we know the relative delays and magnitudes of each path, we can retrieve the original signal. We can create delayed and scaled versions and subtract them from the received signal, thus removing the impact of the additional paths. This approach to channel equalization is practical, but its complexity grows very rapidly with increased

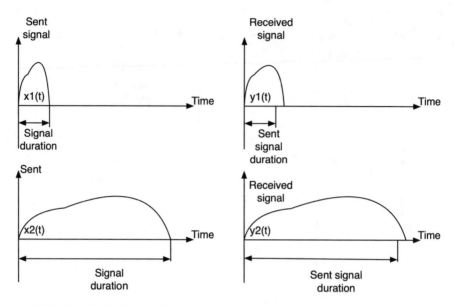

Fig. 4.18 Two signals affected differently by same delay spread

bandwidth. In fact, this approach is just the time-domain equivalent of the frequency-domain equalizer shown in Fig. 4.17.

The ultimate solution to frequency selectivity is orthogonal frequency-division multiplexing (OFDM). We will see this in detail in Chap. 7. But at this point it is important to develop some intuition about one fact: frequency selectivity is not only a property of the channel it is also a property of the signal.

Consider Fig. 4.18. In the upper part of the figure, we observe a signal with a limited duration going through a multipath channel. The multiple paths of the channel significantly affect it, causing it to be distorted and to introduce ISI to the next signal. In the lower part of the figure, we see a signal with a much longer duration suffering from the same channel. Because the arrival times of the multiple paths are insignificant when compared to the signal duration, the delay spread is much less significant as a percentage of the signal duration, and the signal could effectively see the channel as single path.

The same can be seen in frequency domain. The signal with the small duration has a large bandwidth. This is seen in the top of Fig. 4.19, where the signal bandwidth forces it to observe the frequency selectivity of the channel. The signal with the very large duration has a limited bandwidth. As seen in the bottom of Fig. 4.19, this causes the signal to lie within a section of the channel response that for all intents and purposes is flat.

In short, whether the channel is flat or selective will depend on the signal as much as it does channel conditions. We will develop this notion further in Sect. 4.9 once we also have Doppler under our belt.

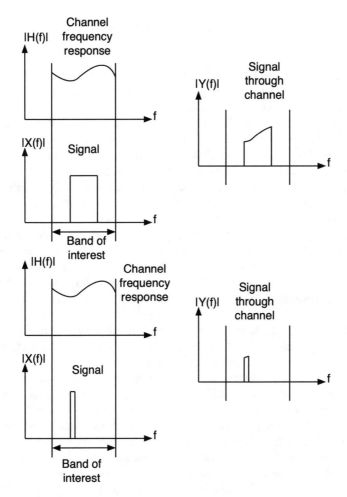

Fig. 4.19 Flat or frequency selective?

4.5 The Cost of Distance, Path Loss

Path loss is an umbrella term used to describe the loss of power as the signal travels from the transmitter to the receiver. The term is used in variable and often confusing ways to describe losses due to a variety of phenomena. Some sources even include the attenuating effects of Rayleigh fading under the term. In this book path loss will only refer to loss of signal strength due to radiation through the medium and miscellaneous losses in the RF front end.

The equation for path loss relates received power to transmitted power, and is very helpful in clarifying what sources of loss are included under the term:

$$P_r = P_t G_t G_r \cdot \frac{1}{d^n} \cdot \frac{1}{L} \cdot \left(\frac{\lambda}{4\pi}\right)^2$$

The equation can also be stated in dB form, although the absolute form is good enough for our current purposes. Now to define each of the terms in the equation and from there to understand what it is telling us.

P_r: Received signal power, this is the received power level at the input of the RF chain. This will determine the SNR of the system and will thus give us an idea of the expected BER. It will set a benchmark for the performance expected from the system. P_r is critical because it represents the highest possible SNR that the rest of the system can achieve. In fact, because noise factors are always positive and nonzero, the SNR observed by the baseband will necessarily be lower than that dictated by received power.

P_t: Transmit signal power. The transmitter pumps a very high-power signal out of its antenna. Signal strength is derived from the power amplifier at the end of the transmitter RF chain. P_t will be a major determinant of battery life in a mobile platform, consuming upward to half of the total power. In this book, we define it as the power emanating from an isotropic transmit antenna in the direction of the receiver.

G_t: The gain of the transmitter antenna. Antennas are passive elements. However, gain is defined as the increase in signal strength above the level that would be expected from an isotropic antenna. An isotropic antenna is one that radiates equally in all directions. Some antennas have the capacity to be directional, for example, through beamforming. Directionality allows the signal to be focused on a certain direction, thus concentrating more of the signal power in a useful way. Notice that P_t was defined as the isotropic power transmitted toward the receiver, and thus $P_t G_t$ is the actual radiated power from the transmitter array toward the receiver. Neither P_t nor $P_t G_t$ is the total power radiated from the transmitter. The first is the power directed at the receiver from a nondirectional transmitter. The second is the power directed at the receiver including directionality of antennas.

G_r: The gain of the receiver antenna. Again, this is not active gain in amplifiers. Instead, it is the beamforming bonus obtained by training the receiver antennas in the direction of the transmitter relative to the case where the receiver is trying to receive in an isotropic fashion.

L: This is a miscellaneous loss factor. It is more useful to define this factor in dB. It includes all frontend circuit-related losses that are not accounted for otherwise. This is typically limited to losses in cables and connectors to and from the antennas. Since P_t is the power radiated out of the transmit antenna, L will typically only account for losses on the receiver side.

λ: This is the wavelength of the carrier. The significant factor is $(\lambda/4\pi)$, which is the antenna aperture. It represents the effective receptive cone size of the antenna. Aperture is a property of the receiver antenna, not the transmitter. Transmitting at lower frequencies requires antennas with much higher aperture, which is the main reason we raise the signal to a carrier frequency. Transmitting at a higher frequency

quickly degrades path loss, which is a good reason for not transmitting at extremely high frequency. To reiterate, high-frequency transmission is promoted by low aperture, but demoted by high path loss.

d: This is the distance between the transmitter and receiver antennas. This is the main factor impacting path loss and is our main concern here.

n: This exponent factor determines how dependent path loss is on distance. In free space the exponent is 2, representing the well-known square root dependence of received power on distance. In the presence of major reflectors and when moving through non-air media, the exponent rises. Transmission in typical terrestrial situations involves a major reflection off the Earth. In such situations the exponent is around 4.

Path loss, when defined as above, is a sheer loss of signal power due to physical impediments in the antennas and in propagating through the air. The equation shows us options to mitigate the effect. Reducing L involves improving connections and cables in the RF front end of the receiver. It is preferable to transmit at a lower frequency to reduce path loss. However, this comes at two costs: first the antenna size must be larger, and second the spectrum is not necessarily free at the desired low center frequency. In the presence of multiple antennas at either or both the transmitter and receiver, better beamforming can increase the amount of power that manages to make it across. However, this comes at the cost of not using the antennas for diversity or spatial multiplexing (Chap. 9).

Overall, we have no good options to manage path loss. Most good radios will have already milked all the options to minimize the drop in power. We are forced to contend with the fact that as the signal travels, it loses power, fast. The most realistic way to address path loss with increased distance is to increase transmit power at the expense of battery life.

4.6 Doppler, Channels That Do Not Stand Still

Figure 4.20 shows a nonstationary transmitter and receiver. The two bodies are moving with a relative velocity and at an angle relative to each other. We can take account of the angle by substituting relative velocities as $v * \cos(\theta)$ rather than v. But for now, consider the case where the transmitter and the receiver are moving either directly toward each other or away from each other. Doppler shift is a

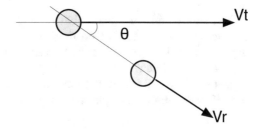

Fig. 4.20 Doppler shift between two moving bodies

phenomenon where the transmitted and received frequencies are shifted due to motion:

$$f_r = f_t \left\{ \frac{c + v_r}{c + v_t} \right\}$$

We can also calculate the *shift* which is the difference between the true and observed frequencies:

$$f_t - f_r = f_t - f_t \left\{ \frac{c + v_r}{c + v_t} \right\} = f_t \left\{ 1 - \frac{c + v_r}{c + v_t} \right\}$$

$$\Delta f = f_t - f_r = f_t \left\{ \frac{c + v_t - c - v_r}{c + v_t} \right\}$$

$$\Delta f = f_t \left\{ \frac{v_t - v_r}{c + v_t} \right\} = \frac{f_t v}{c + v_t}$$

where v is the relative velocity between the transmitter and the receiver. Realizing that transmitter speed is much lower than the speed of light in any medium and expanding transmitter frequency in terms of speed of light and wavelength:

$$\Delta f = \frac{f_t v}{c + v_t} \approx \frac{f_t v}{c} = \frac{c}{\lambda} \cdot \frac{v}{c} = \frac{v}{\lambda}$$

We have been ignoring the angle of the relative velocity so far; we can reintroduce it by multiplying by its cosine:

$$\Delta f = \frac{v}{\lambda} \cos \theta$$

This is the amount of frequency shift that the receiver observes. This shift could vary from nothing when the bodies are relatively stationary or are moving orthogonally to a very high value when the bodies are moving along the same line. The worst case is the latter, leading to a total shift in the carrier of:

$$\Delta f = f_m = \pm \frac{f_c v}{c}$$

The sign of this maximum shift depends on whether the transmitter and the receiver are moving toward or away from each other. Given the fact that most objects on Earth move at a modest speed compared to the speed of light, we can expect this maximum shift in frequency to be rather minor. Let us consider what happens with typical values. If the carrier frequency is 3 GHz and the relative motion is at 36 km/hour, then Doppler shift is:

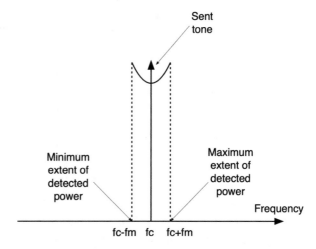

Fig. 4.21 Doppler affecting a single tone

$$f_{\mathrm{m}} = \frac{3 * 10^9 * 36 * 1000}{3 * 10^8 * (60 * 60)} = 100\ \mathrm{Hz}$$

Relatively stationary radios will typically observe Doppler shifts on the carrier in the few Hz, while fast-moving systems may observe Doppler in the low hundreds of Hz. The problem with Doppler shift is that it grossly complicates the task for the receiver. A phase shift does not really affect the receiver much; in fact as we will see in later chapters, it can simply roll it in with channel estimation. Even a frequency shift between the carrier at the transmitter and the receiver is not a big deal; shifts in the few tens of kilohertz can be effectively addressed by frequency offset estimation and correction at the receiver (Chap. 11).

But from Sects. 4.3 and 4.4, we saw that each impulse in the channel is in fact multiple wavefront arrivals coming in at different angles. Doppler shows a strong dependence on the angle of arrival. Because each wavefront hits at a different angle, they will suffer from different frequency offsets. This causes each wave to have a time-varying phase shift that is different from all the other waves. The way these wavefronts interact can thus be difficult to address.

Before trying to solve Doppler shift, we should first consider how to model it. Figure 4.21 shows the PSD of a single tone after suffering from Doppler. The maximum frequency shift introduced by Doppler is f_{m}. The single tone should have arrived as a single tone. Instead, it has spread. It now has a presence with considerable power starting from a frequency f_{m} above the original carrier to f_{m} below it. At f_{c}, where the original signal was sent, power is diminished.

We call this phenomenon Doppler spread, because the original tone seems to have been smeared or spread across a bandwidth it was not supposed to occupy. The same can roughly be applied to more complicated signals. This is shown in Fig. 4.22. The original spectrum is spread by f_{m} above and below the original maximum and minimum frequencies of the signal.

Fig. 4.22 Doppler affecting
a signal PSD

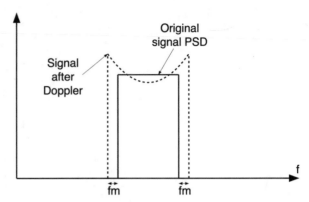

How can we deal with the effects of Doppler? There is basically no good way to address this. The impact is fundamental. However, Fig. 4.22 gives us a clue about what we really need to do. In the figure if the signal had a very large bandwidth, it would not suffer much from the Doppler spread because the amount of spread would be small relative to the original bandwidth. This is starting to sound similar but opposite to how delay spread affected wideband and narrowband signals in Sect. 4.4. We will further discuss this and how it makes sense in the context of packet-based communication in Sect. 4.9.

4.7   Having a Line of Sight, Rician Fading

The Rayleigh fading model discussed in Sects. 4.3 and 4.4 has a major limitation: it fails if one of the paths is dominant. In other words, the Rayleigh fading model assumes that all paths carry powers within the same range. This does not mean the multiple paths are all equal, or even close. It does, however, mean that no one single path consistently dominates a significant proportion of the arriving power.

This assumption is valid for most scattering-rich environments. However, there is one practical situation where it fails: when there is a line of sight (LOS) path between the transmitter and the receiver. The LOS path, as shown in Fig. 4.23, will most likely arrive with a much higher amplitude than all the scattered paths. This leads to an interference pattern at the receiver that does not fit with Rayleigh fading.

LOS-dominated channels are better represented by the Rician distribution. This distribution is fully characterized by two constants K and Ω, where K is the unitless ratio of the average power of the LOS path to the total average power in all the other paths and Ω is the total average power in all paths including the LOS path. We can use these two constants to derive two parameters used in the pdf:

Fig. 4.23 Channel with an
LOS path

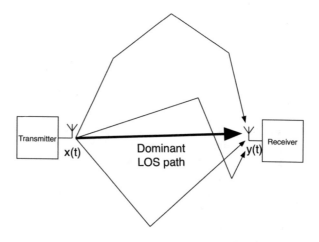

$$\nu^2 = \frac{K}{1+K}\Omega$$

$$\sigma^2 = \frac{\Omega}{2(1+K)}$$

The pdf of the received signal envelope is:

$$f_X(x) = \frac{x}{\sigma^2} e^{\left(-\frac{(x^2+\nu^2)}{2\sigma^2}\right)} I_0\left(\frac{x\nu}{\sigma^2}\right)$$

where I_0 is the zero-order modified Bessel function.

The mean and variance of the distribution are:

$$\text{Mean} = \sigma\sqrt{\frac{\pi}{2}}L_{\frac{1}{2}}\left(-\frac{\nu^2}{2\sigma^2}\right)$$

$$\text{Variance} = 2\sigma^2 + \nu^2 - \frac{\pi\sigma^2}{2}L_{\frac{1}{2}}^2\left(-\frac{\nu^2}{2\sigma^2}\right)$$

where L is the Laguerre polynomial.

Like the Rayleigh distribution, the Rice distribution is nonzero mean, and its mean and variance are not independent from each other. However, by virtue of having a dominant path, the Rician distribution can have a high mean with a relatively low variance when compared to the Rayleigh distribution.

A Rician-faded channel at first seems to be more friendly than a Rayleigh channel. The strong direct path provides a good high mean with less variance. This means a stronger signal received more consistently. But in Chap. 9 we will discover that channels with LOS are devastating to the capacity of MIMO systems.

Thus, it is not always preferable to deal with a Rician channel as opposed to a Rayleigh channel.

4.8 Large Obstacles to Communication, Shadowing

Shadowing is a special form of signal deterioration that occurs due to the presence of a large dominant obstacle between the transmitter and the receiver. This is distinct from fading and the two can occur simultaneously. Fading occurs due to the scattering of the transmitted signal across reflectors from the transmitter to the receiver. Shadowing occurs due to the presence of a large blocking body. Because of the size of shadowing obstacles, shadowing tends to be a more slow-changing effect than fading.

Shadowing affects transmission in open environments when natural objects like mountains or hills are blocking direct communication. But it can also affect urban environments, where the shadow is cast by large buildings. When shadowing occurs, the signal can only arrive through indirect paths. Shadowing is a relatively unimportant effect in indoor environments; thus, we will mostly ignore it for Wi-Fi. However, it must be modeled in cellular systems.

There are many ways to model shadowing. The simplest approach would be to roll its attenuation into path loss. This makes path loss responsible for all slow-changing, large-scale fluctuations in signal power. However, this approach has the disadvantage of "contaminating" the definition of path loss, for path loss is easier to address when it is only concerned with losses due to propagation.

A better approach to model shadowing is to understand that it changes the received power of the signal into a random variable. Specifically, the random variable is the ratio between transmit and receive power. This random variable has a lognormal distribution which is characterized by two parameters μ and σ.

Thus with:

$$x = \frac{P_t}{P_r}$$

The pdf is:

$$f_X(x) = \frac{1}{x\sigma\sqrt{2\pi}} e^{-\frac{(\ln x - \mu)^2}{2\sigma^2}}$$

The mean is:

$$e^{\left(\mu + \frac{\sigma^2}{2}\right)}$$

And the variance is:

$$\left(e^{\sigma^2} - 1\right)e^{\left(2\mu + \sigma^2\right)}$$

Like all propagation envelope distributions, the lognormal distribution is nonzero mean. Like the Rayleigh and Rician distribution, the lognormal distribution has a mean and variance that are not independent.

4.9 How to Determine the Type of Channel

In Sects. 4.2, 4.3, 4.4, 4.5, and 4.6, we discussed the four main channel effects we want to consider for Wi-Fi systems. Path loss and AWGN are both very easy to understand. More of either is bad. More noise means less SNR. More path loss also means less SNR. Less SNR means higher BER, which means less goodput and more energy per good bit received. This is very straightforward, noise bad, signal good!

However, in Sects. 4.4 and 4.6, we introduce two effects that are a little more complicated. Multipath fading and Doppler represent two independent effects. Multipath is a representation of the preponderance and distribution of scatterers (reflectors) in the environment between the transmitter and the receiver.

Doppler is a measure of the total motion in the system. In Sect. 4.6, the suggestion was that this motion is limited to the transmitter and the receiver. But Doppler also accounts for motion of scatterers in the environment. In fact, Doppler is a measure of *total* motion in the system. If a photo is taken of the system and we come again in a second and take another shot, Doppler would be a measure of how different the two shots are.

Figure 4.24 shows the effect of multipath and Doppler shift in time and frequency domains. Assume that the signal we are sending is flat in both time and frequency domains. This is obviously not the same signal, but the flatness will help expose the channel effects. In the top left of Fig. 4.24, we see the received signal after suffering from Doppler. Doppler changes the channel with time, which causes the envelope of the received signal to also vary with time. In frequency domain (top right sub-figure), we see the flat PSD of the signal spreading beyond its original bandwidth.

The bottom of Fig. 4.24 shows the impact of delay spread. In time domain the multiple paths cause the signal to extend beyond the original time of the symbol. In the bottom right, the received signal PSD is still within its original bandwidth, but the frequency selectivity causes the originally flat signal to become variable.

There is an interesting symmetry to Fig. 4.24. Doppler causes the signal to spread in frequency; multipath causes the signal to spread in time. Doppler causes the signal to become variable in time; multipath causes the signal to become variable in frequency. In fact, the two are two sides of the same coin; what one does in time domain, the other does in frequency domain.

Now it is important to revisit an important concept: whether delay spread and Doppler are important is a function of the signal properties as much as the channel. Figure 4.25 shows two signals in both time and frequency domains. $X1$ is a

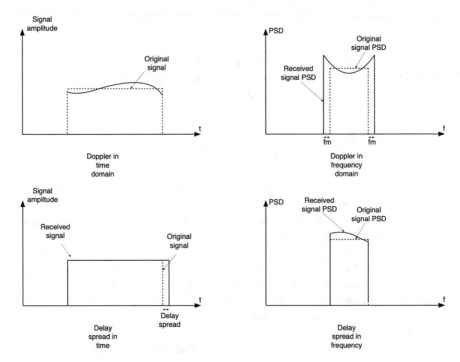

Fig. 4.24 The time-frequency effects of Doppler and delay spread. Shown is the received signal. The sent signal is flat in both frequency and time

narrowband signal with a large symbol duration; $X2$ is a wideband signal with a small symbol duration.

$y1(t)$ and $y2(t)$ show the received signals after $x1$ and $x2$ pass through a channel with the same delay spread. The impact is a lot more pronounced on $y2(t)$ because its duration is comparable to the spread. The same can be seen in frequency domain. $Y2$ (f) has a much wider bandwidth than $Y1(f)$. The channel looks flat for $Y1(f)$, but its selectivity has a detectable impact on $Y2(f)$.

So how do we determine if the channel is or is not frequency-selective? The above discussion shows that we first must define metrics that quantify delay spread and frequency selectivity in the channel before we compare it to the signal.

In time domain the metric is delay spread σ_τ. If the input to the channel is an impulse, the output from a single-path channel is also an impulse. But if the channel is multipath, an impulse input will bleed in time. The delay spread is thus the amount of time that the impulse response of the channel remains above the noise threshold. This is also the amount of time that a signal exceeds its original symbol time at a measurable level. If symbol time is much higher than delay spread, then the channel is essentially single path for that signal.

In the frequency domain, we define coherence bandwidth, B_c. This is the frequency range over which the channel is essentially flat. If the signal bandwidth is much smaller than the coherence bandwidth, then the channel is essentially flat for

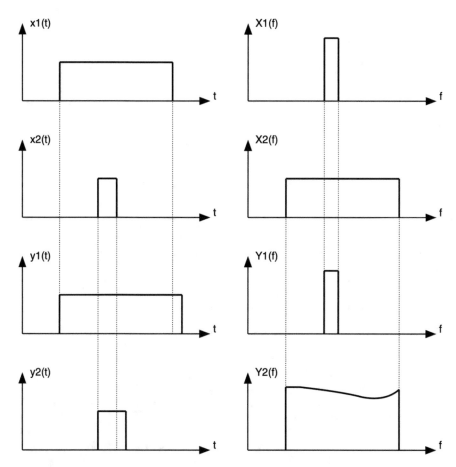

Fig. 4.25 Signal properties that protect from delay spread. $X1$ is a narrowband signal. $X2$ is a wideband signal. $Y1$ and $Y2$ are the signals after passing through the same channel

that signal. Notice that the channel being flat is equivalent to the channel being single path. Thus, tests using delay spread and coherence bandwidth should yield the same conclusion about whether we should consider the channel flat or frequency-selective. In other words, and to be explicit, if your test using coherence bandwidth says the channel is flat, your test using delay spread *must* say that the channel is single path.

   Figure 4.26 shows that like delay spread, the same channel might exhibit Doppler for one signal but not for another. For example, the narrowband signal $Y2(f)$ sees a significant spread in frequency domain and does suffer from Doppler. In time domain, the symbol duration of $y2(t)$ is long enough that Doppler has a chance change the channel before the symbol ends. On the other hand, the wideband signal $Y1(f)$ does not observe a significant enough spread in frequency domain, which can be seen in time domain as $y1(t)$ ends too fast for Doppler to change the channel.

   Along the same lines as delay spread, we define figures of merit and rules of thumb to test if a signal will or will not observe Doppler in the channel. First, we

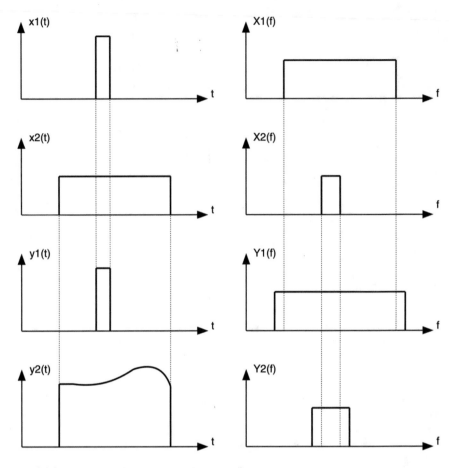

Fig. 4.26 Signal properties that protect from Doppler

define Doppler spread B_d, which is the same as f_m from Sect. 4.6. This is the maximum frequency spread that a tone observes passing through the channel. If the signal bandwidth is much larger than Doppler spread, then the signal will *not* suffer from Doppler.

In time domain, we define coherence time T_c. This is the time over which the channel is essentially unchanged. If the symbol duration is much smaller than coherence time, then the signal will not observe the effect of Doppler. As with delay spread, the tests using coherence time and Doppler spread are looking for the same thing and should yield the same conclusion.

Notice that the words "coherence" and "spread" occur in reference to metrics for both Doppler and delay spread, which can be a source of confusion. But this strengthens the symmetry between the two. Doppler spreads in frequency, while multipath spreads in time. Doppler introduces incoherence in time, while frequency selectivity introduces it in frequency.

Fig. 4.27 The four
"channel types"

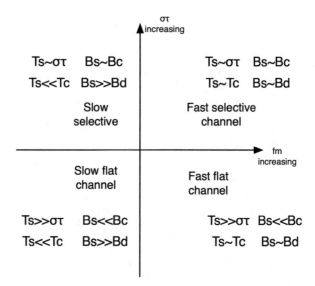

With these figures of merit and rules of thumb, we introduce four broad categories of channels in Fig. 4.27. The four are the permutations of the two sets of signal tests. A signal that suffers from multipath will observe frequency-selective fading. One that does not will suffer from flat fading. A signal that suffers from Doppler will see the channel changing *fast*, while one that does not will observe a slow channel.

Thus, there are four permutations. A channel can be slow-flat, slow-selective, fast-flat, or fast-selective. For example, a calm forest is a slow-selective environment. It is very rich in scatterers, but nothing about their distribution changes. An urban environment is likely to be fast-selective, for it is rich in fast-moving scatterers. A desert on the other hand is a slow-flat environment, devoid of reflectors, and devoid of change. An airplane communicating with another airplane will observe fast-flat fading because they are both moving fast, but there are not many scatterers.

So, which of the four channel types is "best"? First, the four quadrants of channel classifications are completely arbitrary. They are just a general guide to help us predict what to expect. Channels exist on a continuum in the plane of Fig. 4.27. But second, the word "best" is always underdefined. Best for what?

It is easy to argue that a slow-flat channel is *easier* to handle than any other type of channel. Our intuition is also true that a fast-selective channel will require more complicated decoders at the receiver. But this does not mean that selective channels are always "worse." For example, in Chap. 9, we see that multipath channels are necessary for MIMO communication. Having multiple paths introduces spatial diversity to the channel. If one of the paths is being strongly blocked or absorbed, the other paths might get through. The more independent paths there are, the greater the diversity. We still have not described practical ways to benefit from such diversity, and it will take us entire chapters to do so, but the opportunity is certainly there.

Chapter 5
Baseband Modulation

5.1 Sending a Message Over the Air

In Sect. 2.1, we discussed how a message is sent over the air. The baseband message is multiplied by the carrier sinusoid to give the passband signal. Let us begin with a baseband binary signal consisting of 1's and 0's. This produces the baseband analog time-domain waveform in Fig. 5.1. This passband signal has a problem: for a significant portion of time, it has zero energy; this means that the time average of the signal is nonzero. So, the signal contains a DC component. Information-carrying DC values are undesirable because coupling capacitors in the analog sections of both the transmitter and the receiver will reject them.

This problem can be solved by precoding the baseband signal to take the values 1 and -1 instead of 1 and 0. This leads to the time-domain signal in Fig. 5.2. This signal is zero mean. The data produced by the MAC layer is still binary. Transforming these 1 and 0 values to 1 and -1 requires an additional block, albeit a very simple one, at the transmitter side. This block is called the baseband modulator.

The baseband modulator takes in a stream of binary numbers and outputs baseband "symbols." When they are multiplied by the sinusoidal carrier, these baseband symbols produce the passband signal. The modulator corresponding to Fig. 5.2 is trivial. Each bit is transformed into a single symbol, and each symbol corresponds to a single bit. The transformation is 0 to -1 and 1 to 1.

We call this form of modulation binary phase-shift keying, or BPSK. The reason is obvious from Fig. 5.3 which shows the passband signal after multiplication by the carrier. In periods where the symbol is $+1$, the sinusoid is a cosine of the carrier frequency. In periods where the symbol is -1, the sinusoid is a minus cosine of the carrier frequency. This is a cosine function with a phase shift of π. So, specifically:

© The Author(s), under exclusive license to Springer Nature Switzerland AG 2023
K. Abbas, *From Algorithms to Hardware Architectures*,
https://doi.org/10.1007/978-3-031-08693-9_5

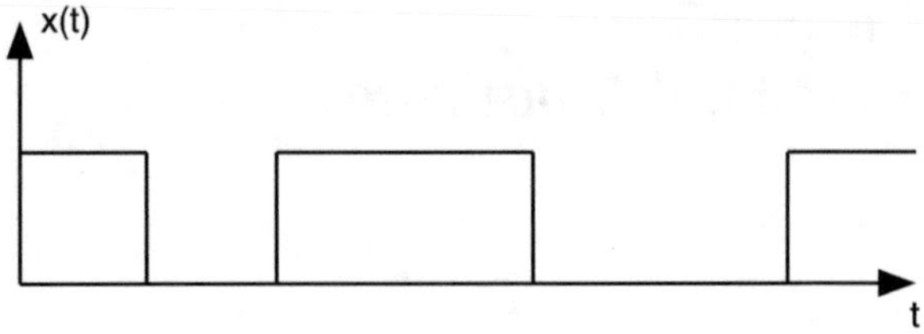

Fig. 5.1 Return to zero passband signal in time domain. This signal corresponds to the binary vector "1011001"

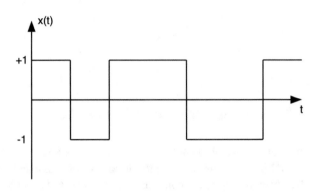

Fig. 5.2 Nonreturn to zero baseband signal in time domain

Fig. 5.3 BPSK passband signal in time domain. This corresponds to the same sequence as the two previous figures. Carrier frequency is unrealistically low for illustration purposes

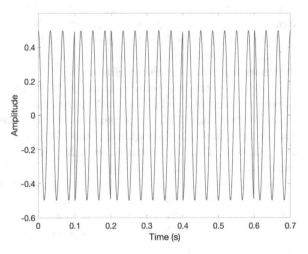

$$\text{bit} = 1$$

$$s(t) = 1.\cos{(\omega t)}$$

$$\text{bit} = -1$$

$$s(t) = -1.\cos{(\omega t)} = \cos{((\omega t + \pi)}$$

Fig. 5.4 Baseband signal constellation for BPSK

The values used for carrier frequency in most examples in this chapter are not typical. We are using unrealistically low carriers so we can see full cycles within a symbol duration. In BPSK symbol duration and bit duration are identical because one symbol carriers one bit. One symbol duration should contain complete cycles of the carrier.

Figure 5.4 shows a useful representation of BPSK. We are not encoding the carrier; instead, we are showing only the baseband symbols. There are two possible symbols: 1 and -1. These represent the magnitudes by which the cosine carrier will be multiplied. They also represent the *phase* of the carrier because that phase is either 0 or π.

Constellations like in Fig. 5.4 show the phases and relative amplitudes of the possible symbols correctly. They are not reflective of the actual amplitude over the air. In other words, the +1 and -1 in Fig. 5.4 are only illustrative. The actual amplitude of the transmitted signal will depend on the power amplifier. Thus, the passband signal is:

$$s(t) = \pm A . \cos{(\omega t)}$$

5.2   The Missing Dimension, QPSK

In Sect. 2.1, we discussed how a sine and cosine are orthogonal to each other. This allows us to extend the number of bits we can transmit simultaneously. The passband signal $x(t)$ is composite and formed by the addition of two independent signals $A(t)$ and $B(t)$ named the in-phase and quadrature components, respectively. The in-phase component is carried on a cosine; the quadrature component is mixed with a sine carrier. The passband real signal is thus:

$$A(t)\cos{(\omega_c t)} + B(t)\sin{(\omega_c t)}$$

And the corresponding phasor equation is:

$$V(t)e^{j\omega_c t + \phi}$$

where the envelope is:

$$V(t) = \sqrt{(A(t))^2 + (B(t))^2}$$

And the phase is:

$$\phi = \tan^{-1}\left(\frac{B(t)}{A(t)}\right)$$

This corresponds to a baseband signal (Sect. 2.1):

$$a + jb, \quad a = A(nT), b = B(nT)$$

where this baseband signal is a series of complex numbers. The real part of every number is the magnitude of the in-phase component, and the imaginary part is the magnitude of the quadrature component.

The main reason we are doing this is to allow us to have more than two symbols in our constellation. This will allow each symbol to carry more than one bit and will improve the spectral efficiency of the system. In Fig. 5.4 we had only two symbols; thus, each of them corresponds to a single bit. But we could in fact have introduced more than two symbols. For example, in Fig. 5.5, we have four symbols arranged on the x-axis. Each pair of symbols will have the same phase and will be distinguished only by their amplitude.

Every "symbol" is going to occupy 1 Hz if it is repeated every second. Thus, if the symbol is repeated at a rate of B times a second, it will occupy a bandwidth of B. In Fig. 5.4, every symbol carries a single bit. This leads to a bitrate of B Mbps in a bandwidth of B. In Fig. 5.5, there are four symbols instead of two. This allows every symbol to encode two bits at a time as shown on the figure. This allows the throughput to jump to $2B$ Mbps in the same bandwidth. This is an increase in spectral efficiency. Spectral efficiency is how many bits per second can be pushed into a single Hz of bandwidth; it strips away the effect of bandwidth, allowing us to compare modulation schemes fairly. Thus, the BPSK scheme in Fig. 5.4 has a spectral efficiency of 1 bps/Hz, while in Fig. 5.5 it is 2 bps/Hz.

However, there must be a price we pay for this increase in bitrate. If we compare Figs. 5.4 and 5.5, we notice that the distances between the symbols have decreased. In Sect. 5.4, we will discover that this will directly lead to a deterioration of BER for the same SNR. A decrease in distance between constellation points means a decrease in the amount of noise necessary for received symbols to cross decision boundaries (Sect. 5.6).

However, we are not using any of the observations about orthogonality of sine and cosine. The fact that we can use a sine wave means that we should have a y-axis

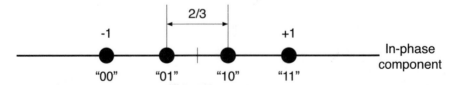

Fig. 5.5 Four symbol cosine-only signal

Fig. 5.6 Constellation of
QPSK with pure real and
imaginary symbols

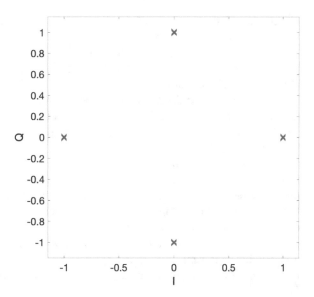

in the constellation diagram. This allows us to draw points on the x-axis as well as
the y-axis. Point on the y-axis will be purely imaginary in the baseband representation. In the passband real signal, y-axis signals are multiplied by the sine carrier.

Thus, in Fig. 5.6, we see a four-point constellation where every point encodes two
bits. The points in the constellation are more distant from each other than in Fig. 5.5.
In fact, let us make a quick calculation. For BPSK in Fig. 5.4, the symbols are at a
distance of 2 from each other. For the constellation in Fig. 5.5, the symbols are at a
distance of 2/3 from each other. In the QPSK constellation in Fig. 5.6, the symbols
are at a distance of $\sqrt{2}$ from each other. This suggests that the probability of error for
BPSK will be the lowest, while it will be higher in QPSK and the highest in the
modified BPSK constellation of Fig. 5.5. This is completely wrong.

There are a couple of creases to the above analysis. First, it must be done for
normalized constellations. We can increase distances between symbols in any
constellation by multiplying it by a factor. This will increase the power of the
transmitted signal and is not a good way to compare apples to apples. In Sect. 2.7,
we reached the conclusion that BER can only be assessed if we know the SNR; thus,
we should only compare modulation schemes once we establish that their constellations have the same SNR; this is equivalent to ensuring that average signal power is
equal given that noise temperature is the same.

All symbols in both Figs. 5.4 and 5.6 have an amplitude of 1, and thus the average
power is also unity. The problem is with Fig. 5.5. In this constellation, the average
power is $\left(1 + 1 + \frac{1}{9} + \frac{1}{9}\right)/4 = \frac{20}{36}$. This means that the signals in Fig. 5.5 must be
normalized by $\sqrt{20/36}$. This leads to a distance between symbols of $\frac{2}{3} *$
$\sqrt{36/20} = 2/\sqrt{20}$. This means the performance of Fig. 5.5 is slightly better than
we expected because its average power as drawn is lower than Figs. 5.4 and 5.6. The

distance between symbols in a normalized Fig. 5.5 would be 0.89, so it is still doing worse than the two other options.

The other crease is that QPSK (Fig. 5.6) has the same BER as BPSK (Fig. 5.4). This is surprising because symbol distances are lower for the same SNR in Fig. 5.6. However, like signals, noise can be orthogonal to itself. Noise in the cosine direction will have no impact on the signal in the sine direction. Noise in the sine direction will have no impact on the signal in the cosine direction. Thus, for QPSK and only for QPSK, we can measure the distance between the j and $-j$ signal instead of that between the j and the -1 signal.

Figure 5.7 shows a time-domain representation of a signal typical of the modulation scheme in Fig. 5.6. In the waveform we see that the amplitude (envelope) of the signal never changes. You can also see this from the constellation in Fig. 5.6 because the magnitude of the phasor vector to each symbol is the same (unity). What changes from symbol to symbol is the phase. You can see this in Fig. 5.7 as a sudden jump in phase every symbol duration. This is also visible in Fig. 5.6, where the phase each symbol vector makes with the x-axis is the same as the phase that the symbol has in time domain. The following are the baseband to passband mappings for Fig. 5.6:

$$x = 1 \rightarrow x(t) = \cos(\omega_c t)$$

$$x = -1 \rightarrow x(t) = -\cos(\omega_c t) = \cos(\omega_c t + \pi)$$

$$x = j \rightarrow x(t) = \sin(\omega_c t) = \cos\left(\omega_c t + \frac{\pi}{2}\right)$$

$$x = -j \rightarrow x(t) = -\sin(\omega_c t) = \cos\left(\omega_c t - \frac{\pi}{2}\right)$$

In Fig. 5.6, we must choose a specific mapping from binary to the QAM symbols to be able to move to the time-domain representation in Fig. 5.7. We will choose the

Fig. 5.7 Time-domain representation of a signal from the constellation in Fig. 5.6

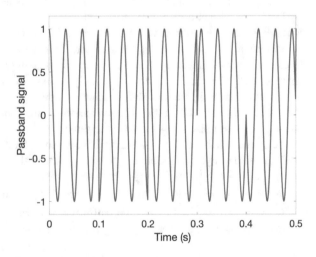

mapping 01 to pure sine, or alternatively in baseband j. 11 is 1, 10 is $-j$, and 00 is -1. We will talk later about why this specific mapping, although it does not matter for the time being since the constellation in Fig. 5.6, is not a valid constellation in any Wi-Fi standard.

In Fig. 5.7, unlike in BPSK, there is a distinction between symbol duration and bit duration. Each *symbol* in Fig. 5.7 occupies a time of 0.1 s. But each symbol carries two bits. Thus, bit duration is 0.05 s. This distinction should carry forward for the rest of this chapter for higher modulation schemes. In general, we care about symbol times when calculating bandwidth. We care about bit durations when calculating throughput. The ratio between the two is the spectral efficiency.

The symbols in Fig. 5.6 do not have to be purely real or purely imaginary. The modulation scheme shown in Fig. 5.6 is known as QPSK, short for quadrature phase-shift keying. This is along the same lines of BPSK. Bits are encoded by variations in the phase of the carrier only. The amplitude of the carrier remains constant. QPSK does not trade off performance for throughput relative to BPSK because we have used the orthogonality of the two carriers in our favor.

The constellation for QPSK shown in Fig. 5.6 is nonunique. In fact, any four symbols spaced at constant phases from each other on the unit circle will give the same performance of the constellation in Fig. 5.6. The constellation in Fig. 5.8 is more typically used for QPSK and is the constellation used in 802.11n. In this constellation the amplitude of the signal is still unity, but the phase belongs to the set $\left\{ \frac{\pi}{4}, \frac{3\pi}{4}, \frac{5\pi}{4}, \frac{7\pi}{4} \right\}$. Note that at the transmitter, we do not generate these phases for the carrier. Instead, they are generated by adding different magnitudes for the in-phase and quadrature components. For example, to generate a phase of $\frac{3\pi}{4}$, we add an in-phase component of $+\frac{1}{\sqrt{2}}$ and a quadrature component of $-\frac{1}{\sqrt{2}}$.

Fig. 5.8 Constellation of QPSK as a special case of QAM symbols. Symbols in this constellation are normalized so that their average power is 1

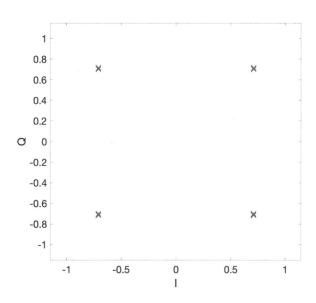

Finally, it is important to understand where the factor $\frac{1}{\sqrt{2}}$ comes in. This factor is used to normalize the amplitude of the carrier to unity. This guarantees that all points in the constellation lie on the unit circle. We do this to normalize the amplitude of the baseband signal. The passband signal amplitude will thus be determined solely by the gain of the PA in the transmitter RF. This allows us to fairly compare the performance of different baseband modulation schemes. When we look at a constellation, distances between points will directly correspond to the amount of noise they can resist because we have already normalized transmit power. This avoids the problem we faced with understanding the performance of the constellation in Fig. 5.5.

5.3 Phase-Shift Keying and the Economy of Empty Spaces

The discussion in Sect. 5.2 has one logical extension. If two bits per symbol double the throughput, why not try more bits per symbol? Figure 5.9 shows just such a constellation. We have eight symbols to choose from, allowing us to encode three bits per symbol. Since we can push one symbol in 1 Hz, then we have a bit rate of $3B$ Mbps, where B is the bandwidth.

But what are we sending here? What distinguishes the eight symbols in Fig. 5.9 from each other? All the symbols have the same amplitude, but they are distinguished by their phases. The passband symbol is:

Fig. 5.9 8-PSK constellation

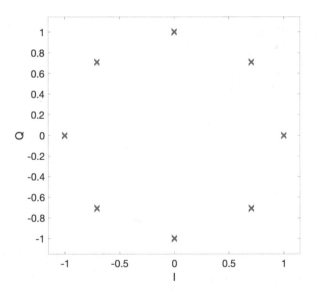

$$x(t) = A \sin (\omega_c t + \phi)$$

where A is constant, and for each symbol duration, the phase may change into any of the following:

$$\phi \in \left\{ 0, \frac{\pi}{4}, \frac{\pi}{2}, \frac{3\pi}{4}, -\frac{\pi}{4}, \pi, -\frac{\pi}{2}, -\frac{3\pi}{4} \right\}$$

Practically, this will be implemented by sending different magnitudes on the in-phase and quadrature components of the mixers in different symbol durations. For example, to send a symbol with phase $\frac{\pi}{4}$, the amplitude of the in-phase component will be $A/\sqrt{2}$, and the amplitude of the quadrature component will also be $A/\sqrt{2}$. To send a symbol with phase 0, the in-phase component will have an amplitude of A, while the quadrature component will have an amplitude of 0.

This is called phase-shift keying (PSK) because the information is carried in the phase of the carrier. This can be seen in any PSK constellation by the fact that symbol radius is constant while symbol phases vary. Taking this further, we get the 16-PSK constellation in Fig. 5.10. 16-PSK carries 4 bits per symbol and has a spectral efficiency of 4 bps/Hz. We can keep increasing the order of modulation, but there must be a limit. There is a price we pay for this spectral efficiency.

The price can be gleaned by comparing Figs. 5.8, 5.9, and 5.10. What we observe here is that the symbols are always arranged in a circle. If the radius of the circle is constant, then increasing the number of symbols will reduce the distance between such symbols. The distance between symbols will directly reflect on the BER. In fact, by observing Fig. 5.11, we find that the BER curves for higher-order PSK constellations will always have a higher BER than lower-order constellations for the same SNR. The only exception is the move from BPSK to QPSK for reasons we discussed in Sect. 5.2.

Fig. 5.10 16-PSK constellation

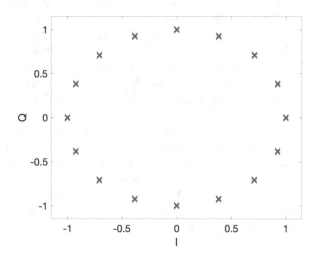

Fig. 5.11 BER-SNR curves for different PSK constellations

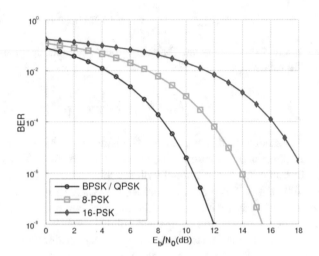

So why are spaces between symbols related to BER? This comes from the method of demodulation. Demodulation is the process of deciding which symbol from the possible sent signals corresponds to the noisy received symbol. We will discuss this in Sect. 5.6, but the main idea is that we find the constellation point to which the received symbol is closest.

Assume, for example, that a symbol with phase 0 was sent. Noise will be added to this symbol at the receiver; this noise will cause the received symbol to move both in the y and x directions. If enough positive noise is added, for example, the received signal may move closer to the symbol with $\frac{\pi}{4}$ phase than the signal with zero phase. This will cause us to make a wrong decision, which will lead to a bit error.

Thus, the distance between symbols is the amount of noise we can withstand before crossing a decision boundary and causing an error. But why not just increase the radius of the circle to restore distances between symbols? We can certainly do this, but the radius is the power of the symbol. Thus, this would just be increasing SNR, and we need to compare BER at the same SNR.

From Figs. 5.8, 5.9, and 5.10, we can see why PSK is generally considered "bad." Notice how the perimeter of the circle becomes progressively more crowded while its interior becomes more glaringly empty. This means we are not distributing the symbols optimally to maximize distances. We will see in Sect. 5.4 how we can best use this empty space.

PSK suffers from another major disadvantage that has to do with how phase changes. The phase of the sinusoid will change every symbol duration. It will make sudden jumps that could be as high as a full 180°. This leads to sudden discontinuities in the waveform of the sinusoid in time domain. In frequency domain, this translates into very high-frequency components, increasing the side lobes of the signal and requiring filtering before we can transmit so that we do not introduce interchannel interference.

So PSK is full of trouble. It is not a modulation technique that scales well for higher orders. But there must be something good about it. This can also be seen from the figures in this section. Notice that for all constellations, the symbols will lie on a circle. This means that the sent waveform has a constant envelope. In Sects. 5.4 and 7.3, we will understand why such a property is very good insofar as the power amplifier is concerned.

5.4 Quadrature Amplitude Modulation, Trading Efficiency for BER

Quadrature amplitude modulation, QAM for short, is a baseband modulation technique that is favored by OFDM systems. QAM has significant advantages when compared to PSK, and its disadvantages will be insignificant when used in an OFDM system.

In QAM, we allow symbols to have different magnitudes as well as different phases. This allows us to change the magnitude of the phasors in the constellation. This in turn solves the problem of empty spaces we observed with PSK constellations and makes BER more optimal. Having variable amplitudes translates into the following:

- a and b in the baseband signal $a + jb$ can take multiple values.
- The envelope (amplitude) of a and b, $\sqrt{a^2 + b^2}$, can also have different values, this matters because it is what sets QAM apart from PSK.
- In both cases above, these variable values are drawn from a discrete set.

Figure 5.12 shows a 16-QAM constellation, while Fig. 5.13 shows 16 PSK versus 16 QAM. Figure 5.13 normalizes both constellations to unity *average* energy.

Fig. 5.12 16 QAM. This constellation is not normalized

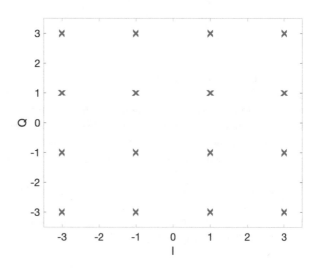

Fig. 5.13 16 PSK versus
16 QAM. The two
constellations are
normalized to have unity
average power

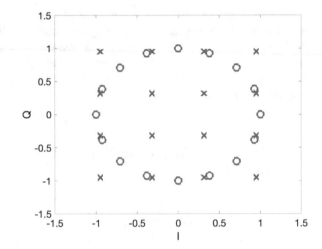

Fig. 5.14 Time-domain
signal for a 16-QAM signal.
The waveform shows three
envelope values
corresponding to three
possible energy values

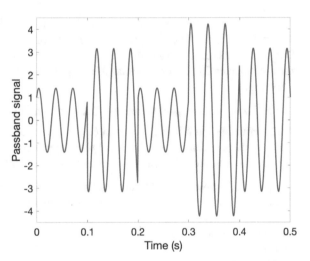

Average energy is the best way to compare apples to apples. For the PSK signal,
average energy is the same as any symbol energy because the signal has a constant
envelope. In QAM signals in general, different symbols will have different energies.
In the case of 16 QAM, there are three distinct values for energy.

A QAM signal in time domain is easy to deduce from the constellation. It will be a
sinusoid where both the phase and the amplitude can vary from symbol to symbol;
this is shown in Fig. 5.14. What is incredible about the QAM signal is how much it
looks like the PSK signal, the only difference being the variability of the envelope.

QAM has a clear advantage relative to PSK. QAM fills the available space better,
maximizing distance between signals, and reducing BER significantly. This can be
seen in Fig. 5.13. The price that QAM pays is better seen from time-domain signals.

Particularly, compare Figs. 5.7 and 5.14. QAM has a nonconstant envelope, while PSK has a constant envelope. A constant envelope is good.

The envelope matters to the power amplifier. In Sect. 1.3, we discussed the PA as part of the transmitter RF chain. It is near the end of the chain, and its only job is to make the magnitude of the signal large so that it can suffer from path loss and still arrive at the receiver with an acceptable SNR.

The power amplifier is a special kind of amplifier that focuses on a trade-off of linearity and efficiency. But at the end of the day, it is still a transistor-based amplifier. Amplifiers need to be biased. The biasing point is the DC operating point around which signal swing occurs at the input and the output. With a constant envelope signal, it is relatively easy to choose an appropriate biasing point. Specifically, choose it so that as the amplified signal swings at the output, its peaks are still within the supply rails.

With a nonconstant envelope, things are not so simple. Figures 5.15 and 5.16 show the conundrum. The signal has large peaks corresponding to large amplitudes, and much smaller swings corresponding to smaller amplitude signals. We have two extreme choices for the bias point: we can choose it so that it accommodates the largest swing or choose it to accommodate smaller swings.

In Fig. 5.16, we accommodate the largest peak. The bias point is chosen around 5 V; the largest peak swings between 0 and 10 V. With a 10-V supply, this large peak is fully taken care of at the output. However, the signal has a much lower swing for a significant proportion of time. During these periods, the DC point is dissipating power corresponding to the 5 V bias, while the signal component is consuming a much lower power. Efficiency is the ratio between signal power delivered at the output and DC power drawn from the supply. Figure 5.16 leads to a drop in PA efficiency which means the battery is going to die fast.

Figure 5.15 shows the other option: we bias closer to the swings of smaller signals. Here we use a supply of 8 V and a bias point of 4 V. This increases

Fig. 5.15 Power amplifier bias point trade-offs. This bias point accommodates low amplitude swings, but the supply (8 V) and ground will clip large swing signals in both directions

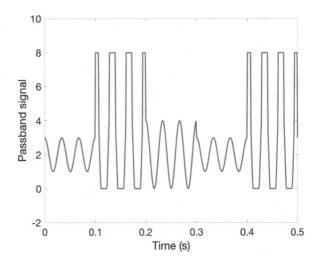

Fig. 5.16 The other side of PA trade-offs. This bias point is set to accommodate large signal swings. There is a lot of unused headroom for small swing signals. For example, the smallest swing here is biased around 5 V when it only needs a bias around 1 V; this leads to reduced efficiency

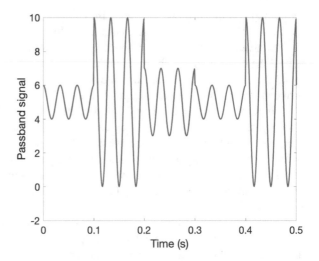

efficiency because DC power is smaller. However, the 10-V swing of the large peak is now clipped for a duration corresponding to 1 V at both ends of the swing. This is an extreme form of nonlinearity. The clipping will lead to an increase in BER which will require retransmissions and thus will also drain the battery faster.

This trade-off between efficiency and linearity is fundamental. The PA is a major power drain in radios, especially when the transmitter and receiver are far apart. PA inefficiency could quickly drain the battery. Extreme nonlinearity could destroy all sent packets, which would also drain the battery through retransmissions.

It sounds like there is a major contradiction in the choice between PSK and QAM. However, things are clear: QAM is better. In Sect. 7.3 we will see that OFDM by its very nature leads to time-domain signals with terrible peak to average power ratios. This is independent of the underlying modulation. In other words, you will have to deal with the PA bias conundrum whether or not you use QAM, so you might as well use QAM. Therefore, QAM is often associated with OFDM.

Figures 5.17 and 5.18 show the constellations used for the allowed QAM modes in 802.11n. Table 5.1 shows the normalization factors used with each constellation. All the shown constellations have non-unity average energy because their real and imaginary parts are integers. To normalize the sent signal to unit energy, we multiply it by the factors in Table 5.1. QPSK is a special case because it is the "four-scheme" of both PSK and QAM.

Symbols in all constellations are not arranged in order, meaning that if we read the BCD equivalent of the symbols left to right top to bottom, we would not be counting in order. Instead, symbols are arranged using Gray encoding.

In Sect. 5.6, we will observe that an error occurs when a received symbol crosses the decision boundary between two sent symbols. The most likely error event involves a received symbol crossing between two adjacent constellation symbols. If we use BCD encoding, then crossings between neighbors will lead to multiple bit

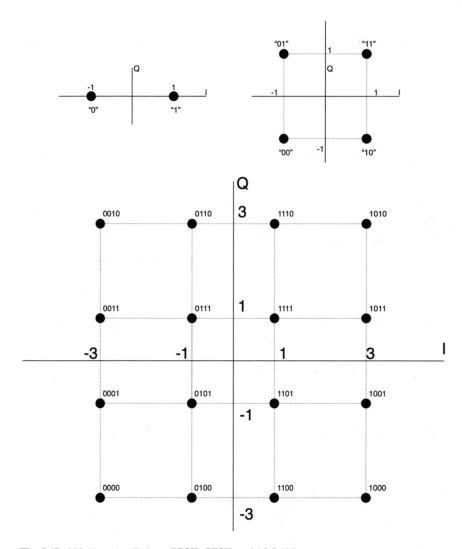

Fig. 5.17 802.11n constellations, BPSK, QPSK, and 16 QAM

errors. Using Gray encoding in Figs. 5.17 and 5.18, crossing between adjacent symbols will lead to a symbol error, but only a single bit error.

In Chap. 6, we find that isolated bit errors can be detected and corrected by forward error correction, whereas multiple bit errors in sequence are a lot harder to correct. Thus, Gray encoding combined with forward error correction can help avert a lot of packet retransmissions.

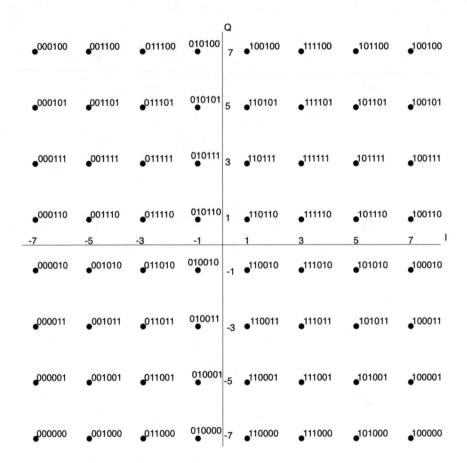

Fig. 5.18 64-QAM constellation in 802.11n

Table 5.1 Normalization factors for QAM constellations in 802.11n

QAM scheme	Symbols	Bits/symbol	Rate in bps/Hz	Normalization factor
BPSK	2	1	1	1
QPSK	4	2	2	$\frac{1}{\sqrt{2}}$
16 QAM	16	4	4	$\frac{1}{\sqrt{10}}$
64 QAM	64	6	6	$\frac{1}{\sqrt{42}}$

5.5 Visualizing Channel Effects Through Constellation Diagrams

Figure 5.19 shows a QPSK constellation as it would be sent from the transmitter. The symbols are missing the normalization factor, but this is fine if we are aware it. We will examine how the constellation looks when exposed to different channel effects

Fig. 5.19 Clean QPSK
constellation

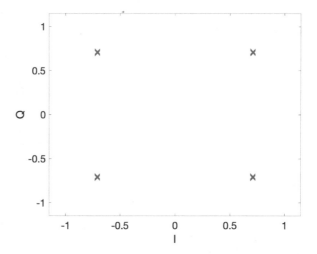

Fig. 5.20 QPSK
constellation suffering
from AWGN

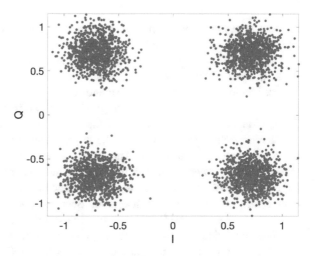

and transceiver non-idealities. Visualizing the impact of these factors will help understand how to counteract them and why they affect the signal the way they do. The constellation in Fig. 5.19 represents hundreds of symbols being transmitted, but because every symbol must fall upon one of the four points $\{1 + j, 1 - j, -1 + j, -1 - j\}$, the constellation looks like it shows only four symbols. When we look at constellations affected by the channel in upcoming figures, we will see hundreds of points as the coincident sent symbols spread out.

Figure 5.20 shows many symbols at a receiver suffering from AWGN. None of the points manage to arrive at the exact points from which they were transmitted in Fig. 5.19. This is because noise moves these sent points in the plane. Motion is in both the x and y directions. Motion in the x direction is due to noise added to the

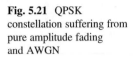

Fig. 5.21 QPSK
constellation suffering from
pure amplitude fading
and AWGN

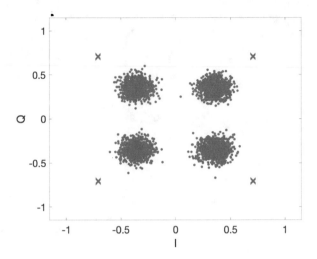

in-phase component, while motion in the y direction is due to noise added on the quadrature arm of the receiver.

The received points form a cloud; the center of the cloud is the correct position of the sent signal. This is because AWGN is zero mean, and thus it will always be centered on the correct symbol. The cloud is denser near the middle and gets thinner the farther away we move from the original sent point; this is a result of the Gaussian distribution of noise. The radius of the cloud is related to the variance of the distribution, which is essentially noise power. Thus, the higher the noise power, the more expansive the cloud. The cloud is circular because for most practical systems, noise power is equal in the in-phase and quadrature directions. If noise is large relative to the signal, some of the transmitted points will cross the axes, leading to an error in estimating the sent symbol (Sect. 5.6), which will lead to bit errors.

Figure 5.21 shows the QPSK constellation suffering from amplitude fading. This is a theoretical channel effect, where h is a pure real number that is usually less than 1. The constellation is shrunk, and all symbols are affected equally. In QPSK, this does not change decision boundaries. But in higher-order QAM, we can see how this kind of fading of signal power will lead to symbol errors.

Amplitude fading also reduces received signal power, leading to more susceptibility to noise. Notice that noise power will remain the same. This means that the radius of the cloud around each point will remain the same. Since the points are now closer to the decision boundaries, this will necessarily lead to more boundary crossings and bit errors.

Figure 5.22 shows the constellation suffering from a phase shift only. The radius of the constellation points remains the same; however, all the points rotate at the same phase. This should not strictly lead to an increase in bit errors, because signal power remains the same. We only need to adapt the decision boundaries. This is contingent on us knowing the phase shift accurately. If we do, we can rotate the

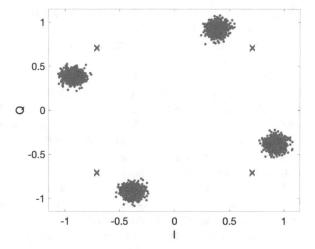

Fig. 5.22 QPSK constellation suffering from phase shift and AWGN

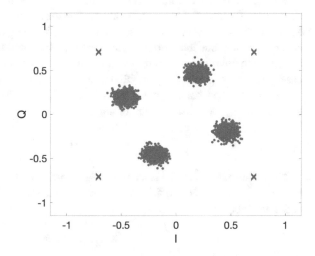

Fig. 5.23 QPSK constellation suffering from a flat-fading channel and AWGN

received symbols back or rotate the decision boundaries forward. Thus, a (known) pure phase shift will not result in a deterioration of BER.

Figure 5.23 combines the amplitude and phase fading effects into one figure. This is what we observe in flat-fading channels. The signal is attenuated by the magnitude of the channel factor h, while all points are also rotated by the phase of the channel. This will require us to either rotate the symbols back and amplify them or distort the decision boundaries the same way before we decide. What we are describing here is called channel inversion, which is the process of inverting the effects of the channel. It is simple in single-antenna systems, but in MIMO it will become a major issue and will consume all of Chaps. 9 and 10.

So, how will a frequency-selective channel change the constellation? It is extremely difficult to represent the impact of a multipath channel on a constellation diagram. The channel will affect different spectral components of the signal differently. However, in OFDM each of the subchannels will itself be flat, allowing us to represent each of them the same way as Fig. 5.23, although each of the subchannels will have its own fading values.

How about Doppler? Doppler is fading that changes with time. We can represent this either using a sequence of figures that look like Fig. 5.23 or using a movie that shows Fig. 5.23 evolving with time. Notice that high Doppler normally describes the rate of change of fading rather than SNR. We should expect noise power to remain the same. Recall that noise is a receiver antenna phenomenon (Sect. 4.2) rather than a channel phenomenon. Thus, the movie should show the centers of the clouds in Fig. 5.23 moving, but their radius remains the same.

In Sect. 11.6, we will spend some time looking into frequency offset estimation and correction. In Sect. 7.3, we will understand why leaving a frequency offset in the baseband signal can be devastating for OFDM signals. Let us take a very basic look at what frequency offset is and what it does to the signal constellation.

Frequency offset occurs because of an unavoidable mismatch between the frequencies produced for the carrier at the transmitter and the receiver oscillators. Manufacturers of crystal oscillators make every effort to guarantee that their product is very accurate, but some offset is unavoidable. This offset is extremely small when compared to the frequency of the carrier. But when compared to the bandwidth of the baseband signal, particularly an OFDM subchannel, this offset is devastating.

Assume the baseband signal at the transmitter is $x(t)$. When raised to the carrier at the transmitter, the passband phasor signal is:

$$x_p(t) = x(t)e^{j\omega_{ct}t}$$

Now assume the channel is ideal; in other words $h = 1$. But the receiver multiplies by a carrier that is a little off from the carrier at the transmitter. The baseband signal at the receiver is:

$$x_r(t) = x(t)e^{j\omega_{ct}t}e^{-j\omega_{cr}t}$$

$$x_r(t) = x(t)e^{j\Delta\omega t}$$

The difference between the angular frequencies of the two carriers is small relative to the carrier but is significant relative to the frequency content of $x(t)$. Assuming again that x is a QPSK signal, what would the received constellation look like? The frequency offset is adding a phase offset that is time dependent.

Thus, the symbols of the QPSK signal will rotate with time as shown in Fig. 5.24. Frequency offsets are expressed as a time-dependent phase shift. Again, this is best shown through a movie. But in Fig. 5.24, the trail of symbols shows how symbols evolve through time. If the symbol duration is small relative to the frequency offset, then the total angle that the symbol moves during the symbol duration will be negligible. This might be so small that we can ignore it.

Fig. 5.24 A relatively short
duration symbol suffering
from frequency offset but no
AWGN. Symbols rotate
only slightly within symbol
duration

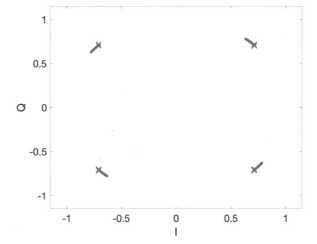

Fig. 5.25 QPSK
constellation suffering from
frequency offset. Symbol
duration is long enough for
the frequency offset to
manifest as a phase offset

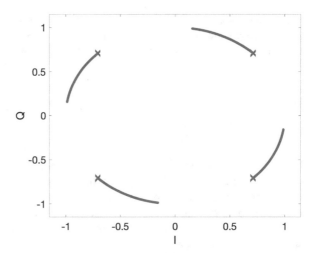

If the symbol duration is large, as in Fig. 5.25, the signal will move substantially during the symbol duration. This will bring it close to the decision boundary and will devastate the BER. When discussing OFDM, we will be particularly concerned about frequency offsets because symbol durations in OFDM are orders of magnitude longer than in single-carrier systems.

Figure 5.26 shows a constellation suffering from a very common transceiver non-ideality called IQ mismatch. This is not a channel effect but rather a defect in the functionality of the transmitter and the receiver. The total gain in the in-phase and quadrature paths of the transmitter, but more often the receiver, are mismatched. This leads to a deformation of the constellation. If gain in the in-phase direction is higher than quadrature, the constellation is stretched horizontally. If gain is higher in the quadrature arm, then the constellation is stretched vertically. IQ mismatches can

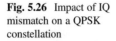

Fig. 5.26 Impact of IQ mismatch on a QPSK constellation

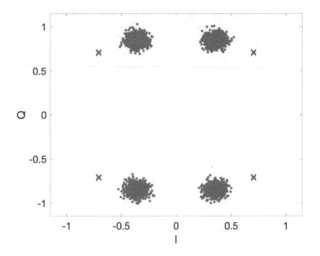

significantly increase BER because they bring symbols closer to decision bound-aries. But the good thing about IQ mismatch is that it is deterministic and thus easy to calibrate using headers and pilots (Sects. 11.1, 11.2, and 11.3).

5.6   How to Do Hard Decision Decoding, the Maximum Likelihood Result

"Decoding" can be a confusing word because it is used differently depending on the context. In the context of baseband modulation, decoding is sometimes used in conjunction with or instead of the term demodulation. In such case, decoding means extracting the bits from the received baseband QAM symbols. This is "done" once we decide which symbol was sent, because translating the sent QAM symbol to bits is a one-to-one mapping.

Figure 5.27 shows a received 16-QAM symbol in an AWGN channel. The symbol (shown as an empty circle) does not coincide exactly with any of the 16 symbols that could have been sent because the received signal has noise added to it. The complex noise will move the symbol on both the x- and y-axes. What is the best guess we can make about which symbol was sent? The answer is intuitive, obvious, and actually optimal: we should pick the nearest constellation point to the received signal. In Fig. 5.27, you would immediately assume that the most likely transmitted symbol is "0011," and you would be exactly right.

Thus, if x is the set of possible symbols we could have sent, $\hat{x}$ is our best estimate of the sent signal, and y is the received signal, then:

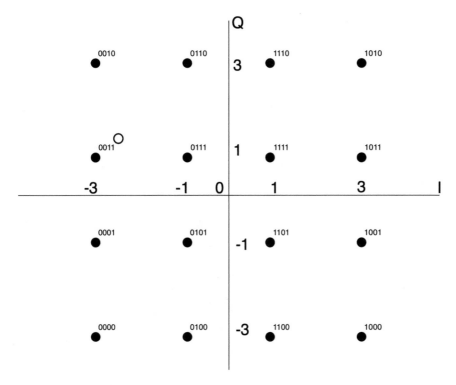

Fig. 5.27 Received signal compared to 16-QAM constellation in an AWGN channel. The received signal is the empty circle, while the solid circles are the possible constellation points

$$\widehat{x} = \min_x ||y - x||^2$$

Thus, the best guess is the one that minimizes the square distance between the observation and the guess. We calculate the square of the distance because the Euclidian distance involves calculating a square root. The x that minimizes the square distance is also the x that minimizes the distance. So, sticking to square distance allows us to avoid using the square root function, which is very costly in hardware. Note that both x and y are complex numbers, and thus the square norm above is calculated as:

$$||y - x||^2 = (y_r - x_r)^2 + (y_i - x_i)^2$$

This solution is called the maximum likelihood solution. It is the most likely signal to have been sent. It is intuitive, simple, and optimal. There is no solution that can provide a better BER for the same SNR. This is the best guess anyone can ever make. Note that this solution is contingent on the zero-mean Gaussian nature of AWGN. Other kinds of noise could have other optimal solutions.

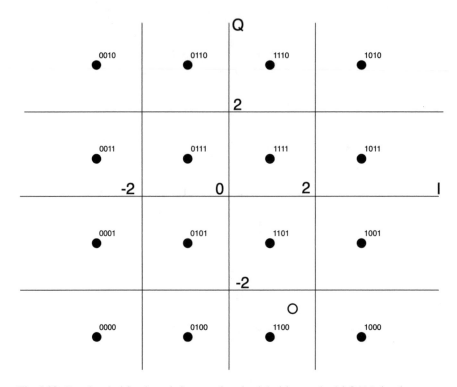

Fig. 5.28 Drawing decision boundaries to make a hard decision on the 16-QAM signal

One problem with the maximum likelihood solution is how its complexity scales. You need to calculate k distances (or rather square distances) and compare them, where k is the number of symbols in the baseband constellation. Thus, in QPSK, we need to calculate 4 distances, whereas in 256 QAM we need to calculate 256 distances. But do we really? By examining Fig. 5.28, all we need to do is decide which symbol in the constellation is closest to the received signal.

We can achieve this result without calculating a single distance. This can be done by drawing decision boundaries between constellation points as seen in Fig. 5.28. We can figure out which symbol the signal is closest to by comparing its real and imaginary parts to the decision boundaries. One way to do this is to successively exclude areas of the constellation to reach a final decision.

For example, if we find that the real part of y is positive, we should only consider symbols in the right half plane. Then, if the imaginary part of y is negative, we only need to consider the lower right quadrant. Then, if the real part is less than 2, we need to only consider symbols $1-j$ and $1 - 3j$. If the imaginary part is less than -2, we decide that the most likely symbol is $1 - 3j$.

We have done this using simple successive comparisons, which in hardware can be implemented using adders. Because the real and imaginary parts of QAM symbols tend to be small integer numbers (Sect. 5.4), these adders can be small.

Using binary elimination as we did above, we need $\log_2(k)$ comparisons to decide which symbol was sent.

The discussion above has one major limitation: it was about an AWGN channel. We must at least extend it to flat-fading channels. In Sect. 7.1, we will find that even frequency-selective channels are treated as a sequence of flat-fading subchannels in OFDM.

With flat fading, the signal is multiplied by a Rayleigh fading coefficient before AWGN is added. Comparing the signal to the constellation, either through distance calculation or successive comparisons, will lead to a BER of ½ because the decision boundaries have moved. We must get rid of the effect of the channel first.

The received signal is:

$$y = hx + n$$

The receiver estimates the channel using the header, pilots, or both. The channel estimate is never equal to the channel because there is noise on the preamble data used in estimation. However, assuming we have perfect knowledge of the channel, then the channel estimate is equal to the channel:

$$\widehat{h} = h$$

Having figured the channel out, we can multiply the observation with the inverse of the channel, to obtain our best guestimate of the sent signal:

$$\widetilde{x} = h^{-1}y = h^{-1}(hx + n) = x + h^{-1}n$$

This is the point in the complex plane which we now plug in to the distance calculation or the successive comparisons. The rest of demodulation proceeds identically to the AWGN case. The problem with this implementation is that it requires us to calculate the inverse of h, or more accurately to divide y by h. In Chap. 3, we found that division is possible in hardware. But it is bulky, ponderous, and has horrible dynamic behavior.

The distance we want to minimize is:

$$\widehat{x} = \min_{x} \left\| h^{-1}y - x \right\|^2$$

But the same x that satisfies the above also satisfies:

$$\widehat{x} = \min_{x} \left\| y - hx \right\|^2$$

So, instead of dividing y by h, we multiply x by h. We have replaced a division by a multiplication, which is worth it in hardware. However, there are two major caveats here. Recall that the distance must be measured for k copies of x. There is only one

y but k copies of x. So, we are replacing a single division by k multiplications! Using symmetries in the constellation, we can reduce the number of multiplications, but it will always be more than a single multiplication.

But more important is what the above insinuates about using decision boundaries instead of distance measurement for decoding. Dividing y by h before doing the comparisons means we are removing the distorting effects of the channel from y. Multiplying x by h instead means that we will be contorting the decision boundaries to match the contortions on y. In Sect. 4.3, we found that the channel rotates and attenuates everything. This causes the decision boundaries to move from those corresponding to x to those corresponding to xh. The way we did successive comparisons above is contingent on the boundaries being horizontal and vertical at the midpoints of constellation symbols, which will not happen if we pre-distort them with the channel. Thus, it is often preferable to swallow the pill of dividing by the channel coefficient, especially for larger constellations.

5.7 Soft Decisions

The way we decide which symbol was sent in Sect. 5.6 is called hard decision decoding. Making a hard decision means that we make a final decision about what value each bit takes. When you decide that a certain symbol was sent in Sect. 5.6, you are also making a final decision about the $\log_2(k)$ bits that this symbol carried.

There is an alternative to this where we make a soft decision about the bits that were sent. This will reflect our level of certainty about the sent bits. As a very simple example, see Fig. 5.29 and Table 5.2. We are considering BPSK, so we are either sending a "0" or a "1." At the receiver, we observe if the received symbol lies to the right or left of the j-axis. If it lies to the left, then we guess a 0 was sent. If it lies to the right, then a 1 was sent. This is a hard decision.

On the other hand, in Fig. 5.28, we divide the range of "0" and "1" into multiple subranges. These subranges reflect our degree of certainty about what was sent. For example, at a certain positive value close to 1, we are fairly certain that the received

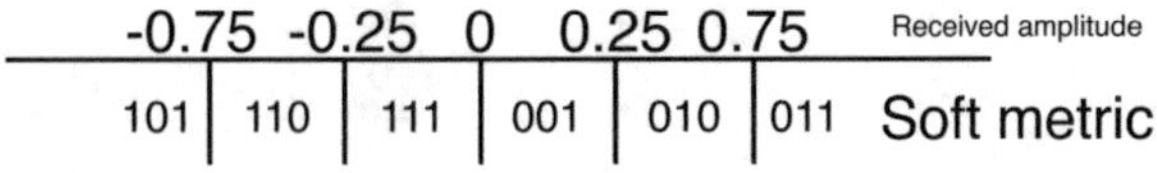

Fig. 5.29 BPSK and soft decisions

-0.75	-0.25	0	0.25	0.75	Received amplitude
101	110	111	001	010	011	**Soft metric**

Table 5.2 Baseband values and soft decision metrics representing degrees of confidence about the received symbol for BPSK

Received baseband value	Soft decision metric
0–0.25	1 = "001"
0.25–0.75	2 = "010"
0.75 and above	3 = "011"
−0.25–0	−1 = "111"
−0.75 to −0.25	−2 = "110"
−0.75 and below	−3 = "101"

signal reflects a sent "1." In this example we take this value to be 0.75. Any higher value would also reflect this same level of high certainty.

At a received value between 0.25 and 0.75, we have an intermediate level of confidence that the sent bit was "1." Between 0 and 0.25, we still think that a "1" was sent, but we can see a very distinct possibility that a "0" was sent and was pushed there by noise. We also have symmetrical certainty levels about a sent "0" in the negative half plane.

Thus, the decision we make about what was sent is not a single bit, nor is it a final decision. Instead, we produce a multi-bit, usually integer, soft decision metric that represents our level of certainty about what was sent. The decisions are listed in Table 5.2 with the corresponding values of the soft metric.

Two questions remain:

- Can this be expanded to larger constellations, and how?
- Why even bother with all these complications, and what gain do we get from making soft decisions as opposed to hard decisions?

Yes, we can produce soft metrics for larger constellations where the metrics encode levels of confidence about each bit in the decoded symbol. In Chap. 6, we will see that a few steps after baseband modulation, the results are fed to the channel decoder. The decoder will use redundancies in these coded bits to detect and correct errors. If the demodulator produces soft metrics instead of hard metrics, the decoder can detect and correct more errors, improving BER for the same SNR.

5.8   Channel Capacity, the Best Anyone Can Do

Capacity is a very important concept in information theory. It is the maximum rate that a transmitter and receiver can reliably share mutual information. When applied to wireless channels, the theory has a fundamental implication. Capacity for AWGN channels is:

$$C = B \log_2 \left(1 + \frac{P_s}{BN_0}\right) \text{ bits/s}$$

Capacity is measured in bits per second. B is the bandwidth of the channel. P_s is the signal power at the receiver. N_0 is the AWGN power spectral density at the receiver antenna. Thus, BN_0 is total white noise at the receiver, and $\frac{P_s}{BN_0}$ is the signal to noise ratio. This allows us to write capacity in a more compact form:

$$C = B \log_2(1 + \text{SNR}) \text{ bits/s}$$

Capacity is the maximum rate that the transmitter can put into the channel and expect the receiver to be able to decode with an arbitrarily small error rate. In

Chap. 6, we will look at error correction codes. These are codes that add redundant bits to allow the receiver to detect and correct bit errors. But to what degree can error correction codes correct errors, and how does redundancy and error correction affect the net throughput we can push through the channel? This is the question that capacity answers.

This is best illustrated by an example. Consider a system where the SNR is 20 dB. Plugging this into the capacity equation, and realizing we are not given the bandwidth:

$$\frac{C}{B} = \log_2(1 + 100) = 6.65 \frac{\text{bits}}{\text{s.Hz}}$$

This is capacity per unit bandwidth. Thus, it measures how many bits per second we can push in 1 Hz. In Sect. 5.4 we saw that depending on the baseband modulation scheme, we can push more than one bit per second in every Hz. Capacity tells us that you can only push 6.65 bits per second in every Hz. If we use 64 QAM, we are pushing 6 bits per second per Hz. Since the channel capacity is 6.65, there exists a channel coding scheme that allows us to retrieve the data we push without errors. This error decoder can be very complex, but it exists.

But what if we try to use 256 QAM for the above SNR? We would be pushing 8 bits per second per Hz. No matter how good error correction is at the receiver, we cannot properly retrieve all this data. In fact, as the message gets very long, our BER will be ½, no matter how much error correction we use.

Channel capacity is so powerful because it provides a theoretical ceiling to our performance. The rate that capacity gives us is the best we can do, not because of limitations in transceiver design but because of fundamental limits in information theory. This is invaluable because it tells us how much data is worth trying to send. Most standards offer multiple modulation schemes to choose from, we need guidance on which to use, and the best guide is capacity.

How about capacity in fading channels? The expression we obtained above included only the effect of white noise. If we include the impact of fading, all it does is reduce the signal power, leading to a lower effective SNR. Assuming the channel coefficient is h, capacity is:

$$C = B \log_2\left(1 + |h|^2 \text{SNR}\right) \text{ bits/s}$$

Fading obviously reduces capacity. Note however that there is a nonzero probability that Rayleigh fading could have instantaneous amplitudes greater than one, leading to short-lived improvements in capacity. The problem with fading is that it is only known at the receiver, while it is the transmitter that makes the decision on what modulation scheme to use. Depending on the Doppler, h will keep changing. Even in low Doppler environments, we can expect h to change from packet to packet. Thus, unless the transmitter always knows channel state information at the receiver

(CSIR), there is no way to prevent a situation where the transmitter tries to send above capacity.

For example, assume the transmitter knows SNR at the receiver is 20 dB. The transmitter will decide it is safe to use 64 QAM and will do so. However, in the presence of fading, the effective SNR ($|h|^2$SNR) at the receiver will sometimes fall below 20 dB. This will happen no matter what because there is a finite probability of this happening in Rayleigh distribution.

Whenever this happens, the channel capacity drops below 6.65 bps/Hz, conceivably even below 6 bps/Hz. The transmitter is still pushing 6 bps/Hz, but this will lead to packets that will certainly be dropped if fading is deep enough. We call such a situation "outage."

The above problem stems from lack of knowledge of channel state information. Particularly, the transmitter is responsible for picking the constellation to use. In Sect. 7.14, we will discuss the possibility that the receiver will send back this CSI to the transmitter. This allows us to use adaptive radios that mold to channel conditions. However, even in cases where the receiver regularly feeds back this information to the transmitter, this would always be average CSI. Fading can be relatively fast changing, and outage will be inevitable. In Sect. 9.5, we will expand our understanding of capacity by considering multiple antennas and by calculating an expected value for capacity in faded channels.

Chapter 6
Forward Error Correction

6.1 Why Do Channel Coding

Channel coding adds error resistance to the message. This is done in two distinct though related ways. First, the code allows us to distinguish messages which are received correctly from messages where an error occurred; we can call this error detection. Second, the code allows us to pinpoint locations where bit errors occurred, and by extension to correct these errors by flipping the bits. We can call this error correction.

Channel coding achieves its error-fighting ability by introducing redundancy. Let us consider a trivial form of coding: simple repetition. Assume we want to transmit a "0." The time it would take to transmit this bit in passband is T_b. With a repetition code, we repeat this bit three times as "000." Thus, we transmit the "0" three times over a period of $3T_b$. The additional two zeros carry no information because they are exact copies of the first bit, yet they take exactly as much time as if they were information bits. The bitrate of the "uncoded" message is $f = 1/T_b$, but the coded message has a throughput of $\frac{f}{3} = 1/3T_b$. This drop in throughput is a fundamental feature of error-correcting codes; the more error correction ability you add, the less the actual data throughput.

Assume noise impacts the three sent zeros at the receiver. The probability that the noise meaningfully impacts only one of the bits is much higher than the probability that it impacts two which is in turn much lower that the noise affects all three values. This owes to the uncorrelated nature of AWGN. If noise is large enough to flip one bit, this tells us nothing about what happened to the two other bits.

We can use majority voting to determine what bit was sent, and we would be able to retrieve the correct value even if noise ruined one of the received waveforms. If we observe "000," then our best guess about what was sent is "0." If we receive "001," "100," or "010," then an error has occurred regardless of what was originally sent. Our best guess in each of the three cases is that a "0" was sent. However, if we receive "110," "011," "101," or certainly "111," then our best guess about what was sent is "1."

One might object by saying that multiple bit errors will lead to a misinterpreted result, for example, how do we know that "011" was originally "111" with the first bit being flipped rather than "000" with the last two bits flipped? This is a realistic scenario, which is why this form of channel coding, or indeed any form of channel coding, will always have nonzero BER. We still make mistakes. However, single bit errors are much more likely than multiple bit errors in sequence under AWGN, which is why channel coding improves BER (but does not nullify it).

But what if we want to be even more resistant, in other words what if I want to protect the system against multiple bit errors in proximity, what if I want to be able to correct the scenario where "000" turns into "110" or even "111." The only way to do this is to increase the order of repetition. Repeating the sent bit five times gives us more protection than three times because the code will be able to correct errors in two bits. Seven is even better; it will be able to correct three bits.

We pick an odd number of repetitions because this allows us to do majority voting at the receiver. In general, using N bits for the coded message allows us to correct against $(N - 1)/2$ errors. So, what is the absolute "best" we can do? What is our utmost error correction ability? If we repeat the bit an infinite number of times, we would naturally be able to correct for an infinite number of bit errors, and yet it would take an infinite time to transmit a single bit of information. So, this is an absurd metric.

So, is there a meaningful upper limit to how much coding can help? In other words, if we use the "ultimate code," do we get zero BER? The answer is no. In Sect. 5.8, we described how a channel can carry no more than a maximum throughput called capacity. There is no error correction code that could allow us to send data higher than capacity. All bits sent above capacity will be erroneously decoded. In fact, channel capacity is a ceiling that can only be achieved by using a perfect error detection code. Therefore, perfection in terms of channel coding does not mean we will achieve zero BER; it means that the receiver can achieve the *useful* goodput (Sect. 2.7) dictated by Shannon capacity.

But there must be a price. The coding scheme that corrects more errors must be paying a heavier price compared to one that corrects fewer errors. If we examine the repetition codes described above, the answer is very simple: the price is useful throughput. At this point, we should be specific about what we mean by throughput.

In Chap. 5, there is a net bitrate that we calculated coming out of the baseband modulator. This is the bitrate that will be radiated out of the antenna. We call this the coded bitrate; it is the rate that already carries within it the "repetition." This is distinct from the useful data bitrate. Data bitrate carries unique information in every bit without redundancies. This is ultimately the bitrate which the MAC layer pushes down to the PHY layer. We will talk more about these distinctions and the trade-off that happens shortly.

Figure 6.1 shows the encoder and decoder in a baseband transmitter and receiver. The encoder is the first block in the transmitter baseband. It receives data bits from the MAC layer. To the physical layer, all these data bits equally carry "information." Each input bit to the encoder must produce more than one output bit because channel coding must introduce redundancy. We will shortly discover that the redundancy

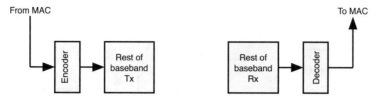

Fig. 6.1 Encoder and decoder in the transmitter and receiver

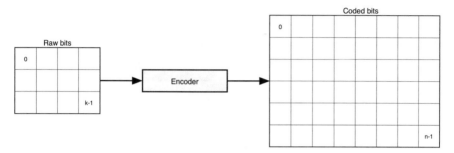

Fig. 6.2 Block codes, shown a rate k/n encoder

introduced does not have to be in whole bits but could be in fractions of bits. Thus, the data rate exiting the encoder is higher than the bitrate entering it.

At the receiver the corresponding block is the decoder, or more formally the channel decoder. The decoder examines the received bit stream with redundancy and then makes a "best guess" about what corresponding data bit was originally sent. The decoder thus detects errors and corrects them if possible. Just like the encoder is the first block in the transmitter; the decoder is the last block. Its output goes up to the MAC layer as the physical layer's best guess about the payload that was sent. The decoder input bitrate is higher than its output bitrate because its input is coded, while its output is uncoded.

The "raw" form of coding we described above by sheer repetition is not the best code we can use. There is no theoretical foundation for how the code is constructed, nor any guarantee on its performance.

Error-correcting codes are classified in a few different ways. The first distinction is whether it is a block code or a convolutional code. The difference between the two is in the nature of the input to the encoder/decoder. In Fig. 6.2, a block encoder works on a block of bits of size k and produces a larger block of n bits at the output. Ideally, each bit in the output block is a function of all the bits in the input block. For example, the value of bit 0 in the output block is a function of the values of bits 0 to $k - 1$ in the input block. Once we have encoded a block, we can start encoding the next block. This also applies to decoding. The values and decisions in one block are completely independent from those in all other blocks. Usually, the mathematical representation of block codes involves matrices of binary numbers.

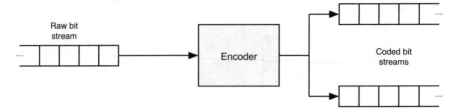

Fig. 6.3 Convolutional codes, shown a rate ½ encoder

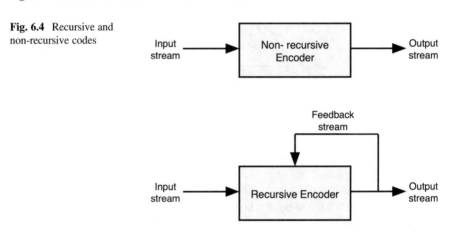

Fig. 6.4 Recursive and non-recursive codes

Convolutional codes are the other class of channel codes (Fig. 6.3). In a convolutional encoder, there is no input block and output block; instead, there are input and output *streams*. A continuous flow of input bits is combined in the encoder to produce an output stream of bits. The input stream bitrate is necessarily lower than that of the output bit stream. The 802.11n encoder is convolutional, and thus most of the rest of this chapter will discuss convolutional encoders in detail. In general, convolutional encoders involve some sort of shift register. This causes each output bit to be a function of several input data bits depending on the length of the register, which determines the "memory" of the encoder.

Codes can also be classified as recursive and non-recursive. This classification generally applies only to convolutional codes. As shown in Fig. 6.4, a recursive encoder is one which has feedback. A non-recursive encoder does not and is purely feed-forward. Stated another way, a recursive code uses the current coded bits as input bits to upcoming encoding operations, whereas non-recursive codes only use incoming uncoded bits as inputs. Recursive codes can be significantly more complicated to decode than non-recursive codes. However, they also deliver the best performance of any channel decoders, pretty much achieving Shannon capacity.

The final method to classify codes is by whether the code is systematic or non-systematic. As shown in Fig. 6.5, in a systematic code, the input bit appears explicitly as one of the coded bits. A non-systematic code is one where all coded bits are processed bits. In a systematic encoder, we usually do not need to look carefully to see the systematic bit at the output because it is directly shorted.

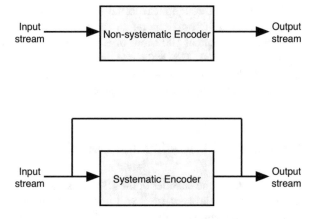

Fig. 6.5 Systematic and non-systematic codes

There is one parameter that matters more than all in evaluating a code, and that is the coding rate. The coding rate, r, is a fraction between 0 and 1. It represents the proportion of the encoder output throughput that is information data. This fraction ultimately represents the proportion of data radiated out of the antenna that is the raw data sent down by the MAC layer, the balance making up the error correction portion of the stream. Consider a transmitter that is radiating data out at a rate of 300 Mbps and has a coding rate of 1/3. This means that unique useful data provided by the MAC layer is being sent out at a rate of 100 Mbps; the remaining 200 Mbps is used for error correction.

The fraction r is defined as $r = k/n$. k is the number of input bits to the encoder and n is the number of outputs from the encoder. Likewise, k is the number of bit outputs from the decoder, while n is the number of input bits to the decoder. Thus, $r = 8/9$, for example, means that of each nine bits transmitted over the air, eight are data bits, and one is an error correction bit.

There is a trade-off between coding rate and bitrate. A higher coding rate means less bits are used for error correction. This means that the throughput is mostly useful data throughput, but there will be less error correction capacity. A lower coding rate means that more bits are used for error correction; this will translate into lower BER and a lower data bitrate. So what coding rate is "best"?

In Sect. 2.7, we discussed the concept of goodput, where goodput = Throughput $(1 - \text{BER})$. Maximizing goodput would be a fair objective for channel coding. Goodput maximizes the benefit to the application by combining both high data rate and low BER.

But again, what will determine what coding rate gives us the best goodput? The answer is channel conditions. A favorable channel has high SNR, high capacity, and few errors. This means the impact of BER on goodput is muted. This promotes a higher coding rate to raise throughput. A bad channel means less SNR and a lot of errors. This magnifies the impact of BER on goodput and encourages us to use a lower coding rate to salvage some of the errors, thus maximizing goodput.

The bitrate or throughput reported for a system is always a point of contention. What is a fair bitrate to report? In the system above, should it be advertised as transmitting at a rate of 100 Mbps or 300 Mbps? The correct and ethical answer is to report the 100 Mbps. This is the rate that the user can use; however, this is not a rule that all vendors will necessarily stick to. We will take a deeper dive into misleading practices in reporting throughput in Sect. 7.14.

6.2 Channel Encoding and Decoding in Wi-Fi

Figure 6.6 shows the encoder of the 802.11n standard. The standard will not directly describe the decoder because it will generally avoid restricting receiver implementation. The standard has loftier aims: to allow *everyone* to get their data over the air reliably and efficiently. Thus, the standard will only tell you what to do at the transmitter. It only cares about two things: that the data you get on the air has a standard form and that the data leaks into surrounding channels only within certain limits. The second aim is self-evident. The standard cares about the overall network, so it must ensure that no terminal is infringing on the channels of other terminals. The first aim is meant to ensure that people can design transceivers independently but that all transceivers can then integrate with each other seamlessly. This is essential to allow the ad hoc nature of the Internet to work.

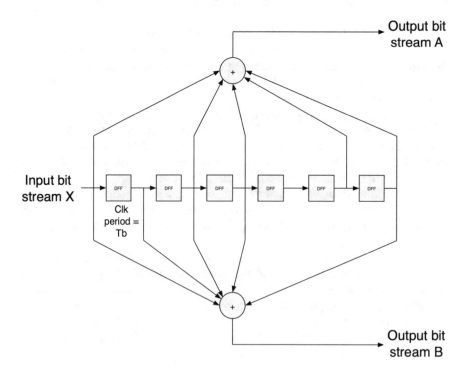

Fig. 6.6 802.11n encoder

So, you should expect the standard to only cover the transmitter in detail. More specifically, the physical layer standard should tell you how each block in the transmitter changes data and the ultimate nature of data before it is radiated out of the transmitter antenna. The designer is free to implement any receiver that can meaningfully interpret such data. This allows individual designers to make improvements at the receiver without adversely impacting interoperability. In practice, because the standard dictates the nature of the message, it implicitly says a lot about the receiver.

Figure 6.6 shows the 802.11n encoder. It is convolutional because it acts on a continuous stream of incoming data (Input bit stream X). Each DF block in the encoder delays the stream by one bit duration. This is implemented using a shift register where the clock period is equal to the duration of a single bit. Both output bit streams produced by the encoder are linear combinations (XOR) of the input bit and the delayed bits. The code is non-systematic because the input bit is not an explicit unprocessed part of either output stream. The code is also non-recursive because it is purely feed-forward.

Thus, the default 802.11n encoder is a convolutional, non-systematic, non-recursive encoder. There are two parallel output streams that are produced at the same rate as the input stream. The two can be combined into a single stream whose bitrate is double that of the input bitrate. The coding rate of the encoder is thus ½. The standard allows for other, higher coding rates. We want to be able to do this without using multiple encoders. In Sect. 6.7, we will see that the structure of the decoder varies dramatically based on the encoder. Thus, having multiple encoders will incur an untenable cost at the receiver. In Sect. 6.4, we will see how puncturing and depuncturing allow us to obtain derivative coding rates without changing either the encoder or the decoder.

We must start discussing the issue of clock speed here, because it can lead to a lot of confusion. Figure 6.6 and the discussion right after suggest that when implemented in hardware, the encoder will have a clock period dictated by the duration of the bit. A simple question is: which bit, the coded bit or the uncoded bit? And what happens if my system clock is different from the bit duration? In Sect. 10.12 we will discuss the issue of clocking in practical implementations and how it relates to system requirements, but for now be aware that the bitrates we are talking about here are not necessarily going to dictate the hardware system clock(s).

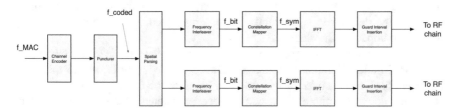

Fig. 6.7 Transmitter architecture for the baseband of 802.11n

6.3   Transmitter Architecture Revisited

Figure 6.7 is the transmitter architecture of the baseband in 802.11n. It shows the flow of data from the MAC layer to the mixed signal and RF chains. The first block in the chain is the channel encoder. It takes its input from the MAC layer. For the physical layer, all the bits incoming from the MAC layer are payload; they all count as data bits regardless of whether they contain headers for higher layers. This is the basis of the layer model; each layer is agnostic to the function of layers above and below it.

The input to the encoder is at a rate f_MAC Mbps. This is the incoming bitrate. The encoder produces a coded message whose bitrate is higher than the input bitrate. The ratio between the two bitrates is the coding rate of the encoder. In fact, from Sect. 6.2, we deduce that $f\_coded = 2 * f\_MAC$ in 802.11n. We also know that the encoder combined with the puncturer produce data at a rate of $f\_coded = f\_MAC/$ coding rate.

Every block from and including the encoder and up to the baseband modulator is said to be a bit-domain block. In other words, it processes data as binary bits. Every block following the baseband modulator acts on QAM symbols and is thus called a symbol-domain block. The modulator works to transform from bit domain to symbol domain. If the input bitrate to the baseband modulator is f_bit, then the output from the modulator is f_sym, where $f\_sym = f\_bit/k$, and k is the number of bits in each QAM symbol.

The puncturer allows us to derive *higher* coding rates than the incoming coding rate without using a different encoder. It is not actually the encoder that we are worried about. In Fig. 6.6, we clearly see that the encoder has a very simple structure. In fact, it costs next to nothing to duplicate the encoder as many times as we need at the transmitter. What we are worried about is the prospect of having to use multiple *decoders* at the receiver.

In Sect. 7.15, we will take stock of a very detailed proposed structure for the Wi-Fi receiver. Everything at the receiver falls under one of the two broad categories: doing the reverse of something that happened at the transmitter or removing the impact of the channel from the received signal. The receiver has a very simple goal: to construct the best *guess* possible about the bits that the MAC gave to the transmitter. It will then give this data to the MAC at the receiver. Thus, the two tasks required from the receiver make immediate sense. To extract raw bit guesses at the receiver, we must counter channel effects and then counter transmitter processing. The channel decoder has a counterpart in the transmitter: the encoder. And yet, the decoder is significantly more complicated, exactly because it must handle the impact of the channel.

In fact, the bit domain in the receiver is pretty much the reverse of its counterpart in the transmitter. Demodulated bits are sent to the depuncturer, where their rate *increases* back to f_coded. These are then given to the decoder, where the input stream has double the bitrate of the output (decoded) stream. This decoded bit stream is handed up to the MAC layer.

6.4 Puncturing and Depuncturing, Obtaining Multiple Coding Rates with the Same Decoder

The 802.11n standard describes four compulsory coding rates that all transceivers must support: 1/2, 2/3, 3/4, and 5/6. There is also an optional mode supporting coding rate 7/8 for very high SNR channels. These higher coding rates are derived from the native rate of ½ at the output of the encoder. The block that transforms to higher coding rates is called a puncture.

A puncturer is a block that accepts an input stream, removing the occasional bit from the stream to increase its coding rate. Two different coding rates derived by puncturing are shown in Fig. 6.8. Gray bits are stolen bits that are dropped to increase the rate.

Figure 6.9 demonstrates puncturing and depuncturing in detail for coding rate 7/8. The input stream to the convolutional encoder is X. The encoder will always produce two coded bits per each input bit. Thus, if we wait for 7 input bits $X0$ to $X6$, the encoder will produce 14 corresponding coded bits. These are shown as $A0$ to $A6$ and $B0$ to $B6$, with each letter denoting an output stream from the encoder. If we transmit all these bits, our coding rate would be ½.

However, as in Fig. 6.9, we do *not* transmit all 14 bits. Instead, we *steal* six bits (shown in gray) and throw them away and transmit only the remaining bits. There are $14 - 6 = 8$ bits remaining. So, we transmit eight bits over the air, of which only seven carry unique data ($X0$ through $X6$). These data bits do not appear explicitly in the sent message because the encoder is non-systematic. However, we know that no more than seven bits carry data because this was the input to the encoder. Thus, we transmit eight bits, of which only seven are data bits, and the coding rate is 7/8.

At the receiver, the decoder is designed for the trellis of a rate ½ encoder (Sect. 6.5). So, we pass the received message stream ($ra0$ through $rb6$) through a depuncturer that returns it to the ½ rate. This is done according to Fig. 6.9 by

Puncturing scheme for coding rate 3/4. Punctured bits in Grey.	A0	A1	A2	A3	A4	A5	A6	A7	A8
	B0	B1	B2	B3	B4	B5	B6	B7	B8

Puncturing scheme for coding rate 2/3. Punctured bits in Grey.	A0	A1	A2	A3	A4	A5
	B0	B1	B2	B3	B4	B5

Fig. 6.8 Puncturing for two different coding rates: 2/3 and ¾

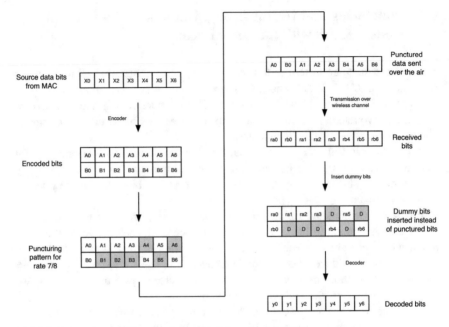

Fig. 6.9 Puncturing and depuncturing in detail for coding rate 7/8

inserting dummy bits in place of the bits that were stolen at the receiver. The dummy bits could be zeros, ones, or any mix of the two. All that matters is that they are inserted in the right location.

There are a few critical points we must note about puncturing. The location at which data bits were stolen is important. These locations are derived by simulation and must be used as stated in the standard. If we steal bits at random locations, there is a chance the message will not be recoverable at the receiver.

Also notice that once we steal the bits and throw them away at the transmitter, they are gone forever. The bits we insert at the depuncturer are dummies. They are constant values that are inserted for any received message. This means they carry zero information, neither about the channel nor about the transmitted message. This will translate into lost ability to detect and correct errors. We no longer have access to information that could have allowed us to improve our estimate of the message that was sent, and thus BER will rise relative to the unpunctured code. This makes sense; we expect a monotonically increasing relation between BER and coding rate. Recall that we also expect a monotonically increasing relation between data throughput and coding rate.

Puncturers and depuncturers seem to be very easy to implement in hardware. All we do is ignore an occasional bit according to a certain pattern. The main problem arises from clock management. The punctured data rate is a very unhelpful fraction of both the rate incoming from the MAC and the rate produced by the encoder. Figure 6.9 suggested something very misleading: that we are going to deal with the

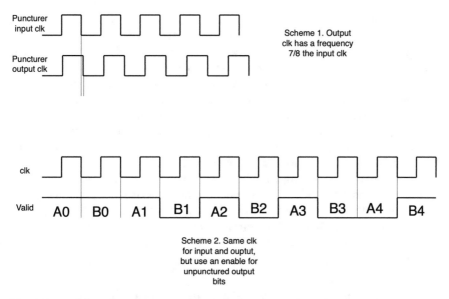

Fig. 6.10 Two strategies to deal with puncturing data rate

code in blocks. This code is convolutional; everyone expects it to flow in a stream. It is helpful to think of the code in blocks for the puncturer and depuncturer, but the output is a continuous stream.

In Sect. 10.12, we will discuss why having multiple (or at least too many) clocks on the same system is difficult. This is significantly more complicated when the clocks are not integer multiples of each other, as is the case for the input and the output of the puncturer and depuncturer.

There are two ways to deal with this problem; they are both shown in Fig. 6.10. The first method is shown in the bottom half of the figure. We use a single clock for puncturer input and output. This greatly simplifies clock design, and there will be a single clock for the entire bit domain, specifically double the MAC throughput.

The problem with this method is that not every period of the output of the puncturer carries unique information. Periods during which the bits are stolen are invalid and should be disabled. This requires the use of a qualifying signal with the output of the puncturer. This signal is shown as "valid" in the figure. It asserts in cycles in which the output is valid. The problem with this is that the remainder of the receiver must heed this enable signal. This significantly complicates control downstream.

The second method (top of the figure) is to use a different clock for the output of the puncturer. This allows us to guarantee that the output is valid for every cycle of the output of the puncturer. The problem with this method is that it requires the puncturer to have inputs and outputs in different and unrelated clocks. This opens the block up for metastability, which requires the use of an asynchronous FIFO (Sect. 10.12).

So, which of the two methods is "best"? Through most of this book, we cannot give decisive answers to questions with the word "best" in them. But this case is an exception; the first method is better. The second method could have been tenable if all we had to do is to transform between the native coded rate and some other rate. This is not the case here; the standard allows multiple punctured rates; thus, it requires us to support multiple possible punctured rates. This means multiple completely unrelated clocks. Clocks are not available for free on a chip. They require specialized buffering and distribution networks. Switching clocks on the fly is much easier said than done. There is no practical chip which can supply and switch between clocks at 2/3, ¾, 5/6, and 7/8 an original clock.

The complexity of dealing with the "valid" signal in Fig. 6.10 is not even that taxing. This signal will only be significant up to the baseband modulator in Fig. 6.7. In the modulator, the bits are transformed into symbols, and the rate changes into either 20 MHz or 40 MHz. Between the puncturer and the modulator, there are only simple blocks which can accommodate the "valid" signal readily.

6.5   Trellis Derivation, Visualizing the Encoder Message

Understanding channel decoding requires expanding possible message streams into a trellis. The size of the trellis is a function of the depth of the encoder. The 802.11n encoder in Fig. 6.6 is six flip-flops deep. Its trellis will be too wide and will need dozens of cycles to reach a steady state.

So, instead we will use a simpler convolutional encoder with fewer states. In Fig. 6.11, the encoder is only 3 DFF deep. This allows us to visualize the encoder and understand the decoder better. There is no loss of generality, and all the conclusions made through the rest of the chapter can be readily applied to deeper codes.

Each convolutional encoder is associated with a trellis. The trellis is a succinct graphical representation of the evolution of the state of the encoder as new inputs are accepted. The trellis is also the best way to understand how and why channel decoders can perform error detection and correction. The encoder in Fig. 6.11 is described by the trellis in Fig. 6.12. For the rest of this section, we will show how Fig. 6.12 is derived from Fig. 6.11.

At the heart of any convolutional encoder are several flip-flops forming a shift register. The contents of the shift register are the "state" of the encoder. As a new bit is shifted in, the contents of the shift register change and we enter a new state. Each branch in the trellis diagram represents a transition from one state to another, and thus a new input bit. There is an isolated bit on each branch representing the input that caused the transition.

The other important aspect of an encoder is the output bits in response to the current input. In the encoder in Fig. 6.11, there are two output bits for each input bit. This means that the native coding rate is 1/2, as is the case with the 802.11n encoder in Fig. 6.6. The two output bits corresponding to each transition are shown on the branches of the trellis within braces.

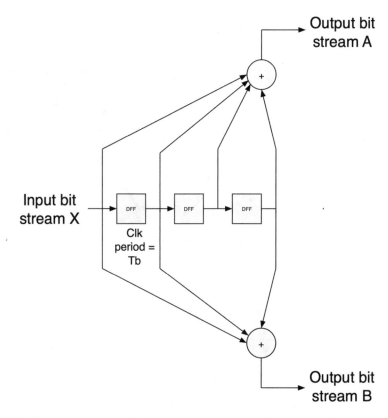

Fig. 6.11 Simple convolutional encoder. This is the case study for the rest of the chapter

The output bits are dependent on the connections in the encoder. This can be seen from the encoder schematic in Fig. 6.11. *Before* the input bit has been shifted in, the shift register contents are $X_0X_1X_2$. The two output bits are:

$$\text{Output bit1} = \text{In} \oplus X_0 \oplus X_1 \oplus X_2$$

$$\text{Output bit2} = \text{In} \oplus X_0 \oplus X_2$$

For example, when we are in state 000 and there is an input bit 0, this causes us to remain in state 000. The output bits in this case will be 00. If the input bit is 1, the next state will be 100; the output bits will be 11. If we are in state 010 and the input bit is 0, the next state will be 001. The associated output bits are 11. But if the input bit is 1, the next state is 101 and the output bits are 00. Similarly, we can derive all the state transitions, output bits, and corresponding input bits. These transitions and associated outputs are listed in Table 6.1. Notice that because we are using the input bit as one of the linear inputs to calculate the output bits, the state used in calculating the output bits is the state *before* transition. This can be seen from the encoder in Fig. 6.11.

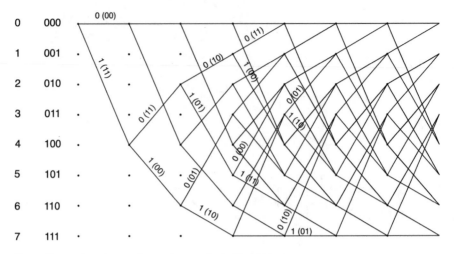

Fig. 6.12 Trellis corresponding to the encoder in Fig. 6.11

Table 6.1 Table summarizing state transitions and outputs for the encoder in Fig. 6.11

Current state	Input bit	Next state	Output bits	Input bit	Next state	Output bits
000	0	000	00	1	100	11
001	0	000	11	1	100	00
010	0	001	10	1	101	01
011	0	001	01	1	101	10
100	0	010	11	1	110	00
101	0	010	00	1	110	11
110	0	011	01	1	111	10
111	0	011	10	1	111	01

We exit every state when there is a new input bit. The input bit can either be 0 or 1. This means that there will be two branches coming out of each state, one corresponding to each of the possible values of the input bit. If we had more input bits, we would have more branches coming out of each state. The number of bits in the braces on branches is the number of simultaneous output streams, which in Fig. 6.11 is two. In short, the number of output branches from each state is $2k$, while the number of bits in the braces is n.

The trellis in Fig. 6.12 will always have the same branches coming out of the same states. The information on each branch will also always be the same. We assume we start from state 000. This is just an assumption, and we could start from any state if both the transmitter and the receiver are aware of it. With each input bit from the MAC layer, we produce two output bits and move a branch on the trellis. In other words, the information on each branch is only dependent on the starting state and ending state.

Table 6.2 Coding the message 1011101

Input bit	Original state	Next state	Output (coded) bits
1	000	100	11
0	100	010	11
1	010	101	01
1	101	110	11
1	110	111	10
0	111	011	10
1	011	101	10

As a result, in Fig. 6.12 after a transient phase, the trellis fills up and starts to repeat. This means that all the information on the trellis could really be summarized in a single transition. This steady state is reached once each state has been visited at least once. Tables (e.g., Table 6.1) are useful to summarize this steady state, although the trellis is better at visualizing transitions both in the initialization phase and in steady state.

Consider the case where we are encoding an incoming message 1011101. Each bit from the input message will lead to a transition in the trellis diagram and corresponding output bits from the encoder. We can quickly figure out the encoded bits from Table 6.1. Of course, we must assume a starting state, which in this case is 000. The results are summarized in Table 6.2.

But it is instructive to see what this looks like on the trellis. This is shown in Fig. 6.13. The first input bit is 1; thus, we move to state 100 and the output bits are 11. The next input bit is 0, we move to state 010, and the output bits are 11. The next input bit is 1 causing us to move to state 101 with output bits 01. We can keep following this and drawing the corresponding path on the trellis to reach Fig. 6.13. The transitions are summarized in Table 6.2. The encoded message will be 11 11 01 11 10 10 10.

6.6 Maximum Likelihood Decoding, the Best Anyone Can Do

Where does the error detection and correction capability of convolutional codes come from? In the trellis diagram in Fig. 6.12, the arrows coming out of each state carry output strings of two bits each. The two output bits have four possibilities: 00, 01, 10, and 11. But only two of these strings occur on the output branches coming out of the current state. We also know which state we start from and can trace the states we pass through from then on. This allows us to trace the expected coded bits. We already did this in Sect. 6.5, where we traced the path that a certain message takes through the trellis.

If the received message contains bit pairs that do not correspond to either valid transition out of a state we know we should originate from, then we know that an error has occurred. By comparing the received signal with valid transitions, we can find the transitions that were most likely to have happened at the transmitter and thus correct the error. This is the main idea behind convolutional codes.

The redundancy in the coded bits means that certain sequences of coded bits are impossible. If we see such sequences, we know an error has occurred. We can then examine the culprit state transitions and figure out which was most likely based on the received message. This is a complicated process that often requires us to examine all possible paths through the trellis before deciding. It is also best exposed through an example.

The input message 1011101 above corresponds to a coded message 11 11 01 11 10 10 10 as show in Sect. 6.5. If at the receiver we get this exact sequence, then we know that no error has occurred or at least no error that we can detect. We know this because we can find and trace a valid path through the trellis corresponding to this coded message as shown in Fig. 6.13. Because we can trace the path, we can also retrieve the original (uncoded) message as the bits outside the braces on the branches.

However, if an error did occur and it happened to corrupt one of the bits, then we will be able to tell that something went wrong. Assume, for example, that we received the message 11 11 *00* 11 10 10 10. There is a bit error in this message (marked in bold italic). The receiver, which is blind to what the transmitter produces, can easily tell that this bit is an error.

We first decode the bit pair 11. We start from state 000 and see two 1's as coded bits; this means we must have made the transition to state 100 and the input bit must have been 1. Because we found a valid transition out of state 000 that produces coded bits 11, we can make this decision with a high degree of certainty. Next, we look at the second pair of bits, which is also 11. There is a valid transition out of state 100 to state 010 with outputs 11, and thus again we can decide with certainty.

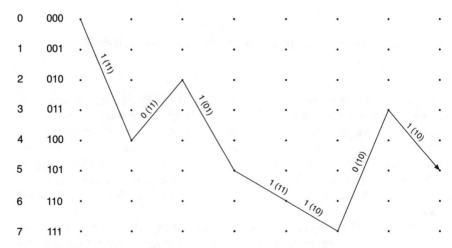

Fig. 6.13 Trellis diagram showing encoding of incoming message 1011101

When we look at the next bit pair, 00, we can tell an error has occurred. There are only two valid transitions out of state 010 as shown in Fig. 6.12. The transition to state 001 produces an output of 01. The transition to state 101 produces an output of 10. Neither produces an output of 00. Thus, an error has occurred, and we have detected it.

This is a fine assessment of the error detection capacity of the code. But it leaves some questions unanswered:

- We can see the error detection capacity clearly above, but error correction is not as intuitive. The erroneous "00" could equally likely have been originally "01" or "10" since both are valid outputs from state 010.
- Is it possible to deal with multiple bit errors, especially in sequence?
- What if a single bit error causes a new valid output to appear? In other words, what if a state has "01" and "00" as valid outputs and an error causes "01" to flip to "00"? In such case you will not even be able to tell that there was an error.

The answer to all three questions is essentially the same:

- Looking at an isolated state transition *might* expose a bit error, but it is highly unlikely to suggest how to correct it. To do that we have to consider the entire message. If the trajectory of the entire message is plotted, we can find the "closest" valid trajectory, and that would be the most likely message to have been sent.
- Most error correction codes and in specific convolutional codes are much less efficient at handling bursts of bit errors than they are at handling isolated errors. This is a fact of life, but again, if we consider the trajectory of the whole message, we might be able to make reasonable conclusions about the most likely sent message even with a couple of consecutive bit errors.
- If the erroneous bit flip causes a switch to a valid transition, the error will still be detected. All this does is push the impact of the error down the message. We might see a valid transition in the current state, but this pushes us to a wrong though valid state. At some point we will hit a state that refuses to give a valid output. Having said that, a properly designed code should maximize the "distance" between its valid outputs. For example, in Table 6.1, you will find that all states produce two outputs which differ in both bit positions. This prevents the situation where the flipping of a single coded bit still produces a legal transition.

What do we mean by the message being "close" to something? We must define distances in such a way that makes sense for received binary messages. To make hard decisions about sent bits, we use either the Hamming distance or the Hamming metric. Hamming distance measures the number of bit locations in which two messages disagree in their bits. Hamming metric is the number of bit positions in which they agree. The sum of the Hamming distance and metric for a message must be the length of the message.

For example, the two bit strings "1100" and "1000" have a Hamming distance of 1 because they disagree in the second bit position only. Their Hamming metric is 3 because they agree in all other bit positions. The sum of the metric and distance is 4, which is the length of either message.

We will now describe a way to guess the closest legal path to the corrupted message 11 11 0*0* 11 10 10 10. Once we figure out this path, its decoded bits allow us to correct the error:

11: There is a valid transition from state 000 to state 100 that could have produced this at the transmitter. We accept this as true and calculate a Hamming metric between the received bit pair and the bit pair that the transmitter had produced of 2. In other words, we see 11, there is a scenario where the transmitter could have produced a 11, and accepting this scenario, the metric is 2 because we agree on both bit positions.

11: We repeat the logic of the first bit pair. We are at state 100, and a transition to state 010 at the transmitter could have produced 11 as coded bits. The Hamming metric of this transition is 2. The entire message so far "11 11" has a total metric of 4, 2 for each bit pair because both have valid transitions in the trellis.

0*0*: There are two possible transitions out of state 010. Transition to state 001 produces an output of 10. Transition to state 101 produces an output of 01. Neither transition matches the received bit pair. Thus, an error has occurred. Both transitions have a metric of 1; neither is more likely. We must calculate the total metric that leads to both states and keep tracing deeper into the message. The received message so far is "11 11 00"; this is impossible. The message could have been either "11 11 01" or "11 11 10." Both possibilities carry a total metric of 5 (4 from sub-message 11 11 and 1 from the last transition).

11: There are four possible transitions that we need to trace here. Two coming out of state 001 and two coming out of state 101. We must calculate the metric on each branch, add them to the paths bringing them in, and keep tracing deeper into the message.

We keep tracking the total metric for all possible transitions from then on. You can see that at the point the error occurs or shortly thereafter, we will start to keep track of many possible paths. By the end of the message, there will be a total path through the trellis that wins because it racks up the highest metric. It will earn this metric by having the greatest number of legal transitions, although if an error has occurred, it will never have no illegal transitions.

In other words, with a rate ½ encoder, an error-free message with an uncoded length of p bits will have a Hamming metric of $2p$. If an error occurs, no paths that we can trace will have a metric of $2p$, but the path with the largest total metric is the most likely.

If we test *every* possible path from the beginning of the message to its end, then the winning path at the end is called the maximum likelihood path, and it represents our maximum error correction ability when using this code. This absolutely does not indicate that every error that occurs can be corrected because with longer messages and more complicated errors, we might end up with a false path winning. But it does represent the maximum theoretical performance of the code.

The complexity of the ML algorithm increases incredibly fast with the length of the message. For example, in the trellis in Fig. 6.12, we have two paths coming out of every state. If the message is deep enough, we will start visiting every transition after a while. So, we must keep track of each of these two paths and any paths that result from them going deeper into the trellis. At each transition, every state has two possible arriving paths meaning that for very long messages, the number of paths that need to be examined is 2^p, where p is the length of the message. The length of each path is itself p. The complexity will quickly become untenable.

But many of the paths do not need to be traced. There are paths that look like they are winning early on and some that we can tell are hopeless. Should we keep some paths and abandon others? Should we drop hopeless paths at every step, and how do we decide that they are hopeless? And if we do, does this not reduce our performance?

6.7 The Viterbi Algorithm, Taking a Calculated Hit

The Viterbi algorithm is the answer to all the questions posed in the last section. It is a systematic way to keep the number of comparisons needed under control. It reduces the number of paths preserved at every step to 2^m, where m is the number of DFF in the encoder, which means only one path survives per state. When we start discussing the algorithm, it might sound obvious. But the main idea behind Viterbi is not which paths are allowed to survive, but rather that there is a ceiling to the performance loss from the abandoned path possibilities.

By examining Fig. 6.12 in Sect. 6.5, the trellis has two stages: a transient stage at the beginning when the trellis is still being filled and a repetitive steady-state stage where all transitions keep repeating. The Viterbi algorithm starts to make sense deeper in the trellis where the possibilities are more numerous. Note that all the transitions that happen in the transient phase are subsets of the transitions in the steady-state stage.

Let us assume that the received message this time around is 11 11 *1*1 11 10 10 10, so that the fifth coded bit is erroneous. At first the Viterbi algorithm behaves exactly like the maximum likelihood algorithm. Figure 6.14 shows the first decision based on the first received pair 11. We start at state 0 and there are only two possible transitions out of it. One transition corresponds to an output of 00, the other to an output of 11. We calculate the Hamming metric for each transition. The transition to state 0 has a Hamming metric of 0 because the output disagrees with the received pair on both bits. The transition to state 4 has a Hamming metric of 2. Obviously, the transition to state 4 "wins." However, we have no way of knowing if this was because we transmitted a 11 or if we transmitted a 00 that was turned by noise into 11. Thus, we keep track of *both* transitions, marking the total metrics for both paths at such a point.

Figure 6.15 shows the second received pair being decoded. There are four possible endpoints corresponding to four possible paths we could have taken from the beginning. Again, on the figure we show the total Hamming metrics that accumulate along the paths. We do this by adding the Hamming metric obtained by comparing the received "11" with the outputs on all the branches to the total path metric we ended up with in Fig. 6.15. For example, the path from state 0 to state 0 to

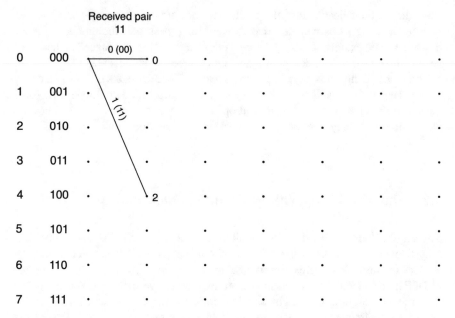

Fig. 6.14 First bit pair according to Viterbi algorithm. Numbers on the nodes are the path metrics

state 4 has a total path metric of 2. This is the sum of 0 from the path metric of state 0 in Fig. 6.14 and 2 from comparing the received pair "11" with the pair on the branch from state 0 to state 4, which is also "11." Notice that the path corresponding to the following state transitions, 0 to 4 to 2, has the maximum possible total metric of 4. This must be true because up to this point there are no bit errors; thus, there must be only one path "winning" and the said path must correspond to the original sent message and have a "perfect" total metric.

The third bit pair, also "11," contains a bit error. It should have been "01." The algorithm does not know when or where an error occurs a priori; thus, it proceeds as it did in the first two steps. We calculate metrics for all state transitions and then add them to the total path metrics we inherit from Fig. 6.15. We end up with the total path metrics shown in Fig. 6.16. By this point there are three coded bit pairs received. An error-free message would have been able to rack up a total metric of 6. None of the paths in Fig. 6.16 have this metric, so you already know that an error has happened.

The trellis in Fig. 6.16 looks very confusing. There is a winning path here with a metric of 4. However, this path is actually erroneous. The correct path now has a total metric of 3 and is tied with two other paths. This is normal. The bit error will throw the winning path out of order. However, we depend on the fact that there are many upcoming transitions to consider. The path that is winning at this step will not be able to keep accumulating more metrics because it will not correlate with the upcoming non-erroneous received bits. The correct path will quickly catch up because all its other transitions will accumulate high metrics.

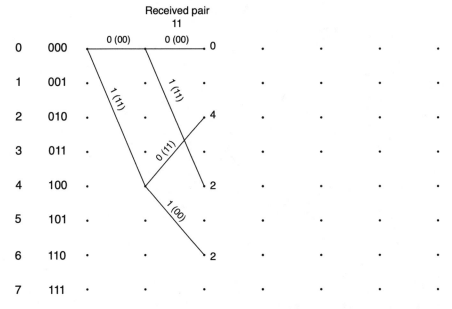

Fig. 6.15   Second bit pair according to Viterbi

The error correction ability of convolutional codes does not only come from the accumulated metrics in transitions after the error. It also comes from the inherited metrics of the paths before the error. This allows us to build a history of good transitions in the past before the error. We will also build forward toward good transitions in the future. Combined, the two will average out the effect of the bit error, allowing us to choose a correct winning path.

Figure 6.17 shows the fourth bit pair decoded using the Viterbi algorithm. Notice that so far, we have been decoding using maximum likelihood. At this step, we start to see Viterbi coming in. The figure shows that for the first time, there are multiple paths incoming to each state. In maximum likelihood, we would preserve all such alternatives and carry them forward with their metrics. In Viterbi, we preserve only one path per state. So, for every state we calculate the total metrics of all incoming paths, compare them, and preserve only the path with the maximum metric. Notice that the correct path starts to accumulate more metrics, although there are still multiple possible paths tying or even surpassing it. We need to go deeper into the message to correct the error.

Figure 6.18 shows the trellis after decoding the fourth pair and pruning all losing paths. Notice that this not only leads to the losing branches being deleted but to some paths earlier in the tree no longer being possible legitimate paths. Is it not possible though that some of the paths we have killed were the winning paths? The answer is yes. Even a path with a very low metric can later accumulate enough metrics to win. This is particularly true when the error happens early in a long message. What this means is that sometimes Viterbi will miss a solution that maximum likelihood

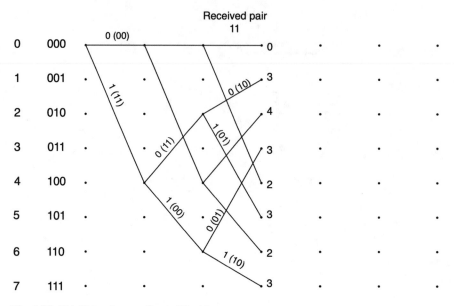

Fig. 6.16 Third bit pair according to Viterbi

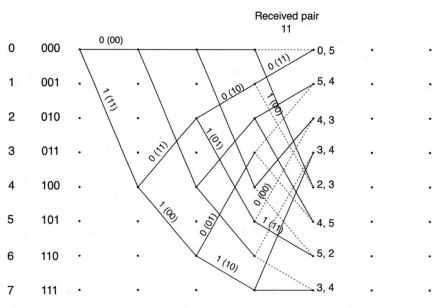

Fig. 6.17 Fourth bit pair according to Viterbi. Dotted transitions correspond to paths that will be killed

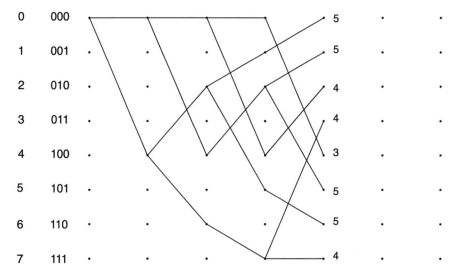

Fig. 6.18 Fourth bit pair with pruned paths

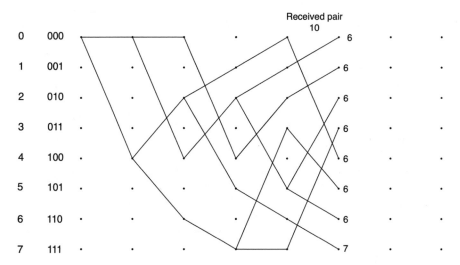

Fig. 6.19 Fifth bit pair being decoded

solution would have found. Thus, Viterbi always has higher BER than maximum likelihood. However, the main genius of Viterbi is that this hit to performance has a well-defined ceiling.

It is at this point that the trellis reaches a steady state and starts repeating. This is also where the algorithm becomes systematic. In Fig. 6.19, we see the fifth bit pair decoded. There are 8 surviving paths from the previous step and 16 transitions into the current step. We calculate the total metrics corresponding to these 16 transitions and again preserve only 8.

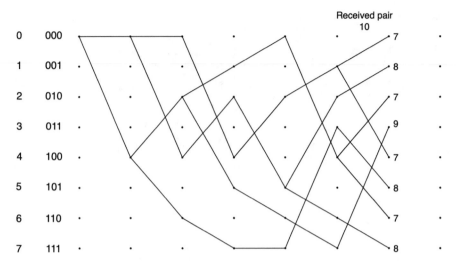

Fig. 6.20 Sixth bit pair

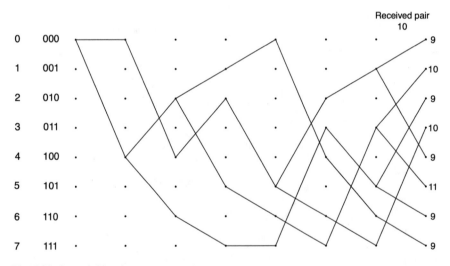

Fig. 6.21 Seventh bit pair

Figure 6.20 shows the sixth bit pair 10 being decoded according to Viterbi. Figure 6.21 shows the seventh bit pair being decoded. We move systematically from now on. Notice that it takes time for the correct path to emerge as a winner. Also notice that at many steps we might have paths with equal metrics entering a state. In such case, pick either randomly.

After the seventh pair has been decoded, the message ends. Only one path emerges as a winner with a total metric of 11. We now perform a task called traceback to finish decoding. Tracing back means moving back over the trellis and

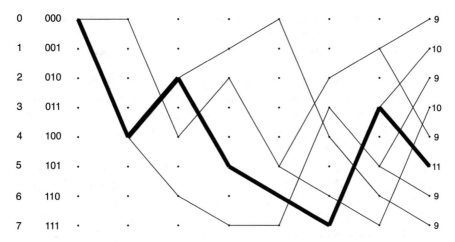

Fig. 6.22 The winning path

figuring out which transitions correspond to the winning path and obtaining their corresponding uncoded bits. This would be the bit outside the braces in the trellis in Fig. 6.12.

Figure 6.22 shows tracing back the winning path in bold. Tracing back might initially seem like an overwhelming task, but in Viterbi it is actually very simple. Going back through the pruned trellis, we find that there is always only one branch available. Notice that in Fig. 6.12, there might be multiple branches coming out of a state, but there is always only one going into a state; thus, tracing back leads to a unique path. Notice that the traced back path in Fig. 6.22 is the same as the message path in Fig. 6.13. Thus, we have retrieved the original message before the error, achieving error correction.

Viterbi is a systematic way to answer the questions posed at the end of the last section. With Viterbi, the number of paths that need to be traced back will always be constant. In fact, we only need to calculate a constant number of path metrics every step. Viterbi provides us with a systematic way to kill highly unlikely paths early in the message while incurring a well-defined hit to performance relative to the maximum likelihood solution.

6.8   Breaking Down a Complicated Solution, Partitioning the Viterbi Algorithm

The Viterbi algorithm is very complicated. When we want to move such a complicated construct to hardware, it is useful to partition it into large blocks. A useful "block" is one that performs an isolated and well-characterized function. It should communicate with other partitions in an easy-to-describe manner. This allows us to design these building blocks in isolation and combine and repeat them in a hierarchical modular design to build up the overall system. In Chap. 8, we will better define these "blocks" as processing units, but for now the mental exercise of breaking down the algorithm into repeatable steps is enough.

Examining the Viterbi algorithm, we find there are three things that happen. We will only consider the trellis once it has reached the steady state, so that the transient transitions have settled. This does not involve any loss of generality because transient transitions are a subset of steady-state transitions.

The three things we do are:

- Every transition out of a state to another state is called a branch. For each branch we calculate the difference (Hamming distance) or agreement (Hamming metric) between the output bits on the transition and the input bits received by the decoder. For example, if we are currently considering a transition out of state 0 and the received bit pair is "11," we need to calculate two metrics. One metric will correspond to the branch from state 0 to state 0, in which case the metric is 0. The other corresponds to the branch from state 0 to state 4, for which the metric will be 2. This function is called the branch metric unit, or BMU.
- At every state we might find multiple branches coming in. For each branch, we find the path that brought it in. We then add the branch metric calculated above to this path metric. The path with the highest total metric or lowest total distance wins and it's the only path preserved. This function is defined as the path metric unit (PMU).
- When we reach the end of the message, we decide a single path that "wins." We then need to trace back this path, deducing the *input bits* corresponding to the transitions that led to this path. This is the decoded message that we need from the decoder. This function is done by a unit known as the traceback unit.

6.9   The Branch Metric Unit (BMU)

The BMU calculates the Hamming metric or distance from a message fragment to a specific state-state transition on the trellis. Thus, if the received message fragment is 01 and the branch output bits are 11, the metric is 1. If the branch is 01, the metric is 2. If the branch carries 10, the branch metric is 0. By examining the trellis, we conclude there are $2 * 2^m$ BMUs. The 2^m is related to the 2^m states because branches come out of each state. There are two branches that come out of each state, because we are dealing with binary numbers and a single input bit.

Figure 6.23 shows the internal construction of a single BMU. It is a simple block. There are as many XNOR gates as there are bits in the output of the encoder, namely, n. In 802.11n, there are 2 output bits; thus, there will be two XNOR gates. The XNOR gates will encode bit positions where the received message and the branch agree. This is not the output that we expect from the BMU; instead we expect a single number that indicates the number of positions where they agree.

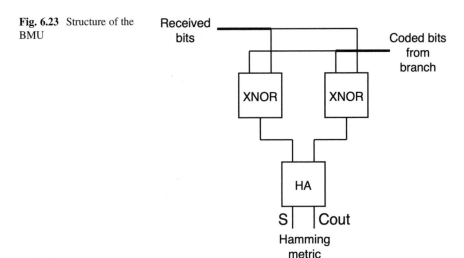

Fig. 6.23 Structure of the BMU

Hamming metric

Thus, the outputs of the XNOR gates are added up to provide the output of the BMU. In encoders with only two output bits, the bits can be added up using a HA. In encoders with three output bits, we must use full adders. In larger encoders we might have to use multiple FA and HA to form an N-bit adder. The adder simply translates the marked bit positions into a number. Thus, if the XNOR gates provide an output of 11, this means the branch agrees with the message in two positions. We cannot output 11 as is, because this would be a three. Instead, we add 1 + 1 in a half adder to produce 10, indicating 2 bit positions of agreement. BMUs that produce Hamming distances instead of Hamming metrics are identical, except that each XNOR is replaced by an XOR.

When we combine the building blocks into a whole system in Sect. 6.12, we will consider an alternative approach toward building BMUs. In general, the computational load of the Viterbi algorithm lies in the PMU rather than the BMU.

6.10   The Path Metric Unit (PMU)

The PMU is significantly more complicated than the BMU. In fact, it forms the core and heart of the decoder. The PMU does one thing: it decides which of the incoming *paths* is a winning path at the current step. There are 2^m PMUs, one for each state. The reason we need only this number of PMUs is thanks to Viterbi, who tells us to preserve only one path for each state at every step.

Figure 6.24 shows the internal construction of a PMU. Understanding how a single PMU is designed is easy once we understand the functions that it performs at every step. The PMU accepts several input path metrics, and an equal number of input branch metrics. The path metrics represent the history of the path, and the

Fig. 6.24 Internal
construction of the PMU,
the ACS

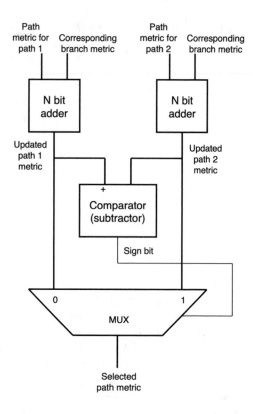

branch metrics represent the latest step along that path. In the example we considered in Sect. 6.7, there were always two paths arriving at each state, although this is not necessarily true for all trellises.

The PMU will then add each branch metric to its corresponding path metric; this calculates an updated path metric, representing the path after the latest input. The PMU will then compare all these updated path metrics. It will pick the largest path metric and will allow that path to survive, killing all the other paths.

Thus, the PMU is often called an Add-Compare-Select unit. It adds the branch metric to the corresponding path metric. It then compares the updated path metrics, and finally selects the surviving path. As shown in Fig. 6.24, the addition occurs in an N-bit adder. We should not ignore what N should be since this adder is probably where the critical path of the decoder lies, and adder delay is often directly proportional to input port length.

The comparison is done using a subtractor, especially when we only have two paths to pick from. The two numbers are subtracted, and the sign bit of the result determines whether the first operand is larger or smaller. Selection is through a multiplexer. The multiplexer picks the path according to the sign bit of the subtraction result. If we are subtracting path metric 2 from path metric 1, then a true sign bit in the result indicates path 2 is the winning path. Thus, this sign bit can be used as the select line of the MUX.

6.11   The Traceback Unit

The traceback unit seems very complicated at first, but a few observations will help us simplify it. At this juncture, it is important to point out that none of the implementations we discuss here are unique. You can feel free to implement any of the blocks in any way you see fit.

Traceback is a memory. This memory will keep track of each transition we pick along each step on the trellis. Once we reach the end of the message, we start tracing back to the beginning of the message to obtain the decoded bits. Recall that the aim of the Viterbi decoder is to obtain the uncoded bits that most closely correspond to the received message. In the trellis diagram, these would be the bits outside the braces on the branches.

The memory is shown in Table 6.3, with contents corresponding to the message in Fig. 6.13. The way the memory is used to trace back is shown in Fig. 6.25. The first column of Table 6.3 is the coded bits of the received message, included for reference. The last column is the address of the row. The concatenation of the rest of the columns is the content of each memory address. Each of these columns corresponds to a state in the trellis. Each row is both a memory address and a step in the trellis.

The content of every column in Table 6.3 (other than the first and last) is the state from which we arrived at the column state. Thus, in the first row corresponding to the first decoded bit, all states other than states 0 and 4 are unreachable. States 0 and 4 are reachable only from state 0. Address 5 corresponds to the sixth received coded bit pair, state 0 is reachable from state 1, state 1 is reachable from state 2, state 3 is reachable from state 7, and so on.

Thus, to reiterate:

- The first column of Table 6.3 is the received bit pair in the current step. In the real memory, these will not be stored. They are included in the table just for reference.
- The last column titled "Add" is the address. There are seven addresses (locations) in this memory; this is the length of the uncoded message.
- The remainder of the columns of the table are numbered by state. Their concatenation in any row would be the contents of the memory location corresponding to address "Add."

Table 6.3 Implementation of the traceback unit

Received bit pair	State 0	State 1	State 2	State 3	State 4	State 5	State 6	State 7	Add
11	0	–	–	–	0	–	–	–	0
11	0	–	4	–	0	–	4	–	1
11	0	2	4	6	0	2	4	6	2
11	1	2	4	7	0	2	5	7	3
10	1	2	5	7	0	3	5	6	4
10	1	2	4	7	1	3	4	6	5
10	1	3	5	7	1	3	5	6	6

Fig. 6.25 Tracing back circuitry

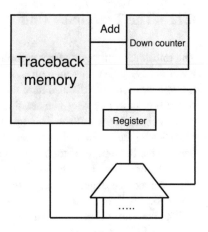

- Each cell includes the state from which we reached the state corresponding to the column. This information can be filled from the trellis diagram of the Viterbi algorithm. For example, in Fig. 6.20, we find that state 110 has a surviving path incoming from state 100. In Add 5 in Table 6.3, under the "State 6" column, we find the number 4.

In this example, there are eight states; each needs to store a state index. This means 3 bits each for 8 states, which makes this a 24-bit memory. We need to store a single state index for each state because for Viterbi, we only keep one surviving branch for every state. While decoding, we store the originating state for each winning path after performing ACS.

Now consider Fig. 6.25. Once we reach the end of the coded message, we know which state contains the winning path. From Fig. 6.22, we know that this state is state 5. Going to the last address of the memory in Table 6.3, we find that the part of the output word corresponding to state 5 contains the number three. We can step back by one address and check the contents of the part of the memory word corresponding to state three to find the number 7, and so on. If we keep doing this till the beginning of the memory, we get the following sequence of states:

$$5 \rightarrow 3 \rightarrow 7 \rightarrow 6 \rightarrow 5 \rightarrow 2 \rightarrow 4 \rightarrow 0$$

This is exactly the path chosen in Fig. 6.22.

In Fig. 6.25, the address bus is obtained from a down counter. It starts with the end position of memory and counts down by one every cycle. Alternatively, it could have been an up counter if we saved by down counting while performing the decoding. Every cycle, the entire memory word is read and split into the state words as in the table. Each state word is fed to a MUX input. In this case, we need a 3-bit 8×1 MUX corresponding to the 8 states of the trellis. The MUX select bus is the state we picked the last cycle. This causes us to read the contents of the state and store that in the register. The content is in turn the pointer to the next state, which is why it is used as a select bus.

Memory size can be a concern if we scale the problem to realistic message length. The number of locations in the memory is the size of the uncoded message in bits, which can be enormous. However, as we will discuss below, there are other limitations imposed on the traceback length due to latency which will cut down this dimension of memory significantly. The width of each output word from the memory on the other hand needs attention. The 802.11n encoder has 64 states; each state needs 6 bits to be encoded. This means the width of the memory data bus must be $64 \times 6 = 384$bits.

However, we only need to store 64 bits. The trace we did in Figs. 6.14, 6.15, 6.16, 6.17, 6.18, 6.19, 6.20, 6.21, and 6.22 is a little misleading in that it mainly exercises the transient part of the trellis. As shown in Fig. 6.12, once the trellis settles, all transitions become repetitive. Also because of the way Viterbi works, each state is only reachable from one of two states from the previous cycle. For example, consider state 6; this state is only reachable from state 4 or 5. Thus, we do not need to store a 3-bit word to encode which of the 8 states we could have come to state 6 from. We only need a single bit to indicate whether we got to state 6 in the current cycle from state 4 or 5.

Naturally, this would require a modification to Fig. 6.25. We need a decoder ahead of the MUX select port. This decoder accepts the current state (which is the contents of the MUX register) as well as the single bit we read to translate them into what the next state should be. For example, if the register contains 6 and the current bit is 0, we translate this into a select bus of 4.

The above discussion misses the point a little. Traceback is about extracting the uncoded bits corresponding to the received coded message. Determining the sequence of states that constitutes the winning path will help us determine these bits, but no part of the discussion so far addresses the uncoded bits. One simple way to do this is to attach a bit to every column (except the first and last) in Table 6.3. This additional bit is the uncoded bit corresponding to the transition that led to that cell. Thus, when tracing back as in Fig. 6.22, we split out this bit from the output of the register and read it as the decoder output.

A major disadvantage of convolutional encoders is their latency. Simply stated, the setup we have described so far will not allow a single decoded bit to be produced by the receiver until the entire message has been decoded. We need to trace the trellis from the very first bit to the very last bit before we can determine the "winning path." Once we do determine such path, we can trace back and only then are we able to start producing decoded bits.

In 802.11n, tracing the trellis through the whole message means tracing to the end of the packet. This is the ideal situation where we obtain the best possible performance. However, this stops the rest of the receiver from working until the entire packet has been decoded. Thus, a compromise is reached, where a number called the "traceback length" is agreed by the transmitter and the receiver. When the traceback length is reached, we truncate the message and traceback, allowing us to start producing output bits sooner. However, this means that less bits are available to average out bit errors. This translates into higher BER. In fact, the shorter the traceback length, the higher the BER.

To be specific, the best performance ever will always be provided by the maximum likelihood path. Viterbi will guarantee a certain maximum performance hit relative to the ML solution. However, this degradation is tightly knit to the length of the message. If you trace back over the whole packet, you will get the best that Viterbi can do. If you divide the packet into smaller messages each of traceback length, you suffer a hit to BER. The shorter the traceback, the worse the hit, but the smaller the memory used and the lower the latency.

Traceback memory is never the limiting factor in the traceback length; it is always the latency which holds us back. Memory is generally cheap and efficient, and the amount of memory used by other parts of the system will cause the traceback memory to pale in comparison. However, latency is a major concern. It could affect quality of service at the user end; any in some cases it may even cause handshaking signals to timeout.

6.12   Bringing It All Together, Building the Viterbi Decoder

We now know how to implement each of the three types of processing units in the Viterbi decoder, but we are still to bring them together to build a full circuit. This is shown in Fig. 6.26. There should be as many PMUs as there are states; in this case there should be 8 PMUs marked 0 through 7 corresponding to the 2^3 states of the encoder in Fig. 6.11. However, Fig. 6.26 shows only the first three states numbered 0 through 2. This is to avoid cluttering and to make the figure more readable. The remainder of the circuit should be very trivial to derive once we understand Fig. 6.26.

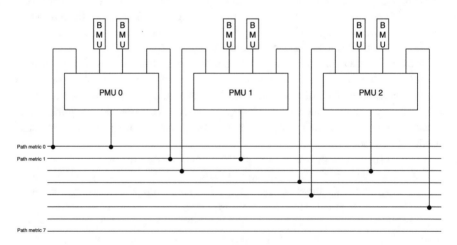

Fig. 6.26 Part of the decoder for states 0 through 2

Each PMU is internally constructed as in Fig. 6.24 and is thus essentially an ACS. The PMU accepts four inputs. According to Fig. 6.24, two inputs are path metrics, and the remaining two are branch metrics. In Fig. 6.26 the two outer inputs of each PMU are path metrics, and the two inner inputs are from BMUs.

Each PMU produces one output which is the path metric of the surviving path. This is fed to one of the horizontal rows in the bottom of the figure marked path metric 0 through path metric 7. Each of the connections on these horizontal lines from a PMU output is a register that stores the surviving path metric. Each PMU feeds its output to the horizontal line with the same number. Thus, PMU 0 stores its output in the row marked path metric 0, while PMU 7 (not shown on the figure) will store its output in the bottom-line marked path metric 7.

The path metric inputs to the PMUs are variable, both in where they come from and in the number of inputs. However, the connections are easily deducible from the trellis. Each connection in Fig. 6.26 is obtained from Fig. 6.12 as the states from which transitions to the current state are incoming in each step. For example, from the trellis we see that path zero will inherit the path metric of either path zero or path one; thus, these are the two lines (horizontal) from which the outer inputs of PMU 0 are obtained. The number of path metric inputs to the PMU will depend on the number of branches incoming to the state in the trellis in the steady state. The specific states from which these inputs are obtained are also directly read from the trellis.

Now let us discuss the BMUs as they are integrated in Fig. 6.26. Each BMU according to Fig. 6.23 accepts two inputs (each two bits). One input is the incoming coded bits. This will depend on the unpunctured rate of the encoder. In our example, our unpunctured rate is ½; thus, we will always accept bits in pairs as we did when discussing decoding in Sect. 6.7. The other input to the BMU is the "correct" output on the branch, also a bit pair in this case, as read from the trellis.

There will be as many BMUs and BMU inputs to each PMU in Fig. 6.26 as there are path metric inputs. Because we only have two incoming branches per state in the trellis in Fig. 6.12, we will always have two path metric inputs and two BMUs per each PMU in Fig. 6.26. The native coding rate of the encoder will determine the internal construction of each BMU but will not impact the number of BMUs.

All the BMUs in the circuit will share one input bit pair, namely, the current message coded bits. The Viterbi algorithm dictates that *all* branches calculate the metric for the current received message. The other input will be a constant for each BMU but will differ from BMU to BMU. For example, the BMUs feeding PMU 0 will have "correct" strings of "00" and "11" for path metrics 0 and 1, respectively. This can be seen from the trellis in Fig. 6.12 where the branch connecting state 0 to state 0 carries an output of "00," while the branch connecting state 1 to state 0 carries and output of "11." In Fig. 6.26 we need $2 \times 8 = 16$ BMUs. And in the general case, we need $2 * 2^m$ BMUs if each state has two incoming branches, or whatever the summation of branches happens to be when moving from one step in the trellis to the next in the steady state.

We can significantly reduce the number of BMUs. This is because all the BMUs share one of the inputs, namely, the received message. For example, for the trellis in Fig. 6.12, we need only four BMUs. There are only two coded bits on each branch;

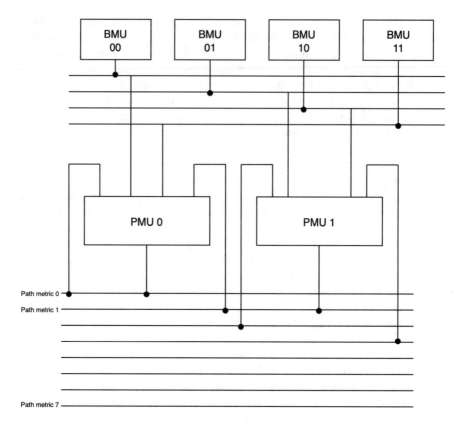

Fig. 6.27 Reducing the number of BMUs

this is a result of the rate ½ coding. Every BMU in Fig. 6.26 will have the same message coded bits every cycle. The other input, denoting the correct bit pair, can only have four possibilities: 00, 01, 10, and 11. Thus, there are only four branch metric values that need to be calculated every cycle, one each for the four possible correct bit pairs.

In Fig. 6.27, the four BMUs are arranged on top, and their registered values run on the top horizontal lines. PMU 0 will need the Hamming metrics for the current message with 00 and 11, and will thus accept the corresponding PMU inputs from these lines. PMU 1 measures the metric between the current message and 01 and 10, and thus accepts these BMU outputs into its PMU.

We clearly see that the hardware center of mass of the decoder is the PMU. The BMUs are simple enough to begin with, and the discussion above reduces our needs to 2^r BMUs. In rate ½ encoders, we only need 4 BMUs regardless of the number of states. So, for example, in 802.11n, we need 64 PMUs and only 4 BMUs. It should be very evident why we should not invest a lot of time trying to optimize the BMU.

The PMU is another matter altogether. We need many PMUs, and the deeper the encoder, the exponentially more PMUs we need. Also, the internal construction of the PMU is relatively complicated. In Fig. 6.24, we notice that the complexity of the PMU lies in the adder. The comparator is also an adder, but it is one that only keeps the sign bit of the output, which allows synthesis tools to optimize it. The selection operation is a MUX, which is extremely cheap in hardware. Thus, the adder in Fig. 6.24 is what we should be worried about.

Particularly, we should be worried about how wide this adder is. Branch metrics have an upper bound. If the coding rate is $1/r$, then the maximum value of the branch metric is r. This is because the incoming string and the branch string, each being r bits long, can agree or disagree on at most r bit locations. This means that the branch metric can be stored in a register that is $ceil(\log_2(r))$ long, and since r is small, the branch metrics are at most a few bits long.

The path metrics are a completely different story. Path metrics can keep growing indefinitely. Their maximum value is only capped by the length of the message or, more specifically, the traceback length. If the traceback length is L, then there can ostensibly be a path metric that is Lr, in which case we need an adder with inputs ceil $(\log_2(Lr))$ One such adder is fine, but recall we need 2^m such adders. When we discuss time-sharing in Chap. 8, we will find that this is not necessarily true, but for the time being, it is.

So, we must figure out a way to reduce the size of the adders if we cannot reduce their number. If you take another look at Fig. 6.24, you will find that the largest path metric keeps growing, but so do all surviving path metrics. At every step, we do not actually care about the value of the path metrics, we care which survives. If the path metrics at a certain step are 15, 13, 7, 11, 10, 8, 11, and 10, all we care about is the relative difference between them. In other words, our zero can be moved to the smallest number here which is 7. To be specific, it makes absolutely no difference to Viterbi if you had the above path metrics or if you had 8, 6, 0, 4, 3, 1, and 4. This allows us to only keep the differences for every step and possibly keep the adders small.

But this is where the trade-off between simplicity and elegance on the one hand and efficiency on the other starts to become challenging. Is the kind of reduction described above worth doing? Do we need to do this? The answer depends on how expensive these adders really are. Let us take a specific example with typical values. In a rate ½ encoder with a traceback length of 100, the adders need to be 8 bits long if we assume unsigned numbers. 8-bit adders are not cheap, but they are not disastrously expensive by any measure.

On the other hand, what kind of spread between path metrics can we observe? The largest path metric is obviously 200. Can a path keep a metric of 0 till the end of the message? This would require that every transition has no agreeing bits. While this is extremely unlikely, there is no fundamental reason that makes it impossible. Thus, even if a path does not have a null metric, it could have a very small metric. The spread between path metrics can thus vary greatly from message to message.

Thus, if we want to reduce adder sizes as we discussed above, we will need dynamic monitoring of metric spread. But what would be the use of any of this if we need to keep the large adders in case of a large spread?

The main takeaway is that hardware optimization is a matter of practicality. Sometimes an idea that might seem brilliant is too complicated and is not worth it once you consider the associated hidden costs.

The traceback unit in Sect. 6.11 is very simple to integrate with the PMU and BMUs. This is because most of the traceback functionality discussed in Sect. 6.11 happens *after* both the BMUs and PMUs have finished their job. This allows the traceback unit to function independently. But in Sect. 6.11, we did not discuss how the memory is filled in the first place. This is where the traceback unit needs to interface with the rest of the Viterbi decoder. During decoding, the memory in Fig. 6.25 is in write mode. The address port of the memory starts with null and is incremented every step in the trellis. The data input of the memory will be the concatenation of all the PMU outputs in Fig. 6.26.

6.13 Frequency Interleaving and Deinterleaving, Allowing Viterbi to Handle Faded Channels

The Viterbi algorithm will throw away a set number of paths every cycle. In a rate ½ decoder with an m bit deep encoder, Viterbi will discard 2^m paths every single cycle. In Sect. 6.7, we discussed how this could lead to Viterbi discarding the ML path by mistake. Viterbi thus takes a performance hit relative to ML, but this hit is quantified well and has a ceiling. This hit is particularly limited when there are isolated bit errors.

With isolated bit errors, the ML path will have accumulated a huge metric, and will continue to accumulate a large metric after the error. This means that Viterbi is very unlikely to discard it. If there is a burst of errors on the other hand, the received path will deviate from the correct path for multiple cycles. This could cause other paths to accumulate a larger metric. With ML, this does not mean an uncorrected error, especially with a long message. This is because the received path still has a chance with upcoming uncorrupted bits to allow the correct path to accumulate a winning metric.

With Viterbi, the burst error is much more likely to cause an uncorrected error than ML. In Viterbi, it is enough for the burst to cause the correct path to lose on only *one cycle* to a competing path arriving at the same state. This will cause the correct path to be permanently discarded. With ML, all paths are preserved to the end.

But maybe this is not a problem we should worry about; maybe errors naturally occur in isolation. Errors occur due to noise. White noise is completely auto-decorrelated. This means that high noise at a certain time does not indicate anything about noise at any other time, not even a femtosecond later. Thus, there is no reason to assume that high noise will sustain over multiple successive bit durations.

This sounds great because it means that bit errors will occur in isolation. This does not inform whether errors are plentiful or rare. Noise power is a property of the variance of its distribution. High-power noise will be strong enough to flip bits more often. However, it is always decorrelated, causing us to see the errors, regardless of how frequent they are, in isolation. And this is all we care about. But this is misleading because we are missing two facts:

- Coding and decoding in 802.11n takes place in the frequency domain.
- We often deal with frequency-selective channels with deep fades, especially in systems that use OFDM.

In Fig. 6.7 the channel encoder and decoder are both in the frequency domain (Chap. 7). In fact, the entire bit domain lies in the frequency domain. At the transmitter the baseband modulator is before IFFT. At the receiver the baseband demodulator is before the FFT. This means that when data is in the form of bits, it must be in frequency domain. Thus, data will always be transformed to frequency domain before it is used for channel coding or decoding.

So, according to Fig. 6.7 and Chap. 7, we should view bits as arranged in the frequency domain. In Fig. 6.28, we see the indices of output bits from the encoder at the transmitter arranged by which subcarrier they will be assigned to. Each QAM symbol is carried on a single subcarrier. In this case, we are assuming 16 QAM, so each subcarrier carries four bits.

If the channel is frequency-selective, then there can be deep fades in some of the subcarriers (Sect. 4.3). This will lead to signal amplitude plummeting for these subcarriers, causing the SNR to decrease and bit errors to become more likely. In fact, even if the fade hits only a single subcarrier, it will cause multiple successive bit errors because the subcarrier carries multiple bits. This suggests that burst bit errors are not only likely in Wi-Fi, but they are also fundamentally inevitable. We must do something about this, or the whole system will consistently fail.

This is where the interleaver and the deinterleaver come in. As shown in Fig. 6.7, the interleaver at the transmitter comes after the encoder and puncturer but ahead of the baseband modulator. This means the interleaver acts on bits rather than symbols. It is also ahead of the IFFT, which means it operates in frequency domain (Sect. 7.5). At the receiver, the deinterleaver is ahead of the depuncturer and the decoder, but after the demapper and the FFT, meaning it also acts in the bit and frequency domains.

Fig. 6.28 Arranging bits in the frequency domain

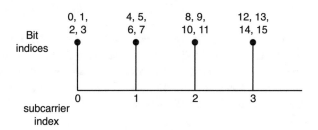

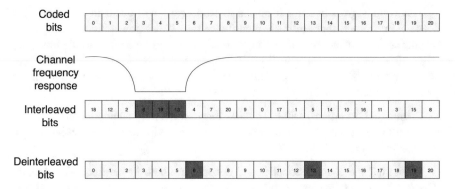

Fig. 6.29 Coded bits interleaved, face a fade, and deinterleaved. Erroneous bits are in red

The interleaver mixes up coded bits to cause burst errors to appear as isolated errors to the decoder. Figure 6.29 shows this through an example. The encoder produces bits with ordered indices. The interleaver takes these bits (shown in the figure arranged in frequency domain) and mixes them up. The order of mixing up is defined by the standard and is thus known to both transmitter and receiver. The particular order shown in Fig. 6.29 is for illustration only.

In the middle of Fig. 6.29, the message experiences a deep fade that destroys the symbol carrying the fourth, fifth, and sixth bits. These bits are shown in red in the figure. At the receiver, the deinterleaver returns the bits back to their original order. Recall that the ordering and jumbling is known to everyone through the standard. This leads to the serially arranged bits at the bottom of Fig. 6.29, which are handed to the Viterbi decoder. The erroneous bits are now isolated, even though over the channel, they were killed in a burst. This allows Viterbi to still do its magic.

802.11n uses a class of interleavers called block interleavers. Block interleavers are a very good fit for OFDM systems; however, they introduce an inevitable latency that further compounds latency issues introduced by traceback (Sect. 6.11) and FFT (Chap. 7). In a block interleaver, a block of bits is written in order and then read in the interleaved order. Most physical layer standards describe this in terms of index equations.

Define the following:

$$N_{\mathrm{BPSC}} = \text{Bits per subcarrier}$$

$$N_{\mathrm{CBPS}} = \text{Bits per OFDM symbol}$$

To understand how the block interleaver in 802.11n works, we introduce a few constants in Table 6.4. The three constants differ based on bandwidth. If we visualize the block of bits being interleaved as a rectangle, then the constants describe the number of rows and the number of columns in the block as well as an optional frequency rotation that we will describe below.

Table 6.4 Interleaver parameters in 20 and 40 MHz in 802.11n

	N_{column}	N_{row}	N_{rot}
20 MHz channel	16	$3N_{\text{BPSC}}$	11
40MHZ channel	18	$6N_{\text{BPSC}}$	29

0	1	2	3	4	5	6	7	8	9	10	11	12	13	14	15
16	17	18	19	20	21	22	23	24	25	26	27	28	29	30	31
32	33	34	35	36	37	38	39	40	41	42	43	44	45	46	47
48	49	50	51	52	53	54	55	56	57	58	59	60	61	62	63
64	65	66	67	68	69	70	71	72	73	74	75	76	77	78	79
80	81	82	83	84	85	86	87	88	89	90	91	92	93	94	95
96	97	98	99	100	101	102	103	104	105	106	107	108	109	110	111
112	113	114	115	116	117	118	119	120	121	122	123	124	125	126	127
128	129	130	131	132	133	134	135	136	137	138	139	140	141	142	143
144	145	146	147	148	149	150	151	152	153	154	155	156	157	158	159
160	161	162	163	164	165	166	167	168	169	170	171	172	173	174	175
176	177	178	179	180	181	182	183	184	185	186	187	188	189	190	191

Fig. 6.30 Bits arranged in a 20-MHz 16-QAM scheme. Original indices before interleaving

N_{BPSC} is the number of bits per subcarrier, which is also the number of bits per QAM symbol. In BPSK this is 1, QPSK 2, 16QAM 4, and so on. N_{CBPS} is the number of bits per OFDM symbol (Sect. 7.13). So, in a 64-subcarrier frame using 16 QAM, this would be 4×64. Notice that in Sect. 7.13, we will refine our understanding of N_{CPBS} based on data subcarriers versus pilot and null subcarriers (Fig. 6.30).

The 802.11n interleaving equation works in three steps. Each step takes indices of bits and transforms them into interleaved indices. The first step takes indices k covering the entire OFDM symbol and transforms them into indices i:

$$i = N_{\text{row}} * (k \bmod N_{\text{column}}) + \text{floor}\left(\frac{k}{N_{\text{column}}}\right)$$

$$k = 0, 1, \ldots, N_{\text{CBPS}} - 1$$

While we can just apply this equation in hardware to transform indices, this is not an efficient hardware implementation. This equation is trying to tell us something very simple, which we can see if we take a specific example. For 20-MHz 16-QAM and 64 subcarriers, $N_{\text{column}} = 16$, $N_{\text{row}} = 12$, and $N_{\text{BPSC}} = 4$. By consulting Sect. 7.13, we find that only 48 of the 64 subcarriers carry data; thus, $N_{\text{CBPS}} = 48 * 4 = 192$. Also notice that in this case $N_{\text{column}} * N_{\text{row}} = 192$. This means that a block $N_{\text{column}} * N_{\text{row}}$ can accommodate the entire OFDM symbol. Now running indices $k = 0, 1, 2, 3, \ldots, 191$ through the transformation above will give us indices:

$$i = 0, 12, 24, 36, \ldots, 191$$

This is equivalent to writing in the rectangular block with 12 rows and 16 columns but then reading column by column instead of row by row as shown in Fig. 6.31.

The equation for the second step is:

$$j = s * \text{floor}\left(\frac{i}{s}\right) + \left(i + N_{\text{CBPS}} - \text{floor}\left(N_{\text{column}} * \frac{i}{N_{\text{CBPS}}}\right)\right) \text{mod } s$$

$$i = 0, 1, \ldots, N_{\text{CBPS}} - 1$$

0	12	24	36	48	60	72	84	96	108	120	132	144	156	168	180
1	13	25	37	49	61	73	85	97	109	121	133	145	157	169	181
2	14	26	38	50	62	74	86	98	110	122	134	146	158	170	182
3	15	27	39	51	63	75	87	99	111	123	135	147	159	171	183
4	16	28	40	52	64	76	88	100	112	124	136	148	160	172	184
5	17	29	41	53	65	77	89	101	113	125	137	149	161	173	185
6	18	30	42	54	66	78	90	102	114	126	138	150	162	174	186
7	19	31	43	55	67	79	91	103	115	127	139	151	163	175	187
8	20	32	44	56	68	80	92	104	116	128	140	152	164	176	188
9	21	33	45	57	69	81	93	105	117	129	141	153	165	177	189
10	22	34	46	58	70	82	94	106	118	130	142	154	166	178	190
11	23	35	47	59	71	83	95	107	119	131	143	155	167	179	191

Fig. 6.31 First step of interleaving visualized

0	13	24	37	48	61	72	85	96	109	120	133	144	157	168	181
1	12	25	36	49	60	73	84	97	108	121	132	145	156	169	180
2	15	26	39	50	63	74	87	98	111	122	135	146	159	170	183
3	14	27	38	51	62	75	86	99	110	123	134	147	158	171	182
4	17	28	41	52	65	76	89	100	113	124	137	148	161	172	185
5	16	29	40	53	64	77	88	101	112	125	136	149	160	173	184
6	19	30	43	54	67	78	91	102	115	126	139	150	163	174	187
7	18	31	42	55	66	79	90	103	114	127	138	151	162	175	186
8	21	32	45	56	69	80	93	104	117	128	141	152	165	176	189
9	20	33	44	57	68	81	92	105	116	129	140	153	164	177	188
10	23	34	47	58	71	82	95	106	119	130	143	154	167	178	191
11	22	35	46	59	70	83	94	107	118	131	142	155	166	179	190

Fig. 6.32 Second step of interleaving visualized

where $s = \max\left(\text{floor}\left(\frac{N_{BPSC}}{2}\right), 1\right)$. s is simply $N_{BPSC}/2$ for any modulation scheme other than BPSK and is equal to 1 for BPSK. This second equation operates on the output indices from the first transformation equation to move indices to j from i. However, let us consider them operating on ordered indices; in other words consider i coming in as $0,1,2, \ldots.,191$ and let us inspect the output j in Fig. 6.32.

The equation has exchanged the indices across every other column, further introducing randomization to the order of interleaved bits. The third equation is shown below:

$$r = \left(j - \left((2 * i_{ss}) \bmod 3 + 3 * \text{floor}\left(\frac{i_{ss}}{3}\right)\right) * N_{rot} * N_{BPSC}\right) \bmod N_{CBPS}$$

$$j = 0, 1, \ldots., N_{CBPS} - 1$$

i_{ss} is the index of the spatial stream, so for SISO systems it is always 1. This third equation performs a rotation of indices in the frequency domain. It has no impact on the first stream and is thus only significant in true MIMO systems. In Fig. 6.33 we show the indices for stream 2 in a 2×2 system to examine what this equation does.

The equation has rotated indices both along the columns and the rows. These rotations will be different for each spatial stream, thus introducing spatial diversity. This helps the antennas perform beamforming.

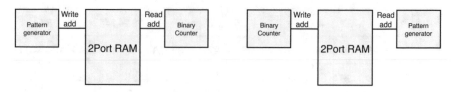

Fig. 6.33 Third step of interleaving visualized. Frequency rotation for the second spatial stream in a 2 × 2 system

The deinterleaver performs the reverse equations of the above to retrieve the original indices:

$$j = \left(r + \left((2 * i_{\text{ss}}) \bmod 3 + 3 * \text{floor}\left(\frac{i_{\text{ss}}}{3}\right) \right) * N_{\text{rot}} * N_{\text{BPSC}} \right) \bmod N_{\text{CBPS}}$$

$$r = 0, 1, \ldots, N_{\text{CBPS}} - 1$$

$$i = s * \text{floor}\left(\frac{j}{s}\right) + \left(j + \text{floor}\left(N_{\text{column}} * \frac{j}{N_{\text{CBPS}}} \right) \right) \bmod s$$

$$j = 0, 1, \ldots, N_{\text{CBPS}} - 1$$

$$k = N_{\text{column}} * i - (N_{\text{CBPS}} - 1) * \text{floor}\left(\frac{i}{N_{\text{row}}}\right)$$

$$i = 0, 1, \ldots, N_{\text{CBPS}} - 1$$

The interleaver and deinterleaver equations are extremely challenging to implement in hardware. They include many multiplications as well as challenging functions such as integer flooring. However, ultimately these equations are only trying to describe how the indices change in a mathematical notation that can be easily examined by designers. There is no intention to insinuate the implementation. In fact, this is a general property of wireless standards; they are not trying to describe specific implementation, just how the result should look like.

We do not need to implement these equations or even look at them to implement the interleaver in hardware. We just need to be able to write to a memory in order and read from it in another order. In fact, we do not need to implement interleaving in three steps. While the three equations clarify why the pattern ends up as it is, we only need to transform from the normal order of bits in Fig. 6.30 to the order in Fig. 6.33 without passing through Figs. 6.31 and 6.32.

In Fig. 6.34 the interleaver is implemented using a two-port RAM. One port is the write port, and another is a read port. In the right implementation, the write address is obtained from a counter. This means that bits are written in binary order along the lines of Fig. 6.31 but in a linear rather than a rectangular array. Each memory location in this case is a single bit. The read address is obtained from an address generator that is a counter containing the counting pattern in Fig. 6.34. The left-hand side is the opposite implementation where writing takes place using the pattern in Fig. 6.33 and thus reading is in order using a binary counter.

148	161	172	185	4	17	28	41	52	65	76	89	100	113	124	137
149	160	173	184	5	16	29	40	53	64	77	88	101	112	125	136
150	163	174	187	6	19	30	43	54	67	78	91	102	115	126	139
151	162	175	186	7	18	31	42	55	66	79	90	103	114	127	138
152	165	176	189	8	21	32	45	56	69	80	93	104	117	128	141
153	164	177	188	9	20	33	44	57	68	81	92	105	116	129	140
154	167	178	191	10	23	34	47	58	71	82	95	106	119	130	143
155	166	179	190	11	22	35	46	59	70	83	94	107	118	131	142
156	169	180	1	12	25	36	49	60	73	84	97	108	121	132	145
157	168	181	0	13	24	37	48	61	72	85	96	109	120	133	144
158	171	182	2	15	26	39	50	63	74	87	98	111	122	135	146
159	170	183	3	14	27	38	51	62	75	86	99	110	123	134	147

Fig. 6.34 Two implementations of interleaver/deinterleaver using block RAM

The block interleaver has one major disadvantage: latency. We must write the entire block of bits before we can read the first bit out. While you can improve the efficiency a little by reading earlier bits as they become available, there is a limit on performance imposed by the block nature.

On the plus side, we do not actually need to use a two-port memory for the interleaver. Since we are writing the bits only as they become available, and since we can only start reading out once all the bits are written, we can use a single port for reading and writing. We will thus multiplex the address bus between the two counters based on whether we are reading or writing. Bits are written at the rate that they are produced out of the puncturer, and after a latency equal to the puncturer clock period multiplied by block length, we can start reading out.

In fact, the address bus of the memory used for the interleaver needs to be highly programmable for more reasons than the one discussed above:

- If we are going to use the same memory for the interleaver and the deinterleaver, the address must be switched between interleaving and deinterleaving patterns.
- The pattern in MIMO systems is dependent on the number of spatial streams.
- The QAM constellation used defines the number of bits in the block and thus impacts both the amount of used memory and the pattern of the counter.

The interleaver and deinterleaver can also be used to accommodate the rate change issue we discussed in Sect. 6.4 for puncturers. The valid signal produced by the puncturer can be used to enable or disable the write enable for the interleaver memory. This allows us to use the same clock at the input and the output of the puncturer; it also allows the valid signal to become irrelevant to all blocks downstream from the interleaver.

The deinterleaver can also implicitly absorb the depuncturer. While reading out of the deinterleaver, we can have an enable signal that stops reading in stolen bit locations. This will cause the value of the last read deinterleaver bits to be held for stolen bits. Since we wanted dummy bits to be inserted at these locations, the depuncturing operation will essentially be done.

Chapter 7
OFDM

7.1 Multipath and Frequency Selectivity: The Conundrum of Equalization

In Sect. 4.4, we discussed how a multipath channel has a destructive impact on a signal. Figure 7.1 shows such a channel. The impact on the signal can be seen in both frequency and time domains. In the frequency domain, the frequency-dependent channel will impact different frequency components of the signal differently. In the time domain, the multiple impulses of the channel impulse response create multiple copies of the sent signal, distorting its shape and causing it to smear beyond its original symbol time. Frequency-selective and multipath channels are synonymous. They are just time-domain and frequency-domain manifestations of the same phenomenon.

In Chap. 5, we assumed that the channel model while finding the maximum likelihood solution is:

$$y = hx + n$$

If this is a time-domain equation, then it assumes that the channel is a single scalar value. This would be correct if the channel is single path in the time domain, or equivalently flat-faded in the frequency domain. Otherwise, this model does not work in either domain. In the frequency domain, we would have to include a dependency on frequency:

$$y(\omega) = H(\omega)x(\omega) + n$$

Note that we are not including a frequency dependency for noise since it is white. In time domain the multiplication does not hold and must be turned into a convolution:

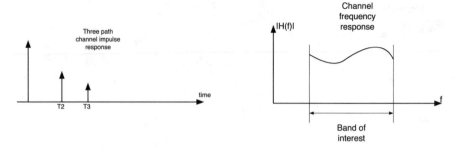

Fig. 7.1 Multipath channel in frequency and time domains

$$y = h \otimes \mathrm{x} + \mathrm{n}$$

Whether a channel is single or multipath and whether it is flat or frequency selective is not a property of the channel alone but also the signal under consideration. In other words, the same channel could be single or multipath depending on the signal passing through it. A channel can be considered single path if the signal passing through it has such a long symbol duration that the delay spread of the channel is negligible in comparison. Another signal with a short symbol duration will observe the same channel impulses as representing distinct paths and thus will see the channel as multipath.

The same can be observed from the frequency domain in Fig. 7.2. In fact, the information conveyed by the frequency domain is the same as that conveyed by the time domain. A signal with a wide enough bandwidth will observe variability in the channel frequency response, while another signal with a much narrower bandwidth will observe the channel to be essentially flat within its band.

The opposite is also true: the same signal may sometimes appear to be wideband or narrowband. In other words, it will sometimes suffer from frequency selectivity and multipath and sometimes it will not. This depends on the impulse response of the channel and if the channel has enough variation within the signal's band.

A frequency-selective channel will cause a susceptible signal to suffer from intersymbol interference. This is shown in Fig. 7.3 replicated from Chap. 4. The delay spread causes all symbols to bleed beyond their original symbol duration, thus interfering with coming symbols. This is an independent problem from the fact that the multipath channel mutilates the symbol itself. In other words, if a symbol is sent in isolation, it will be deformed but will not suffer from ISI. When symbols are transmitted in a train, we also observe ISI.

A frequency-selective channel will affect different frequency components of the signal by different channel coefficients. Thus, extracting the estimated of received signals from the observation signal requires dividing two frequency functions. In other words, $\widehat{x}(\omega) = y(\omega)/H(\omega)$. This is significantly more complicated than equalizing the channel for a narrowband channel, where it was just a scalar division.

Inverting a frequency-selective channel is done in time domain. We can model the channel as multiple impulses, creating multiple copies of the signal. If we can

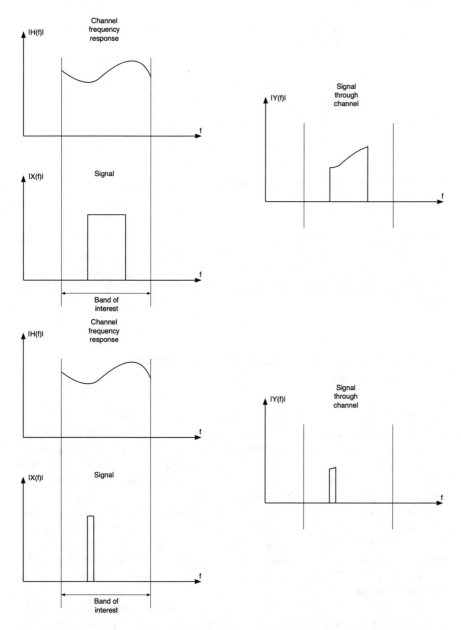

Fig. 7.2 The same channel, different signals. The top signal is wideband and observes the channel as selective. The bottom signal is narrowband and observes the channel as essentially flat

"guess" the times and amplitudes of these impulses, we can retrieve the signal and even remove its interference with the next symbol. This will result in an FIR filter that performs a frequency response that closely matches the inverse of the channel.

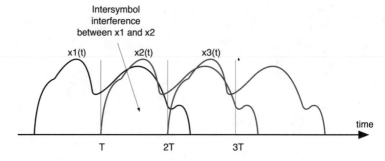

Fig. 7.3 Intersymbol interference. Each of the symbols should have a duration of T, but they extend due to delay spread

The problem with this approach is that the number of "taps," or copies of the signal, that we need to create grows rapidly as the bandwidth increases. In fact, the number of multipliers needed to invert the channel grows quadratically with the bandwidth, assuming channel selectivity remains the same. This is prohibitively complicated.

7.2 Narrowband Transmission and Maintaining Throughput

OFDM is the most ingenious answer to equalization in wideband channels. OFDM makes a very simple proposition. If performing wideband communication is prohibitively complex, don't do it! In other words, always limit yourself to a bandwidth where the channel is essentially flat and delay spread is negligible. If the channel becomes more frequency selective, just reduce your bandwidth. Thus, you can keep using trivial equalizers that are just scalar divisions.

This argument has a glaring problem: throughput. We wanted wideband transmission in the first place because it is the most direct way to increase throughput, and here we are decreasing bandwidth again. But what if we restore throughput by sending multiple such narrowband channels in parallel? If we send as many parallel channels as the ratio by which we decrease bandwidth, then we restore throughput to our desired levels.

Let us explore this with a specific argument. Assume that we find that the throughput we get from a bandwidth of B is enough for our application using our selected MCS. However, this bandwidth B exposes us to a channel that is frequency selective. In other words (Chap. 4), the symbol duration of $1/B$ is comparable to the delay spread of the channel. We find that if we decrease the bandwidth to B/N, the channel will be seen as essentially flat, and the delay spread will be much shorter than the new symbol duration of N/B.

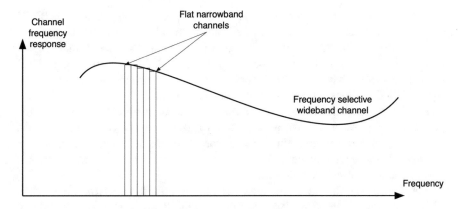

Fig. 7.4 Channel divided into multiple essentially flat sub-channels

This solves our equalization problem, and we can invert this much narrower channel through the simple multiplicative equation we saw in Chap. 5. However, this also leads to a decrease in throughput by a factor of N. But what if, instead of sending a single such narrowband signals, we send N of them simultaneously? In such case our total throughput goes back up to the throughput we get from a bandwidth of B because we have N streams, each with a throughput carried by a bandwidth of B/N.

But we are sending these N symbols simultaneously. This means that each of them will interfere with all the $N - 1$ other symbols for their entire duration of N/B seconds. In frequency domain we can ensure the subchannels do not interfere by doing frequency-division multiplexing. In time domain, we can get rid of the resulting cataclysmic form of ISI by ensuring that the carriers of the subchannels are all mutually orthogonal. As seen in Chap. 2, orthogonal carriers are decorrelated from each other. Thus, what we need to do is *orthogonal frequency-division multiplexing* (OFDM).

Now to make the example we discussed above more specific, consider the relatively wideband channel shown in Fig. 7.4. The bandwidth of the channel is 20 MHz. This is the bandwidth of the main mode of 802.11n; using this practical example is helpful because it carries on to many upcoming discussions. Assume also that the channel we deal with is frequency selective. How would we know that it is? In Chap. 2, we defined one parameter as the frequency-domain determinant of frequency selectivity: coherence bandwidth. Now assume that the coherence bandwidth of the channel is 1.8 MHz. Coherence bandwidth was defined as the bandwidth over which the channel is *essentially* flat. This does not mean that the channel is completely flat within the coherence bandwidth. In fact, there is some degree of variation allowed within the coherence bandwidth. However, in a fraction of coherence bandwidth, the channel can be considered flat. Let us assume, as a rule of thumb, that the channel is flat in one fifth a coherence bandwidth. Thus, the channel is flat in a bandwidth of $1.8/5 = 360$ kHz.

Now, let us assume that the 20 MHz channel allowed a coded throughput of 20 Msps. We will discuss throughput calculations in detail in Sect. 7.14. However, for now the 1–1 correspondence between bandwidth and throughput suggests BPSK as a baseband modulation scheme. If we instead use only a bandwidth of 360 kHz, we will be using a flat channel. This is equivalent to the time-domain signal observing no delay spread. This is because the bandwidth of the narrowband signal is 55 times smaller than that of the wideband signal and the period of the narrowband signal is 55 times longer than that of the wideband signal.

But the narrowband signals still cause ISI among themselves. To solve the problem of ISI, we will leave a guard interval (GI) between each two symbols. The GI is, for now, empty. Thus, the GI forms an overhead for communication because it is a time interval during which we are transmitting no information. But it is also essential because it removes ISI.

But if inserting GI solves ISI, then why do we need to subdivide the original 20 MHz channel in the first place? We have, multiple times, stressed that solving issues of wideband communication in frequency domain solves them in time domain. This is true, however, solving ISI only addresses the impact that consecutive symbols have on each other; it does not address the issues of frequency-dependent equalization within the same symbol that we discussed in Sect. 7.1.

But more importantly, using guard intervals without doing OFDM is not practical (Fig. 7.5). The figure contrasts the impact of GI on a single carrier and an OFDM system. Assume a 20 MHz channel and delay spread of 50 ns. The symbol duration for the single carrier is 1/20 μsec, which is 50 ns. Thus, to transmit a single symbol, it takes 100 ns, of which only 50 ns contain useful information and the rest are GI for ISI. The overhead is 50% and the realized throughput is halved. In other words, the 20 Mbps we expected will only be received as 10 Mbps because we are only transmitting half the time.

Now, if we use the narrowband signal, its symbol duration is 55 times longer than the original signal, specifically 2750 ns. The delay spread is still 50 ns. Thus, leaving an empty GI is a very practical proposition because the overhead on throughput is only 1.8%. This single sub-channel is transmitting 20/55 Mbps = 360 kbps at 98.2% efficiency, for a raw data rate of 353.52 kbps. But remember that we would have 55 such subchannels transmitting in parallel; their total bitrate will be 19.44 Mbps.

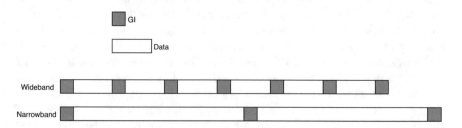

Fig. 7.5 GI as an overhead in long and short duration symbols. Not to scale

Fig. 7.6 Wideband
transmission in frequency
and time

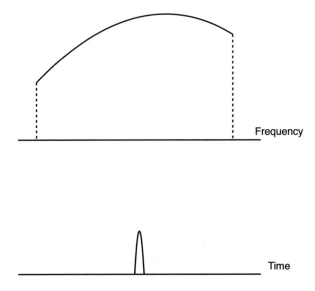

To sum the reasoning behind OFDM:

- Wideband signals are bad (Fig. 7.6) because they have a short signal duration that exposes them to multipath and a wide bandwidth that exposes them to channel selectivity.
- So, we send on a much narrower channel:

  - This solves frequency selectivity because the channel is essentially flat.
  - It solves multipath because symbol duration is huge relative to path delays.
  - It allows us to use GI to negate ISI without much loss of throughput.

- The narrowband channel leads to a proportionate drop in throughput; to restore it we send multiple such narrowband signals in parallel:

  - To stop these multiple simultaneous signals from interfering in the frequency domain, we perform frequency-division multiplexing.
  - To stop them from interfering with each other in time domain, we make sure all the subcarriers are mutually orthogonal.

This is shown in Fig. 7.7. All we have done is to divide the wideband channel into N narrowband channels. Each of the narrowband channels will be effectively flat if their bandwidth is much smaller than the coherence bandwidth. In time domain this is equivalent to the symbols for each of the narrowband channels being long enough that delay spread barely has an impact.

In the example we discussed above, the 20 MHz channel was divided into 55 narrowband channels. We call each of the narrowband channels a subchannel. For reasons that have to do with implementation (see Sect. 7.11), we want the number of subchannels to be a power of 2. If we take the next highest power of

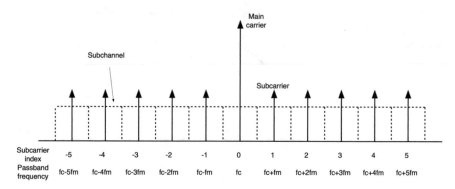

Fig. 7.7 Carriers and subcarriers

2, this will guarantee that modulation still deals with flat subchannels. Thus, in this case instead of using 55 subchannels, we will use 64 subchannels.

When talking about subchannels, it is also useful to talk about subcarriers. Subcarriers are conceptual carriers that offset each subchannel from the main carrier. Let us again consider the 20 MHz single carrier channel example. We divided this channel into 64 subchannels, so each subchannel will be 20/64 = 312.5 kHz wide. Note that this is smaller than the 360 kHz we obtained as one fifth of the coherence bandwidth, which means the subchannels are safely essentially flat.

Now if you imagine the entire wideband channel being moved to baseband as in Fig. 7.7, we start to see what we mean by "subcarriers." The middle subchannel is centered around DC. In passband, this subchannel will be at the main carrier frequency. Any other subchannel will be offset from DC in baseband and thus will be offset from the main carrier in passband. The amount of offset is the subcarrier frequency. In this case subcarrier frequencies are multiples of 312.5 kHZ. We can also number the subcarriers, which would also be numbering the subchannels. There are two alternative approaches to numbering. The first numbers them from 0 to 63. The second recognizes that in baseband some of the subcarriers are negative. This is the numbering scheme used by most Wi-Fi physical layer standards and is thus the method used in Fig. 7.7.

There is a problem we have been kicking down the road. To restore throughput, we were forced to use multiple subchannels in parallel. In parallel means that their time-domain symbols are sent at the same time. Thus, according to Fig. 7.8, we are sending *N* symbols each with a very long duration all at the same time. Whatever loss in throughput we had to contend with because we increased symbol duration is made up for by the fact that we send multiple such symbols at the same time.

But sending multiple signals at the same time causes each symbol to interfere with all other symbols. We seem to have created a monstrously extreme form of ISI that is much worse than what the wideband signal ever had. It is not that the symbols are bleeding into each other's symbol duration; it is that every one of the *N* symbols is interfering with *all* the others for *all* their durations.

Fig. 7.8 Multiple
orthogonal time-domain
signals sent simultaneously
are flat in frequency

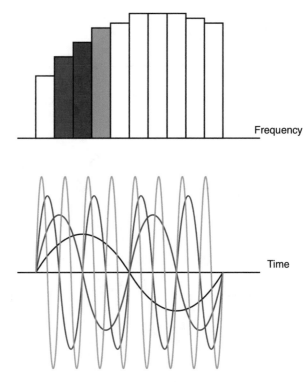

But symbols do not *have to* interfere with each other. Specifically, if the symbols are orthogonal, they will have zero interference. After all, recall that sine and cosine never interfered with each other and that we could always send data on the in-phase and quadrature rails simultaneously and retrieve each perfectly at the receiver side.

So, orthogonality is key here. If each subcarrier is orthogonal to *all other subcarriers*, then each of the signals can be retrieved at the receiver. Note that we can ignore the main carrier in this case because the orthogonality we care about occurs in the baseband. How can we guarantee this form of global mutual orthogonality? The perfect way to do this is if each subcarrier is at $\frac{m}{T_s}$ where T_s is the symbol duration and m is an integer multiple. What this does is that it guarantees two things: that each subchannel produces a symbol with duration T_s and that each subcarrier is orthogonal with all other subcarriers.

Orthogonality can be observed in both time and frequency domains. In Fig. 7.9, we notice that in the time domain, each subcarrier is a harmonic of the other subcarriers. This guarantees that there are whole periods of each subcarrier when trying to correlate with the other subcarriers. This is a result of our insistence that subcarrier frequencies be multiples of the reciprocal of symbol duration.

In frequency domain, if we assume that each time-domain information symbol is a rectangular pulse multiplied by the subcarrier sinusoid, then in frequency domain the pulse is a sinc that is moved to center frequency of the subcarrier. This is shown

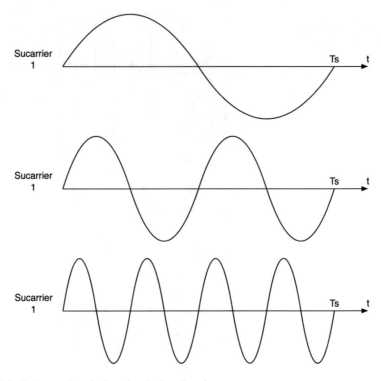

Fig. 7.9 Orthogonality of subcarriers in time domain

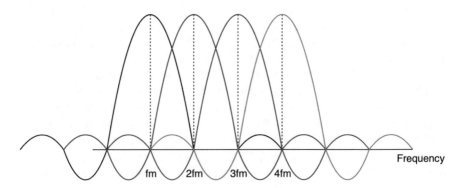

Fig. 7.10 Orthogonality in the frequency domain

in Fig. 7.10. The choice of subcarrier frequencies as multiples of each other leads to the situation shown in the figure. At the center frequency of each subcarrier, there are nulls from all other subcarriers. This is a manifestation of orthogonality, where the effect of each subcarrier on each other subcarrier is null.

OFDM sounds extremely good. We just need a way to implement it. Do we need N correlators at the receiver to decorrelate the subchannels from each other? Do we need N mixers to mix each of the subchannels with its subcarrier at the transmitter? This would be prohibitively costly, especially given the fact that N is typically 64 or more for practical systems.

Another interesting question about OFDM is why it is not used to multiplex the channel between different users. Multiple access is a challenging problem that requires multiple layers to deal with. The problem is that multiple users are all trying to access the same channel at the same time, and while the answer is to divide resources between them, having a specific algorithm to handle this is challenging.

OFDM seems to provide a great answer to this problem: Everyone can send at the same time if their subcarriers are orthogonal and there will be no interference. But in Sect. 7.3, we will find that OFDM is not without faults and one of its major issues is sensitivity to frequency offsets. Addressing frequency offset for a single-user system is feasible and will be done in Sect. 11.6. When multiple users are involved, offsets can be very hard to address.

But using OFDM in multiple access is a practical proposition. It is called OFDMA, or orthogonal frequency-division multiple access. It is a method of choice for upstream communication in asymmetric systems like cellular systems (Sect. 12.5) where a central authority can handle resource assignment, frequency offsets, and power amplifier bias. However, it is not practical in ad-hoc networks like Wi-Fi where there is a fair degree of symmetry between the upstream and downstream.

7.3 OFDM: The Good and the Ugly

OFDM has multiple advantages and a couple of disastrous disadvantages, and we must be aware of both. Dealing with the drawbacks of OFDM is challenging and requires significant changes to the receiver design. If we are not fully motivated by its advantages, we might start to wonder why we even bother handling its problems. One curious thing about OFDM is that its advantages are clearest in the time domain while its disadvantages are clearest in the frequency domain.

The advantages are:

- Insensitivity to timing offsets
- The ability to use GI to mitigate ISI
- Insensitivity to delay spread

The disadvantages are:

- Sensitivity to frequency offsets
- Terrible peak-to-average power ratio (PAPR)

Insensitivity to Delay Spread
An OFDM symbol is N times longer in duration than a wideband signal that gives the same throughput. This means that the signal can tolerate proportionately more delay spread before failing. For example, assume that the wideband signal can

tolerate a delay spread of σ_τ before it is distorted beyond repair. The OFDM signal can tolerate delay spread of $N\sigma_\tau$.

This is a very compelling advantage and is one of the main reasons we use OFDM. We have been talking about this since the beginning of the chapter, except in frequency domain. Coherence bandwidth is the frequency-domain correspondent of delay spread. If the wideband signal needs a coherence bandwidth of B_c to see a flat channel, the OFDM signal can deal with a coherence bandwidth of B_c/N. This might be a little confusing compared to Sects. 7.1 and 7.2, but we must understand that signal and channel properties are interchangeable. The decision on type of channel is informed by both the signal and the channel.

The Ability to Use GI to Mitigate ISI
The above considers the impact that delay spread has on the symbol itself, in other words the kind of equalization needed. But delay spread also leads to the signal bleeding into adjacent symbols, which leads to ISI. Because in OFDM the symbol is longer, we can insert a guard interval (GI) between symbols in time domain. The GI allows each symbol to bleed into the empty duration without impacting adjacent symbols. The GI should cover the delay spread.

We have so far been using the terms narrowband and wideband to distinguish the two cases of signals. At this point, we will introduce the terms single carrier and multi-carrier. A single-carrier system sends the information on the entire bandwidth at once. A multi-carrier system, like OFDM, divides the bandwidth into multiple subchannels. Both systems will achieve the same throughput because they are using the same modulation on the same *total* bandwidth. Single-carrier systems can also use GI to mitigate ISI. But as discussed in Sect. 7.2, a single-carrier symbol has a very short duration, and the GI will destroy its throughput.

The ability to use GI is related to resistance to delay spread but is independent from it. The fact that the symbol is not mutilated by multipath allows us to perform equalization by a scalar division. The fact that symbol duration is much longer than delay spread allows us to practically use GI to fight ISI. Both derive from the long symbol duration, but neither is dependent on the other.

Insensitivity to Timing Offsets
One of the main advantages of OFDM is its lack of sensitivity to timing offsets. Timing offset is a concept that we will talk about in detail in Chap. 11. However, the short version is, once the receiver understands that there is a packet incoming over the air, it does not really know where in the packet we are. The packet edge detection operation (Sect. 11.8) is very coarse. It recognizes a packet is being sent, but there is a very wide window in which this recognition happens.

As shown in Fig. 7.11, a packet consists of samples received serially in time. A timing offset means there is a discrepancy between what the receiver thinks is

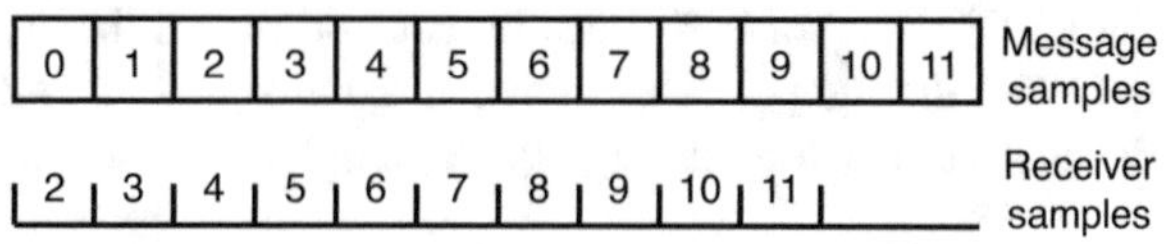

Fig. 7.11 Timing offsets

| 0 | 1 | 2 | 3 | 4 | 5 | 6 | 7 | 8 | 9 | 10 | 11 | Message samples |

| 2 | 3 | 4 | 5 | 6 | 7 | 8 | 9 | 10 | 11 | Receiver samples |

sample 0 and what is actually sample 0. In a single-carrier system, the symbol duration is very small, and even small timing offsets can be devastating.

In OFDM, timing offsets are much less important. We still need to correct them (Sect. 11.7), but the process is very tolerant. In OFDM, the symbol duration is very long. An unfixed timing offset means that the N samples we feed to IFFT/FFT will not be from a single OFDM symbol (Sect. 7.6) but will contain a few samples from an adjacent symbol.

The GI contains a cyclic prefix (CP) instead of nulls (Sect. 7.10). If we miss timing offset in OFDM by as many samples as there are in the GI, we will still perform demodulation correctly. The CP causes the system to still perform circular convolution correctly, and IFFT/FFT will not fail. So why even bother to do any form of timing offset estimation? Because any samples from the GI that are eaten up by timing offset are samples that are no longer available for guarding against ISI.

Extreme Sensitivity to Frequency Offsets
There are two main drawbacks to OFDM. Both can be devastating if not properly addressed. The first is extreme sensitivity to frequency offsets. This is the counterargument to insensitivity to timing offsets. We will discuss frequency offsets in detail in Sect. 11.6. The transmitter and the receiver use commercial oscillators to generate the carrier sinusoids they use to mix the signals. There is an inevitable offset between any two commercial oscillators even if they have the same nominal frequency. This leads to a "frequency offset" as the receiver tries to bring down the passband signal to baseband.

In Sect. 5.5, we see that frequency offsets cause the received signal constellation to rotate with time. In a single-carrier system, the symbol duration is so small that the constellation barely has the time to rotate significantly before the symbol ends (Fig. 5.24). Thus, frequency offsets in single-carrier systems appear as a phase offset from symbol to symbol. This can be rolled into channel estimates as a constant phase offset within the symbol. OFDM symbols are much longer in time; thus, the received constellation will revolve significantly before the symbol ends.

This can also be seen in the frequency domain. Consider Fig. 7.10. The subcarrier frequencies shown are in baseband and assume that the carrier frequency has been subtracted from all subcarriers. When we do not have frequency offset, this brings subcarrier 0 down to DC (zero frequency). With an unfixed frequency offset, the whole of Fig. 7.10 will be shifted slightly to the left or to the right. This will destroy an OFDM system.

In a single-carrier system, a slight frequency offset is not a big deal. This is because the signal is wideband, and thus the offset is small relative to the signal bandwidth. In an OFDM system, the same offset will be significantly more impactful. Using the same numbers we used in Sect. 7.2, assume there is a frequency offset of 300 kHz. In the single-carrier system, this offset is $0.3/20 = 1.5\%$ of the total baseband bandwidth. In the corresponding OFDM system, the offset is still 1.5% of the *total* bandwidth. However, it is $300/(20000/64) = 96\%$ of the bandwidth of each subchannel! You can clearly see why this is a huge deal.

But things are even worse. Even if the offset is much smaller than the subchannel bandwidth, an OFDM signal will still fail. Back to Fig. 7.10, an unfixed frequency offset means we are sampling each subcarrier in frequency domain slightly off from its peak. This is bad because off-peak the signal has less power. But it is terrifying because the peak point was also the *only point* where all the other subcarriers had nulls. This means that sampling off-peak means we are sampling a degraded version of the target subcarrier *and* sampling interference from the side lobes of *all* the other subcarriers.

This is a special form of interference called inter-carrier interference (ICI). ICI is a devastating issue in OFDM. As discussed above, we can be casual about correcting timing offsets in OFDM. But we must be super diligent about correcting frequency offsets.

ICI seems like something we should also care about in single-carrier systems. But single-carrier systems are wideband. This means that we can introduce guard bandwidths between wideband channels to protect them against inter-channel interference, and the overhead of the guard would be small relative to the large channel bandwidth. In OFDM guard bands are impractical because we are not actually talking about inter-channel interference. We are talking about inter-subcarrier interference. Each subchannel is very narrow, and to protect against ICI, we would need a guard band around *each* subcarrier. This will destroy the throughput of the system as most of its bandwidth would be consumed in guards.

This is a similar though complementary argument to the one we made while discussing the viability of GIs in OFDM versus single carrier. This is yet another manifestation of how frequency- and time-domain phenomena tend to be similar but opposite.

Terrible Peak-to-Average Power Ratio (PAPR)
The other major disadvantage of OFDM is the high peak-to-average power ratio. PAPR is measured in time domain, and it is the ratio between the maximum power of the signal and its average power:

$$\text{PAPR} = {}^{\max}\{x^2\}\big/_{E(x^2)}$$

PAPR is a measure of how much bigger the biggest peaks of the signal are relative to the "normal" level of the signal. Signals with high PAPR show large spurious peaks in time domain and a generally moderate amplitude most of the time. Signals with low PAPR do not have large occasional peaks but rather a generally moderate level of envelope variability.

OFDM signals have very high PAPR. This is because the time-domain signal consists of many narrowband signals sent simultaneously and added together. Each of the narrowband signals is carried on a different subcarrier, and all the subcarriers are orthogonal to each other. Each subcarrier carries a QAM symbol. The result is a lot of decorrelated signals added to each other. These signals sometimes add constructively, but often do not. This leads to occasional very large peaks and a moderate average, thus a high PAPR.

Another signal that has very high PAPR is AWGN. OFDM signals are in fact, very "noise-like." This means that if you are not aware in advance that you are looking at a signal, you might be forgiven for mistaking the OFDM signal for AWGN.

This can also be seen in frequency domain. In Fig. 7.10, we saw how the spectrum of an OFDM signal consists of multiple sincs next to each other. These sincs sum up to a nearly flat spectrum over a relatively wide bandwidth. This is exactly the kind of spectrum you would expect to see from white noise.

We have shown why OFDM signals have high PAPR, but why is this a bad thing? In Sect. 5.4 we discussed how a non-constant envelope signal poses a challenge to the design of power amplifiers. OFDM takes the non-constant envelope problem to a whole new level. In Sect. 5.4, we talked about how QAM constellations challenge PA design relative to PSK. When OFDM is used, PAPR is disastrous regardless of the underlying baseband modulation, which is why OFDM is usually paired with QAM modulation rather than PSK.

There are two ways we can adjust the PA to deal with the nature of OFDM PAPR:

- Adjust the operating point to accommodate the occasional maximum swing in the signal. This means that most of the time the DC signal of the PA is burning off power needlessly. Most of the time, signal swing is much lower than the operating point. PA efficiency plummets.
- Adjust PA operating point closer to the average signal rather than the peak. Efficiency is restored. However, we end up clipping the peaks of large signals. Clipping is an extreme form of nonlinearity. Although these clipped peaks are occasional, their loss will certainly lead to a degradation of BER.

So, which should we choose? The ultimate decision will be something in between and it will always be decided based on the tradeoffs that the designer is ready to make. How much BER are you ready to give up, and how much excess power will you burn in the PA? Perhaps the power that you can save by improving PA efficiency can be used to boost signal power, thus overcoming the hit to BER that you suffer by boosting SNR. As with any practical design decision, this requires experience, trial and error, and a familiarity with the QoS requirements of the system.

The above only considers challenges posed by PAPR at the transmitter. PAPR also causes serious problems at the receiver, particularly as far as the ADC is concerned. This is shown in Fig. 7.12. The input signal to the ADC first passes through a VGA (variable gain amplifier). The VGA will amplify this signal by a variable gain. The ADC will digitize a range of voltages produced by the VGA between 0 and V_{max} into 2^{bits} levels, where bits is the number of output bits from the ADC.

The question is, what is V_{max} or rather what should it be? V_{max} is obviously the maximum voltage output of the VGA. We can adjust the gain of the VGA so that V_{max} at its output corresponds to the peak of the OFDM signal at its input as shown in Fig. 7.12. In such case we will not be missing any of the data in the peak signal.

Fig. 7.12 ADC dynamic range when accounting for peak signals

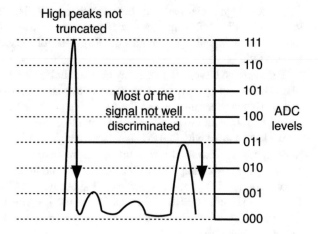

Fig. 7.13 ADC dynamic range when accounting for average signal

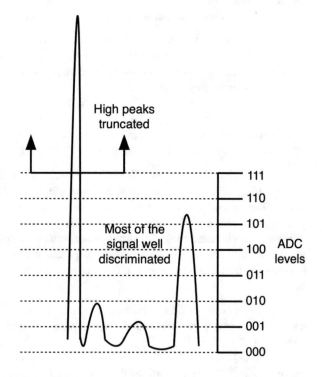

However, the signal is, most of the time, much lower than the peak because of the high PAPR. Thus, we are squeezing most of the signal variations into fewer quantization levels just to accommodate the occasional peak. This reduces resolution for most of the signal.

If we adjust V_{max} to be closer to the peaks of most of the signal (i.e., closer to average peaks) as in Fig. 7.13, we increase our resolution for most of the signal. But we do end up clipping the large peaks. As with the PA design decisions, VGA adjustment and ADC resolution are not easy and require simulations. Unlike the PA, however, efficiency is not a primary concern here. There are no power hungry essentially inefficient analog blocks like the PA; thus, the ultimate decision depends on a compromise that minimizes the distance between the actual BER curve and the theoretical BER curve.

7.4 Avoiding the Use of Multiple Correlators, OFDM in 802.11n

So far, we have dodged a fundamental question about OFDM: how to do it? By saying we will transmit on multiple subcarriers, we are insinuating that we need multiple RF chains to send and receive the data. Each RF chain will raise the subchannel to its subcarrier frequency. This is too many mixers, power amplifiers, low noise amplifiers, and filters.

An OFDM system uses a single RF chain at the transmitter and the receiver. To allow this, we use FFT/IFFT to transform between time and frequency domains. We arrange the subcarriers in frequency domain and then combine them into a single time domain which can be transmitted on a single main subcarrier, thus achieving OFDM.

In Sect. 7.5, we will discuss various time-domain to frequency-domain transform pairs in detail. We will conclude that the transform of choice is FFT/IFFT. But for now, accept that FFT transforms a time-domain signal into its frequency-domain counterpart, while IFFT returns a frequency-domain signal back to time-domain.

The transmitter in Fig. 7.14 contains an IFFT block. The output of this block is thus in time domain. This is the signal that will be sent over the air, because the process of sending happens one sample at a time. Everything that comes before the IFFT is thus in frequency domain. This includes the channel encoder, puncturer, interleaver, parser, and baseband modulator.

Unencoded bits descend from the MAC layer and pass through the encoding combo (encoder, puncturer, interleaver). These coded bits are then fed to the baseband modulator. Each k-bits will form a single QAM symbol. Each QAM symbol will occupy a subchannel of the OFDM signal. Each subchannel has a bandwidth of $\frac{B}{N_{sc}}$ MHz, where B is the channel bandwidth and N_{sc} is the number of subcarriers. This is also the rate at which this particular QAM symbol will be transmitted.

The OFDM "symbol" in frequency domain is thus a set of N_{sc} QAM symbols arranged in a vector. Each QAM symbol is a complex number. In Sect. 7.13, we will discuss how the actual 802.11n OFDM symbol looks like, and we will find that a lot of subcarriers are "special" in that they do not contain normal data. The frequency domain of the OFDM symbol is thus an arrangement of single QAM symbols on

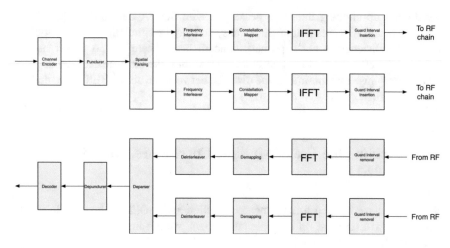

Fig. 7.14 Transmitter (top) and receiver with FFT/IFFT inserted

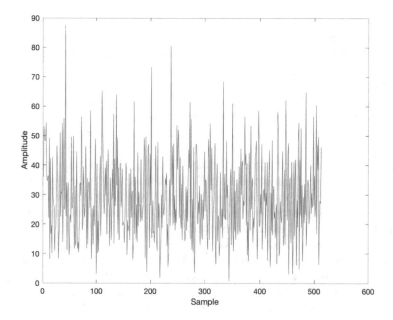

Fig. 7.15 A 512-point FFT QPSK signal in time domain. The x-axis is time-domain discrete samples (integer indices), and the y-axis is the amplitude of the signal

each subcarrier. Subcarriers are indexed by integers rather than by their specific subcarrier frequencies.

Figure 7.15 shows the time-domain amplitude of an OFDM signal. This signal comes from a 512-point OFDM system with QPSK modulation. The x-axis in Fig. 7.15 is discrete time sample indices; in other words it is the n in $x(nT)$. When the signal is sent over the air, these samples are sent out sequentially. The time difference between samples will be $\frac{1}{B}$.

We should also discuss what the integer indices in the frequency domain Fig. 7.7 mean for the band of the signal. When we apply IFFT in the transmitter, it transforms the N_{sc} samples in frequency domain to N_{sc} samples in time domain. The rate at which we produce outputs will determine the specific bandwidth of the subchannels, thus replacing the integer indices of the array of QAM samples with frequency offsets that can be measured in Hz.

Thus, if we produce outputs from the IFFT block at a rate of B MSps, then the overall signal has a bandwidth of B, but each subcarrier will have a corresponding sample in time domain every N_{sc} samples. Thus, the subcarrier has a bandwidth of only B/N_{sc}. Thus, each subcarrier occupies a bandwidth of B/N_{sc} and each occurs at integer multiples of said bandwidth. The time-domain signal produced by the IFFT is mixed with the main carrier (2.4 GHz or 5 GHz in 802.11n) to take the overall OFDM signal to passband. We end up using a single RF chain at both the transmitter and the receiver.

7.5   Time-Domain to Frequency-Domain Transforms

The key to OFDM is the ability to go from frequency domain to time domain and vice versa. This allows us to do OFDM without using multiple mixers. OFDM divides mixing into two steps: In baseband, the frequency-domain subcarriers are mixed into a single time-domain signal, and in passband this time-domain signal is mixed with the main carrier. So, how do we go from one domain to the other? We have already alluded to the fact that we will be using IFFT and FFT as our transform pairs, but it is time to understand why.

The Fourier series (FS) is an ingenious way to represent signals in the frequency domain, but it has one limitation: it only applies to periodic signals. The genius of the Fourier series is that it represents any periodic signal as the summation of sinusoids whose frequencies are harmonics of the frequency of the signal at hand. This applies regardless of how long or short the period of the signal is and how "weird" it looks. If it is periodic, there is a Fourier series for it. In fact, the FS has a general form where the basic functions are not even sinusoids, but it is only the sinusoid transform pair that is of interest to us.

The FS of a periodic signal is:

$$s(t) = \sum_{k=-N}^{N} C_k e^{-jk\omega_0 t}$$

$$\text{Where } C_k = \frac{1}{T} \int_0^T s(t) e^{-jk\omega_0 t} dt$$

How is this a frequency domain transformation? After all, we are still representing the signal $s(t)$, so we are still stuck in time domain. But if you consider the coefficients C_k, these are power levels that describe how this time-domain signal can be broken down among the sinusoids $e^{-jk\omega_0 t}$. This is a frequency-domain analysis. It is a spectral breakdown of the different power levels contained in the signal. So, the coefficients are, themselves, the frequency-domain representation of the signal.

The equation still has plenty of undefined variables; understanding these will help us form a more complete picture of the FS. C_k are the coefficients of the sinusoids. k is the index of the sinusoid taking integer values from $-N$ to N, where N is a maximum value of your choice. The higher the N, the more accurate the representation of the signal. ω_0 is the fundamental frequency of the periodic signal. In other words, ω_0 is the reciprocal of the signal period. Thus, the signal is represented as a summation of the harmonics of its fundamental frequency.

So, what does "harmonics" mean? If the signal we want to breakdown has a period T, then it has a fundamental linear frequency $f_0 = 1/T$ and a fundamental angular frequency $\omega_0 = 2\pi f_0$. We call this the fundamental frequency. The sinusoids that form the Fourier series will thus have frequencies starting at f_0 and repeating at integer multiples of f_0. Each of the components at the multiple frequency is called a harmonic.

Can we use the Fourier series for our transform pair in OFDM? It has one thing going for it that we will always need: discrete components in frequency. But otherwise, nothing about this transform works. Looking back at Sect. 7.4, we see that at the transmitter and the receiver we need a transform that can go back or forth between N_{sc} samples in time domain and N_{sc} samples in frequency domain. Thus, we need a transform that can handle discrete finite samples in both domains.

The Fourier series has discrete components in the frequency domain, but they are not finite. Sometimes, it is possible to stop at a certain harmonic and accept the remaining terms of the summation as an error; this is particularly possible because Fourier coefficients in FS are generally decaying. But the problems of the FS do not stop there:

- It can only handle signals that are periodic in time domain; our baseband modulated signal is most definitely not periodic.
- It can only handle continuous-time signals; we have discrete-time signals.

Which brings us to the Fourier transform (FT). The Fourier transform is what happens to the FS if you set the period T infinitely large. This essentially makes the signal aperiodic. It also reduces the fundamental frequency to an infinitesimally small amount, leading the frequency components of the FS to turn into a continuum. Thus, the FT is a transform pair which moves between a continuous representation of the signal in frequency domain and a continuous representation in time domain as shown below:

$$X(\omega) = \int_{-\infty}^{\infty} x(t)e^{-j\omega t}\,dt$$

$$x(t) = \frac{1}{2\pi} \int_{-\infty}^{\infty} X(\omega)e^{j\omega t}\,d\omega$$

This frees us from the restriction of periodicity of the FS. However, it sets us a step back by making both the time-domain and the frequency-domain signals continuous. While this might seem like a no-win situation, we are getting closer to what we need.

We should now try to "make" the FT work on discrete finite signals in both time and frequency domains. This sounds like a tall order, but we can start by making the time-domain signal discrete. We can do this by sampling the signal. According to Nyquist sampling theory, a signal that has a finite bandwidth can be sampled and reconstructed perfectly if we sample it at twice the rate of its highest frequency content. This means that the only signals that can be perfectly reconstructed are those that are infinite in duration and thus finite in frequency.

Figure 7.16 shows what happens when you sample a time-domain signal and why Nyquist rate applies. When the signal is sampled, its spectrum becomes periodic with a (linear frequency) period of $1/T$ where T is the sampling time. Therefore, we must sample at more than twice the highest frequency content, otherwise the copies of the spectrum at the bottom of Fig. 7.16 will overlap each other, a phenomenon known as aliasing.

To reconstruct the continuous time signal from the sampled signal, you filter out a single period of the spectrum. This filtering has the effect, in the time domain, of reconstructing the inter-sample values by interpolation. Said interpolation will be

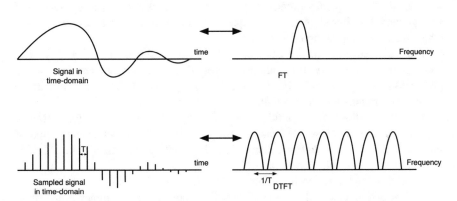

Fig. 7.16   If we sample a time-domain signal, its spectrum becomes periodic. Applying this to FT, we get the DTFT

perfect if the sampling rate respected Nyquist criteria and the reconstruction filter is ideal.

The situation in Fig. 7.16 describes a transform pair known as the discrete-time Fourier transform (DTFT). It is crucial to distinguish DTFT from DFT (discrete Fourier transform). In the DTFT as the name clearly indicates, it is the time-domain signal that is discrete. The frequency-domain signal is still continuous, in addition to being periodic. The DTFT equation is:

$$X(\Omega) = \sum_{n=-\infty}^{\infty} x[n]e^{-j\Omega n}$$

where Ω is the normalized frequency based on the sampling time T_S of the signal; thus, $\Omega = \omega T_S$. If you want to write the DTFT equation without normalization, then:

$$X(\omega) = \sum_{n=-\infty}^{\infty} x[nT_s]e^{-j\omega n T_s}$$

And the inverse equation is:

$$x[n] = \frac{1}{2\pi} \int_0^{2\pi} X(\Omega)e^{j\Omega n} d\Omega$$

The DTFT has discretized the time-domain signal, but we still have multiple issues. The frequency-domain signal is continuous and both signals are infinite in their extent. These issues preclude us from using these transforms to perform OFDM modulation and demodulation.

Notice that there is a correspondence and symmetry between the Fourier series and the discrete time Fourier transform. In FS, the time-domain signal is periodic, and the frequency-domain signal is discrete. In DTFT the time-domain signal is discrete, and the frequency-domain signal is periodic. This is yet again a reflection of the symmetry of time-domain and frequency-domain phenomena we saw through Chap. 4 and the beginning of this chapter.

Thus, discretization in one domain leads to periodicity in the other. It is also worth noting that all the information in a periodic signal is contained in a single period. For example, with DTFT, filtering out a single period allows us to obtain the original continuous time signal with all its information intact.

The above allows us to finally reach the discrete Fourier transform (DFT). Figure 7.17 shows one way we can picture the DFT. In this figure, one period of the DTFT is windowed and sampled. All the information contained in the DTFT is preserved in this single period, and if we sample at a high enough rate, we still preserve all the information in the bottom left signal in Fig. 7.17.

The time-domain signal remains discrete. If there were N_{sc} samples in the time-domain signal in the DTFT, it is enough to sample N_{sc} samples in the frequency-

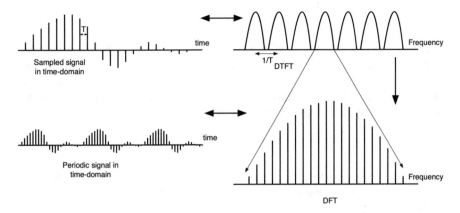

Fig. 7.17 DFT can be obtained from DTFT by sampling one period of the latter

domain period to respect Nyquist criteria. Notice that Nyquist is now operating in reverse since we are introducing periodicity in the time domain by sampling the frequency domain.

The DFT is obtained by sampling one period of the DTFT spectrum. This is kind of hackneyed and the DFT is an ad-hoc solution for transform pairs made specifically for computers. For example, in Fig. 7.17, the time-domain signal corresponding to the DFT should actually be a periodic signal. In a way, the DFT is like the FS we started this section with. However, we will "window" a single period of the time-domain signal in the bottom left of Fig. 7.17 and thus allow an N_{sc} sample to N_{sc} sample transformation.

The foundation of DFT is messy, but it works. It can uniquely transform between finite discrete frequency and time domain signals, meaning that signal A in time domain will always give signal B in frequency domain but more importantly that B will always yield A with the inverse transform, and only signal A. This means that all information is preserved by DFT and IDFT.

We now have a fully discrete and fully finite transform pair that goes back and forth between time and frequency domains. This addresses everything we needed for OFDM. The equation for DFT is:

$$X_k = \sum_{n=0}^{N-1} x_n e^{-2\pi jkn/N}$$

where X_k are the frequency-domain samples and x_n are the time-domain samples. The phasor factor $e^{-2\pi jkn/N}$ is called the twiddle factor. N is the total number of samples in the time-domain signal. This results in N frequency domain signals; thus, we can expect k to run the range:

$$k = 0, 1, 2, \ldots, N - 1$$

It is not strictly "wrong" to try to run k over $N - 1$, but this transform will only yield N unique samples in the frequency domain and running over will cause a repetition of samples. From Fig. 7.17, this implicit periodicity can be seen in the time-domain signal. The periodicity is also clear from the equation because the twiddle factors are sinusoids. In fact, it is trivial to show that:

$$X_{k+N} = X_k$$

This frequency-domain periodicity is also logical because the time-domain signal is discrete; thus, the frequency-domain signal must be periodic.

The IDFT equation is:

$$Nx_n = \sum_{k=0}^{N-1} X_k e^{2\pi jkn/N}$$

And again, this would only yield N unique samples in time domain; thus, we should only attempt to retrieve time-domain samples for the range:

$$n = 0, 1, 2, \ldots, N - 1$$

And again, running over is not wrong but will give us redundant information because the time-domain signal is periodic given that the frequency domain signal is discrete.

Important points to notice about DFT:

- To allow preservation of all information of the spectrum, we must sample at least N samples in one period of the frequency domain.
- N samples are necessary and sufficient, sampling more than N samples in one period will not yield any more information in the time domain but will give us extra null samples.
- The DFT is like sampling the 2π spectrum of the DTFT at spaces of $2\pi k/N$, where k is an integer.
- The period of the DTFT in normalized frequency was 2π and in non-normalized frequency $\frac{2\pi}{T_s}$. Thus, the DFT is sampling at frequency intervals of $\frac{2\pi}{NT_s}$.
- Assume that we are in a 64 subcarrier 20 MHz system, the samples come out in time-domain at $T_s = 0.05$ µs, which means that sampling occurs in frequency at $\frac{1}{64 \times 0.05} = 312.5$ kHz. This is the subcarrier spacing we obtained earlier.
- The DFT being discrete means it is a representation of a periodic signal in the time domain.
- The time-domain signal being discrete should also correspond to a periodic signal in the frequency domain.

- In fact, both signals in the DFT are periodic, in the sense that if you extend either transform beyond N samples, you will get a periodic repetition of the first N samples.
- These periodic repetitions carry no additional information, and thus we can think of the two signals as windowed to yield only a single period.

The DFT is exactly what we need. It transforms between the time domain and the frequency domain and vice versa. It takes N samples from one domain and transforms them into N samples in the other. And the transforms are always unique, meaning that information is preserved.

It is important to distinguish what we mean by "discrete" here because a signal can be analog and discrete. Discrete considers how sampling happens in time or frequency, in other words how sampling occurs on the "x-axis." The values of each of the samples can be continuous or discrete. This y-axis "discretization" is what makes the signal analog or digital. If the values of the signal can only take discrete values, then the signal is digital. Thus, a signal can have discrete samples, but each of the samples can be analog. This happens, for example, inside the ADC after sampling and holding but before discretization.

7.6 FFT as an Implementation of DFT

Let us look at the DFT equation with an eye for its computational complexity. We will soon find out that the same hardware that performs DFT can perform IDFT with hardly any changes; thus, the complexity of both is the same. To measure complexity, we will count the number of complex multiplications and the number of complex additions. We will often ignore the complex additions, especially when the scaling of multiplier count is more than linear.

To calculate one sample in DFT, we perform N complex multiplications with N exponents. Some of the exponents will reduce to simpler multiplications, but we will ignore this for now because it has little impact on transforms with a large N. We need to calculate N samples in frequency domain, and if each sample needs N multiplications, then we need N^2 multiplications to find the entire transform. Each sample in frequency domain also needs N additions in the summation, for a total of N^2 complex additions.

This should be devastating news according to Sect. 7.1. We abandoned normal (single carrier) equalization in favor of OFDM because we found that complexity of equalizers rose quadratically with bandwidth. Because N rises as the bandwidth (or frequency selectivity) of the channel increases, we also have a quadratic rise in complexity here. So OFDM does not seem to have done anything useful.

This is where FFT comes in. FFT is simply a method to implement DFT. In floating point, the results of the two are identical. But FFT reduces the complexity of DFT to logarithmic instead of quadratic. It does this by utilizing symmetries in the exponential function of the DFT equation and the periodic nature of DFT.

FFT weaponizes the devastating quadratic complexity above against the problem. It does this by breaking the DFT down into progressively smaller DFTs. These will then have dramatically less complex DFTs. To complete the picture, FFT uses periodicities in DFT and symmetries in the twiddle factor to efficiently combine these small DFTs into the original sequence.

7.7 Decimation in Time

Let us consider the easiest and most intuitive FFT algorithm. The Cooley-Tukey decimation in time radix-2 FFT. By decimation in time, we mean that sequences are split in time domain. The radix refers to how many sequences we break into at every step. Recall that FFT is not an independent transform, just an efficient implementation of DFT. Also, without needing to constantly restate, we must recall that both time-domain and frequency-domain sequences run only in the range of indices 0 to $N - 1$. Recall also that an N-point DFT is periodic in frequency domain with a period of N. The DFT equation is:

$$X_k = \sum_{n=0}^{N-1} x_n e^{-2\pi jkn/N}$$

$$k = 0, 1, 2, \ldots, N - 1$$

Now split this summation into two summations, one covering the even indices of n and one the odd indices of n:

$$X_k = \sum_{n=0}^{N/2-1} x_{2n} e^{-2\pi jk2n/N} + \sum_{n=0}^{N/2-1} x_{2n+1} e^{-2\pi jk(2n+1)/N}$$

$$k = 0, 1, 2, \ldots, N - 1$$

The first summation will calculate for time indices:

$$2n = 0, 2, 4, 6, \ldots, N - 2$$

While the second summation will calculate for time indices:

$$2n + 1 = 1, 3, 5, 7, \ldots, N - 1$$

There is an implicit assumption here, which is that we have an even number of samples N. This means the last index $N - 1$ is odd. We will soon find out that for any efficient implementation of FFT, the total number of samples must be a power of 2 and thus even, so this assumption is valid.

Because DFT equations feature the twiddle factor extensively, we define a symbol for twiddle factors as a shorthand:

$$W_N^{lk} = e^{-2\pi jkl/N}$$

The twiddle factor allows notation of time and frequency domain indices, as well as the total number of samples N. It spares us having to write the constant $2\pi j$. Note that the negative sign in the exponent is in the definition of the baseline twiddle factor.

We can thus rewrite the DFT equation as::

$$X_k = \sum_{n=0}^{N/2-1} x_{2n} W_N^{2nk} + \sum_{n=0}^{N/2-1} x_{2n+1} W_N^{(2n+1)k}$$

$$X_k = \sum_{n=0}^{N/2-1} x_{2n} W_N^{2nk} + W_N^k \sum_{n=0}^{N/2-1} x_{2n+1} W_N^{2nk}$$

$$k = 0, 1, 2, \ldots, N - 1$$

Define the even indices transform as E and the odd indices transform as O. This allows us to define sample k of signal X as:

$$X_k = E_k + O_k W_N^k$$

$$k = 0, 1, 2, \ldots, N - 1$$

But both E and O are only $N/2$ samples long; when we force their k indices to go over $N/2 - 1$, they start repeating periodically. In other words, we know the following periodicities for X, E, and O:

$$X_k = X_{k+N}$$

$$E_k = E_{k+N/2}$$

$$O_k = O_{k+N/2}$$

So, while calculating samples of X from $k = 0$ up to $k = N/2 - 1$, we can just calculate:

$$X_k = E_k + O_k W_N^k$$

$$k = 0, 1, 2, \ldots, N/2 - 1$$

But for samples that run over $N/2$, we can start to use periodicity:

$$X_k = E_{k-N/2} + O_{k-N/2} W_N^k$$

$$k = \frac{N}{2}, \dots, N-1$$

And we should also note that:

$$W_N^k = W_N^{k-N/2} W_N^{N/2} = e^{-2\pi j \left(k - \frac{N}{2}\right)/N} e^{-2\pi j \frac{N}{2}/N} = W_N^{k-N/2} e^{-\pi j}$$

$$W_N^k = -W_N^{k-N/2}$$

And thus:

$$X_k = E_{k-N/2} - O_{k-N/2} W_N^{k-N/2}$$

$$k = \frac{N}{2}, \dots, N-1$$

Figure 7.18 illustrates what we just did if $N = 8$. We broke down the 8-point DFT into two 4-point DFTs E and O. We then combine (add) the outputs of two (with the odd sequence multiplied by the twiddle) for the four lower indices to obtain X-(0) through $X(3)$. Then combine them again, except with a negative sign next to the twiddle multiplied by the odd sequence to obtain samples $X(4)$ through $X(7)$.

Had we implemented the N-point DFT directly, we would have had N^2 complex multiplications. Now we have two sequences of $N/2$ points each, their combined complexity is $2 \times \left(\frac{N}{2}\right)^2 = \frac{N^2}{2}$. There is also the complexity of combining them to obtain samples of X. This is $N/2$ multiplications of O by the twiddle factors. But the twiddle factors for the even and odd indices of X are identical, so these are the only

Fig. 7.18 The 8-point FFT is broken into two 4-point FFTs recombined using symmetries

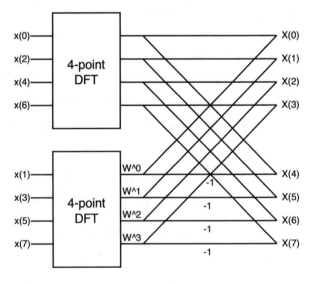

$N/2$ multiplications. Thus, the complexity of Fig. 7.18 as measured in multiplications is $\frac{N^2}{2} + \frac{N}{2}$ which is less than N^2.

We are obviously not going to stop there. Since breaking down the sequences once gave us this reduction in complexity, breaking them down further will certainly be to our benefit. The next step is to break down the E sequence into even of even and odd of even sequences. Also, the O sequence will be broken down into even of odd and odd of odd.

$$X_k = \sum_{n=0}^{\frac{N}{4}-1} x_{2\times 2n} W_N^{2\times 2nk} + \sum_{n=0}^{\frac{N}{4}-1} x_{2\times(2n+1)} W_N^{2\times(2n+1)k} +$$

$$W_N^k \sum_{n=0}^{N/4-1} x_{2\times 2n+1} W_N^{(2\times 2n+1)k} + W_N^k \sum_{n=0}^{N/4-1} x_{2\times 2n+3} W_N^{(2\times 2n+3)k}$$

$$k = 0, 1, 2, \ldots, N-1$$

We now have four smaller sequences each of $N/4$ samples and thus repeating over four periods for the N samples of k. If we get bogged down in the equations, we will quickly lose insight into how this works. Thus, it is better to think of this in terms of the transformation from Fig. 7.18 to Fig. 7.19.

In Fig. 7.19 we see each of the 4-point DFTs in Fig. 7.18 is broken down into two 2-point DFTs. The right-hand side of the figure where the 4-point DFT outputs are

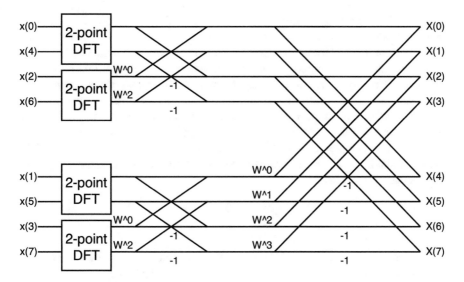

Fig. 7.19 8-point DFT broken down further into 2-point DFTs

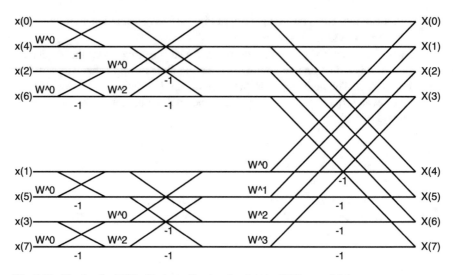

Fig. 7.20 The 8-point DFT with the realization that 2-point DFTs are trivial

combined to form the 8-point DFT remains unchanged. Figure 7.19 contains only 2-point DFTs. We will generally consider a DFT "done" once we reach this stage.

A two-point DFT is very simple. It takes two samples in time and transforms them to two samples in frequency. It multiplies the odd sample by a twiddle factor, combines the even sample and the odd sample with a twiddle to obtain index 0 in frequency, and does the same except with a negative sign to obtain index 1. Thus, Fig. 7.20 shows the 8-point DFT in Fig. 7.19 with these 2-point operations plugged in.

Figure 7.20 is the signal flow graph (SFG) of FFT. It contains only multiplications and additions. Every node is an addition. Every factor on an arrow is a multiplication. If you look at the extreme left column, you will notice each element in the column forms a butterfly shape to combine two samples in time into two in frequency. The remaining columns are also butterfly shaped but operating on farther away samples. Thus, the basic building block of FFT is called a butterfly. The butterfly performs a complex multiply-accumulate on two samples to produce two samples. Any FFT can be formed from a repetition of this basic PU.

But back to complexity. After the second splitting above, the complexity of the two $N/2$ point DFTs is further divided into four $N/4$ sequences. Thus, the complexity of the quadratic part becomes:

$$\frac{N^2}{2} \longrightarrow 4\left(\frac{N}{4}\right)^2 = \frac{N^2}{4}$$

This requires us to do $N/2$ twiddle multiplies in addition to the $N/2$ we had to do to combine the two $N/2$ sequences, and thus the total complexity is:

$$\frac{N^2}{4} + 2\left(\frac{N}{2}\right)$$

In the 8-point example, we stopped after two splits. The reason is that our mini-transforms were now two-point each. If you examine the leftmost column in Fig. 7.20, you will notice that the two-point transforms are just single multiplications by twiddle factors, which means they are like the "remainder" $N/2$ multiplications we did after every splitting, which is why we do not further split after reaching 2-point DFTs.

Starting with $N = 8$, we reached 2-point sequences after 2 steps. Starting with N points, we will reach 2-point DFTs after $\log_2(N) - 1$ splits. Therefore, it is useful to start with an N that is a power of 2, because it means we do not have wasted complexity. So, for an N-point FFT, the complexity is:

$$\left(\frac{N}{2}\right) \times \log_2(N)$$

where $N/2$ multiplies come from the last column of 2-point sequences and the remaining $(N/2) \times \{\log_2(N) - 1\}$ come from the twiddle combinations at every step. Thus, we have reduced the complexity from quadratic to logarithmic. Notice that the outputs of DFT and FFT are identical in floating point. In fact, there is no transform called FFT. FFT is just an umbrella term for algorithms used to implement DFT efficiently.

We have been ignoring complex additions in the calculation of complexity because additions are significantly less complicated than multiplications. The number of complex additions can be seen from Fig. 7.20, where each node is an addition. There are N additions in every column and $\log_2(N)$ columns. Thus, there are $N\log_2(N)$ complex adds. This is also much better than addition scaling in vanilla DFT.

The scaling of complexity in FFT versus vanilla DFT is astounding. For $N = 4$, normal DFT has a complexity of 16, while FFT has a complexity of 4. For $N = 32$ the spread increases to 256 versus 32. And for $N = 32$ The spread widens further to 1024 versus 80.

7.8   Decimation in Frequency

Decimation in frequency is an algorithm to implement FFT. Here we divide the frequency domain signal into even and odd indices. This is similar and opposite to decimation in time. We will shortly find that the signal flow graph of decimation in frequency is the reversal of decimation in time. In general, there is no computational advantage to either algorithm. Both are equal in complexity and produce identical outputs. If you are implementing your FFT in ASIC or FPGA, you should generally pick whichever algorithm you understand best. Some general-purpose processors may be more optimized to one or the other, but in such cases a compiler and a library usually take care of picking the appropriate algorithm.

Let us start again with the DFT equation:

$$X_k = \sum_{n=0}^{N-1} x_n e^{-2\pi jkn/N}$$

$$k = 0, 1, 2, \ldots, N-1$$

The even indices of the DFT are:

$$X_{2k} = \sum_{n=0}^{N-1} x_n e^{-2\pi j2kn/N} = \sum_{n=0}^{N-1} x_n W_N^{2nk}$$

$$k = 0, 1, 2, \ldots, \frac{N}{2} - 1$$

We can split this summation into two summations, one running over the lower half of the n-indices and the other over the upper half:

$$X_{2k} = \sum_{n=0}^{\frac{N}{2}-1} x_n W_N^{2nk} + \sum_{n=0}^{\frac{N}{2}-1} x_{n+\frac{N}{2}} W_N^{2\left(n+\frac{N}{2}\right)k}$$

And noting that:

$$W_N^{2\left(\frac{N}{2}\right)k} = W_N^{Nk} = e^{-2\pi jkN/N} = 1$$

$$X_{2k} = \sum_{n=0}^{\frac{N}{2}-1} x_n W_N^{2nk} + \sum_{n=0}^{\frac{N}{2}-1} x_{n+\frac{N}{2}} W_N^{2nk}$$

$$X_{2k} = \sum_{n=0}^{\frac{N}{2}-1} \left\{ x_n + x_{n+\frac{N}{2}} \right\} W_N^{2nk}$$

So, we must calculate a new sequence that is the sum of the even and odd time indices separated by $N/2$. The above summation is a DFT of length $N/2$ whose complexity is $\left(\frac{N}{2}\right)^2$. We have only calculated the even ordered frequency domain samples. To calculate the odd indices:

$$X_{2k+1} = \sum_{n=0}^{N-1} x_n e^{-2\pi j(2k+1)n/N} = \sum_{n=0}^{N-1} x_n W_N^{(2k+1)n}$$

$$k = 0, 1, 2, \ldots, \frac{N}{2} - 1$$

And again splitting:

$$X_{2k+1} = \sum_{n=0}^{\frac{N}{2}-1} x_n W_N^{(2k+1)n} + \sum_{n=0}^{\frac{N}{2}-1} x_{n+\frac{N}{2}} W_N^{\left(n+\frac{N}{2}\right)(2k+1)}$$

$$X_{2k+1} = \sum_{n=0}^{\frac{N}{2}-1} \left(x_n + W_N^{\frac{N}{2}} x_{n+\frac{N}{2}} \right) W_N^{(2k+1)n}$$

$$X_{2k+1} = \sum_{n=0}^{\frac{N}{2}-1} \left(x_n - W_N^n x_{n+\frac{N}{2}} \right) W_N^{(2k+1)n}$$

This is another $N/2$ point DFT. The two have a complexity of $\left(\frac{N}{2}\right)^2$ for a total complexity of $\frac{N^2}{2}$. Calculating the interleaved sequences requires $N/2$ complex multiplies because there is a twiddle factor involved in calculating $x_n - W_N^n x_{n+\frac{N}{2}}$. The block diagram of the algorithm after one step is shown for an 8-point FFT in Fig. 7.21.

It can be confusing to understand how Fig. 7.21 corresponds to the equations above. The summations and the W_N^{2kn} factors are included in the $N/2$ point DFT blocks. The butterflies only calculate the interleaved time-domain signals. For the top half of the graph, the two time-domain samples are added "as is" to calculate $\left\{ x_n + x_{n+\frac{N}{2}} \right\}$. For the lower half of the graph, a twiddle factor is multiplied by the bottom half to yield $\{x_n - W_N^n x_{n+\frac{N}{2}}\}$.

The natural next step is to expand each of the $N/2$ DFTs into $N/4$ sequences and so on until we end up with 2-point DFTs. The full butterfly diagram for an 8-point DFT

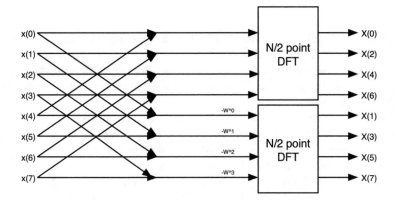

Fig. 7.21 8-point DIF FFT after one step

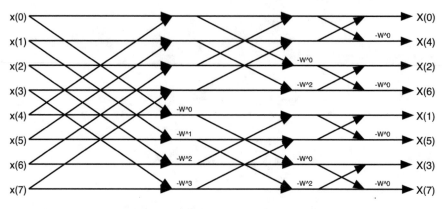

Fig. 7.22 Full DIF FFT butterfly diagram

is shown in Fig. 7.22. Comparing Figs. 7.20 and 7.22 shows that DIT and DIF diagrams are essentially reversals of each other. The tight butterflies are on the right in DIF and on the left in DIT. Indices are ordered at the input in DIF and the output in DIT. The disorder of indices in frequency domain in Fig. 7.22 matches the disorder in time domain in Fig. 7.20.

The total operation count in DIF can be calculated from the SFG or the equations as::

$$\frac{N}{2} \log_2(N) \text{ complex multiplies}$$

$$N \log_2(N) \text{ complex adds}$$

This is the same complexity as DIT.

7.9   Higher Radix FFT

Dividing the DFT into even and odd sequences in every step led to the multiplication count we derived in Sects. 7.7 and 7.8. These algorithms are called radix-2 algorithms because they split into two sequences every step. Would it be possible to split into more sequences per step? The answer is yes; the number of splits per step is the radix of the algorithm.

Splitting by any number is possible, but a radix K algorithm will only be efficient if it is used to handle a K^m-point DFT. Using a higher-order FFT will generally reduce the number of operations from radix 2, but again it is critically important that the number of points in the FFT fits with the radix.

The radix-4 algorithm is the most often used higher radix FFT. It comes in both DIT and DIF flavors. In this section, we will consider the first step of radix-4 DIT; derivation of a full butterfly diagram should then become obvious.

Consider the original DFT equation:

$$X_k = \sum_{n=0}^{N-1} x_n W_N^{kn}$$

$$k = 0, 1, 2, \ldots, N-1$$

We split it into four summations, expanding on the even-odd split we did in radix 2:

$$X_k = \sum_{n=0}^{\frac{N}{4}-1} x_{4n} W_N^{4kn} + \sum_{n=0}^{\frac{N}{4}-1} x_{4n+1} W_N^{(4n+1)k} + \sum_{n=0}^{\frac{N}{4}-1} x_{4n+2} W_N^{(4n+2)k} + \sum_{n=0}^{\frac{N}{4}-1} x_{4n+3} W_N^{(4n+3)k}$$

$$X_k = \sum_{n=0}^{\frac{N}{4}-1} x_{4n} W_N^{4kn} + W_N^k \sum_{n=0}^{\frac{N}{4}-1} x_{4n+1} W_N^{4nk} + W_N^{2k} \sum_{n=0}^{\frac{N}{4}-1} x_{4n+2} W_N^{4nk} + W_N^{3k} \sum_{n=0}^{\frac{N}{4}-1} x_{4n+3} W_N^{4nk}$$

Each of the summations above is an $N/4$-point DFT. These are combined to form the N-point DFT according to Fig. 7.23. The natural next step is to split each of the $N/4$-point DFTs into further four DFTs. As we did with radix 2, we will only stop once we reach 4-point DFTs.

The operation counts for radix 4 are as follows:

$$\frac{3}{8} N \log_2(N) \text{ complex multiplications}$$

$$N \log_2(N) \text{ complex additions}$$

This is the same number of complex additions as radix 2, but only 75% the number of the complex multiplications. By examining Fig. 7.23, we might be surprised by this because the butterflies for radix 4 have more wings than in radix 2. This means we are doing more complex multiplications per stage. You can also see this from the equation at the beginning of this section where three twiddle factors resulted from the first split as opposed to a single twiddle for 2-point. However, notice that we are going to go through far fewer stages in radix 4 than we did for radix 2.

7.10   What Does the Guard Interval Really Contain?

So far, we have assumed the GI contains null samples. We have also ignored where exactly the GI is located relative to the symbol. In 802.11n, the content and location of the GI are very specific. The guard interval is a prefix, meaning that it comes before the symbol rather than after it. An 802.11n packet consists of OFDM symbols of duration 3.2 μsec. Each symbol's guard interval is the GI that comes before it

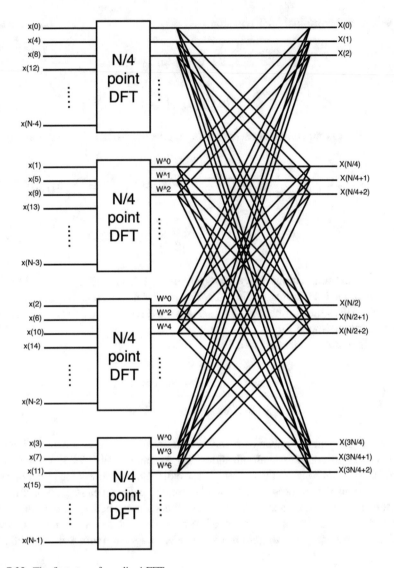

Fig. 7.23 The first step of a radix-4 FFT

rather than after it. You might wonder what difference it makes. After all, we only care that the OFDM symbols are separated by GIs; we do not care which GI belongs to who. But because the GI is non-null, it does matter.

The guard interval in 802.11n is a cyclic prefix (CP). This means that the guard interval contains a repetition of samples from the symbol that follows it. It is also a cyclic repetition, meaning that it is a copy of the end of the symbol (Fig. 7.24). Let us talk specifics. In 20 MHz with a long guard interval (Sect. 7.14), the GI duration is

Fig. 7.24 The GI as a cyclic prefix. The 800 ns of the GI are a copy of the last 800 ns of the OFDM symbol

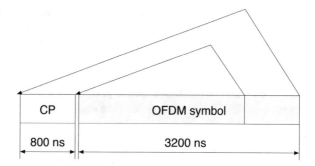

800 ns; this means $800/50 = 16$ samples in time. So, we take the last 16 samples from the following OFDM data symbol and use them as a GI.

So why should we not use null GIs? The reason has to do with the way convolution happens in DFT. We have been programmed by the continuous-time Fourier transform to think of the following transform pair as self-evident:

$$x[n] \times y[n] \leftrightarrow X(\omega)Y(\omega)$$

where the asterisk denotes a convolution. This relation between multiplication in frequency domain and convolution in time domain is one we have constantly used when developing the channel model. We always assume that the channel equation in the frequency domain is a multiplication of the channel response and the input signal spectrum, which translates into convolution in the time domain.

However, this is not true for DFT, or at least not necessarily true. Let us consider two sequences $x[n]$ and $h[n]$ in time domain, each is N samples long. This means their DFTs $X(k)$ and $H(k)$ are also N samples long. Multiplying X and H will yield a signal $Y(k)$ which also contains N unique samples. But a basic knowledge of convolution tells us that convolving x and h will yield more than N unique samples. This is because as we convolve the two signals there will be a tail as one of them starts encroaching into the other and as it exits. This is a fundamental property of convolution.

So, this means that this form of normal convolution, which we will now call linear convolution, can never be represented by multiplication in frequency domain. Why? Because there are only N samples in frequency domain and more than N samples in time domain. In other words:

$$Y(k) = X(k)H(k)$$
$$y[n] = x[n] \times h[n]$$
$$\mathrm{DFT}(y[n]) \neq Y(k)$$

But it is crucially important that we preserve this multiplication in frequency. This can happen in DFT if we consider the signals to be infinitely periodic. Thus, we extend the symbol with a GI containing data, and this data comes from the end of the symbol. This makes the overall symbol look like it is repeating periodically. This transforms the linear convolution into a circular convolution if we are doing it within the GI; if the channel impulse response extends further than the GI, we would be facing ISI anyway and the operation should fail. So, we now define a new operation called circular convolution which is performed using periodically extended signals in time domain. This will translate into multiplication in frequency. In other words:

$$y[n] = x[n] \otimes h[n]$$
$$\text{DFT}(y[n]) = Y(k)$$

7.11 Using the Same Hardware for IFFT and FFT

We have seen that IFFT is necessary at the transmitter while FFT is necessary at the receiver. Do we need a separate FFT and a separate IFFT in every radio? This question is related to whether a platform will work as a transmitter or as a receiver or both. In most Wi-Fi setups, everyone needs to be able to act as a transmitter and a receiver. There is no clear distinction between downlink and uplink. If you want to receive something, you first ask for it, which means you must transmit to the access point. And if the access point wants to send you something, it needs to listen to you saying that you received it.

So, the answer is that every platform in Wi-Fi needs to be both a transmitter and a receiver. This might indicate we need both an FFT and an IFFT. FFT is pretty resource intensive, so we should avoid having to duplicate it. So maybe we can reuse the same block to do both FFT and IFFT. For this to work, two conditions must exist:

- FFT and IFFT are similar enough that it is easy to reconfigure the hardware between the two. We will show that the two are pretty much the same hardware. But you can already see this by examining the equations for DFT and IDFT.
- The system does not need to use FFT and IFFT simultaneously.

The second point above could be addressed by asking whether Wi-Fi is a simplex, half duplex, or full duplex standard. We can clearly see it is *not* simplex because we have established everyone needs to occasionally transmit and receive. The concept of full duplex versus half duplex is also not relevant to modern digital baseband. But the simple answer is that most modern digital communication systems can be considered half duplex. In other words, they are either transmitting or receiving at a certain time. But even if they need to do both simultaneously, if the FFT is fast enough to absorb the extra data, we can do both with a single block (see Chap. 8 for hardware reuse).

So, we only need to figure out how to reuse the same hardware for both FFT and IFFT. Recall that FFT (and thus IFFT) are just mathematical implementations of DFT and IDFT. So, we examine the equations of DFT and IDFT:

$$X(k) = \sum_{n=0}^{N-1} x[n] e^{-2\pi jkn/N}$$

$$k = 0, 1, \ldots, N - 1$$

$$x(n) = \frac{1}{N} \sum_{k=0}^{N-1} X(k) e^{2\pi jkn/N}$$

$$n = 0, 1, \ldots, N - 1$$

The similarities between the two equations are uncanny. We can show with a bit of algebra that:

$$\text{IFFT} = (\text{FFT}(x^{\times}))^{\times}$$

Meaning that according to Fig. 7.25, we can use the same hardware of FFT to implement IFFT. Or to be more specific, we can use any DFT hardware to produce IDFT if we conjugate both the input to the block and its output. Since complex conjugate is a very simple arithmetic operation (negative sign), the hardware cost is negligible.

7.12 Bit Reversal

By inspecting Sects. 7.7 and 7.8, particularly Figs. 7.20 and 7.22 as well as the 16-point butterfly diagram in Chap. 8 (Fig. 8.2), we notice that the indices of the outputs in decimation in time FFT are ordered, while the indices of the inputs are disordered. The "disorder" at the input follows a pattern that arises from taking even and odd indices, then even of even and odd of even, and so forth till we reach 2-point DFTs. When the size of the DFT grows, we need a systematic way to figure out the order of these input indices. The same also applies to DIF, although the disorder in that case affects the outputs.

The indices can be obtained by bit-reversing the binary representations. Thus, in an 8-point DIT DFT, the output indices are {000, 001, 010, 011, 100, 101, 110, 111}. By reversing the bits, we obtain {000, 100, 010, 110, 001, 101, 011, 111} and in decimal {0, 4, 2, 6, 1, 5, 3, 7}, which is the order of time-domain inputs in Fig. 7.20. If we are dealing with a 4-point DFT, the first four indices would be {0000, 0001, 0010, 0011, ...} and in reverse {0000, 1000, 0100, 1100,...} and in decimal {0, 8, 4, 12, ...}. The same can also be applied to decimation in frequency.

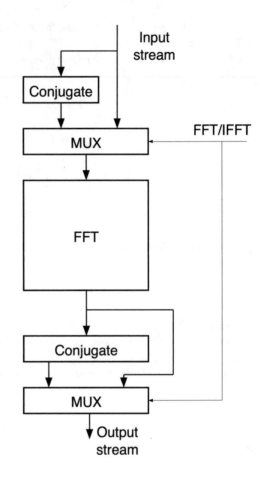

Fig. 7.25 FFT/IFFT switching on the same hardware

7.13 Populating the OFDM Symbol

Table 7.1 lists values that will help us understand how an OFDM symbol is built in 802.11n. In 20 MHz there are 64 subcarriers; thus, the subcarrier spacing is 20/64 = 312.5 kHz. In 40 MHz mode, there are 128 subcarriers; thus, subcarrier spacing is 40/128 = 312.5 kHz. This shows us that the availability of 128 or 64 subcarriers in 802.11n is not to address more multipath in the channel but to support 40 MHz bandwidth. The 312.5 kHz subcarrier spacing is also the width of subchannels, using the rule of thumb from Sect. 4.9; this means we can address channels with coherence bandwidth of about 1.56 MHz, which is fine for indoor environments typical of Wi-Fi.

Subcarrier indices run from -32 to 31 in 20 MHz and from -64 to 63 in 40 MHz. Of the 64 subcarriers in 20 MHz, 48 carry data. This means they carry QAM symbols from the baseband modulator. We already used this number in Sect. 6.13 while

Table 7.1 Important values for OFDM in 802.11n

Symbol	Name	20 MHz	40 MHz
N_{SD}	Number of data subcarriers	48	108
N_{SP}	Number of pilot subcarriers	4	6
N_{SN}	Center null subcarriers	1	3
N_{SR}	Subcarrier index range	$-26, \ldots, 26$	$-58, \ldots, 58$
Δ_F	Subcarrier spacing (kHz)	312.5	312.5
T_{FFT}	Symbol period (μsec)	3.2	3.2
T_{GI}	GI interval (μsec)	0.8	0.8
$T_{short\,GI}$	Short GI interval (μsec)	0.4	0.4

counting the number of bits at the input of the block interleaver. Similarly, for 40 MHz 108 of the 128 subcarriers are data-bearing. There is a specific mapping for which subcarrier indices carry data which we will consider shortly.

In 20 MHz 16 subcarriers do not carry data. According to Table 7.1, subcarrier index range runs from -26 to 26. This means that indices -32 through -27 inclusive are null. Also, indices 27 through 31 are null. According to the table, subcarrier 0 is also a null. This is a total of $6 + 1 + 5 = 12$ null indices. Together with the 48 data subcarriers from Table 7.1, they add up to 60 subcarriers. This means we are still missing 4 subcarriers. In fact, the range -26 through 26 is 53 subcarriers, of which 1 is null at index 0, leaving us with 52 populated subcarriers of which only 48 carry data.

The four remaining non-null subcarriers are at indices $-21, -7, 7$, and 21. These subcarriers are called pilots. They carry known information of low modulation order. Because they carry known information, they do not carry data. Pilots are used (Chap. 11) to perform frequency offset estimation and to refine channel estimation.

Back to the null subcarriers from earlier, why are they needed? The null at index 0 is needed because when moved to baseband, index 0 is at DC. Since a lot of systems use AC coupling, we cannot rely on DC to carry any information. The other nulls are at the periphery, they are there to allow filter roll-off. Every transmitter must have a filter that masks the spectrum of the output signal to ensure that it will not interfere with neighboring channels through its sidelobes. The roll-off of the filter will attenuate peripheral subcarriers, so we reserve these subcarriers as non-data-bearing.

To summarize, for 20 MHz:

$$\text{Null tones} = \{-32 \ldots -27, 0, 27 \ldots 31\}$$

Which leaves populated tones at:

$$\text{Populated tones} = \{-26 \ldots -1, 1 \ldots 26\}$$

Of which pilots are at:

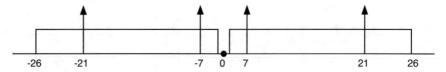

Fig. 7.26 Null, data, and pilot subcarriers in 20 MHz. Pilots are shown as impulses

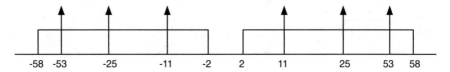

Fig. 7.27 Null, data, and pilot subcarriers in 40 MHz

$$\text{Pilot tones} = \{-21, -7, 7, 21\}$$

And thus, data is at:

$$\text{Data tone} = \text{Populated tone} - \text{Pilot tones}$$

This is also summarized in Fig. 7.26.

Similarly, and for identical reasons in 40 MHz:

$$\text{Null tones} = \{-64 \ldots -59, -1 \ldots 1, 59 \ldots 63\}$$

$$\text{Populated tones} = \{-58 \ldots -2, 2 \ldots 58\}$$

$$\text{Pilot tones} = \{-53, -25, -11, 11, 25, 53\}$$

$$\text{Data tone} = \text{Populated tone} - \text{Pilot tones}$$

Again, the results are summarized in Fig. 7.27.

Samples in an OFDM symbol are produced at intervals of $\frac{1}{20} = 50$ns. There are 64 samples in an OFDM symbol in 20 MHz, for a total symbol duration of 64×50 ns $= 3.2$ μsec which we can confirm from Table 7.1. Note that in 40 MHz the intervals between samples are halved but the number of samples is doubled, leaving OFDM symbol duration the same.

In normal circumstances a GI duration of 800 ns is used, leading to an overall symbol duration of 4 μsec. 802.11n also offers a short GI duration of 400 ns for an overall symbol duration of 3.6 μsec. This can be used in channels with low delay spread. As we will see in Sect. 7.14, GI is an overhead on throughput since it does not carry additional information. The shorter the GI, the more data throughput you get. The longer the GI, the more protection against delay spread and ISI you get.

7.14   MCS, Useful Throughput, and Cognitive Radios

In this section, we discuss what we mean by "throughput." In other words, if an access point advertises that it delivers 300 Mbps, what does this mean? Does the user get to stream video at 300 Mbps? The physical layer standard has a section that lists throughputs that can be achieved for different MCSs. These throughput figures are excellent benchmarks, but first we must understand how they are calculated.

MCS stands for modulation-coding scheme. Any standard allows for multiple QAM constellations as well as multiple coding rates. However, not all combinations are allowed. Any allowed combination of coding rate, modulation order, and number of antennas is called an MCS.

The tables below show the rates that the 802.11n standard lists for different MCS. Table 7.2 covers MCS for single antenna setups. Table 7.3 covers MCS with two spatial streams, Table 7.4 three spatial streams, and Table 7.5 four spatial streams.

The rates listed in the tables above are the rates observed at the input of the MAC layer at the receiver, or the output of the MAC layer at the transmitter. These are the pure data rates that the PHY layer manages to deliver to higher layers. This is still not

Table 7.2 802.11n MCS for single spatial stream. Rates are measured in Mbps

MCS (index)	Modulation	Coding rate	Rate (20 MHz) GI = 800 ns	Rate (40 MHz) GI = 800 ns	Rate (20 MHz) GI = 400 ns	Rate (40 MHz) GI = 400 ns
0	BPSK	½	6	13.5	6.67	15
1	QPSK	½	12	27	13.33	30
2	QPSK	¾	18	40.5	20	45
3	16-QAM	½	24	54	26.67	60
4	16-QAM	¾	36	81	40	90
5	64-QAM	2/3	48	108	53.33	120
6	64-QAM	¾	54	121.5	60	135
7	64-QAM	7/8	63	141.75	70	157.5

Table 7.3 802.11n MCS for two spatial streams

MCS (index)	Modulation	Coding rate	Rate (20 MHz) GI = 800 ns	Rate (40 MHz) GI = 800 ns	Rate (20 MHz) GI = 400 ns	Rate (40 MHz) GI = 400 ns
8	BPSK	½	12	27	13.33	30
9	QPSK	½	24	54	26.67	60
10	QPSK	¾	36	81	40	90
11	16-QAM	½	48	108	53.33	120
1	16-QAM	¾	72	162	80	180
13	64-QAM	2/3	96	216	106.67	240
14	64-QAM	¾	108	243	120	270
15	64-QAM	7/8	126	283.5	140	315

Table 7.4 802.11n MCS for three spatial streams

MCS (index)	Modulation	Coding rate	Rate (20 MHz) GI = 800 ns	Rate (40 MHz) GI = 800 ns	Rate (20 MHz) GI = 400 ns	Rate (40 MHz) GI = 400 ns
16	BPSK	½	18	40.5	20	45
17	QPSK	½	36	.81	40	90
18	QPSK	¾	54	121.5	60	135
19	16-QAM	½	72	162	80	180
20	16-QAM	¾	108	243	120	270
21	64-QAM	2/3	144	324	160	360
22	64-QAM	¾	162	364.5	180	405
23	64-QAM	7/8	189	425.5	210	472.5

Table 7.5 802.11n MCS for four spatial streams

MCS (index)	Modulation	Coding rate	Rate (20 MHz) GI = 800 ns	Rate (40 MHz) GI = 800 ns	Rate (20 MHz) GI = 400 ns	Rate (40 MHz) GI = 400 ns
24	BPSK	½	24	54	26.67	60
25	QPSK	½	48	108	53.33	120
26	QPSK	¾	72	162	80	180
27	16-QAM	½	96	216	106.67	240
28	16-QAM	¾	144	324	160	360
29	64-QAM	2/3	192	432	213.33	480
30	64-QAM	¾	216	486	240	540
31	64-QAM	7/8	252	567	280	630

the data rate that the radio delivers to the user because there are some unaccounted-for overheads.

In a QAM modulation scheme with M symbols, there are $k = \log_2(M)$ bits per symbol. Each Hz of bandwidth can carry a single QAM symbol. In other words, the symbols are sent at the rate of the bandwidth of the channel. Thus, a modulation scheme with k bits should be transmitting kB Mbps. By examining Table 7.2, this is obviously not the case. For example, consider MCS 0. This uses BPSK and for 20 MHz should deliver a bitrate of 20 Mbps; the table lists its throughput at 6 Mbps. We must find out why.

In the MCS table of the 802.11n standard, there are two main columns for each MCS, one for 20 MHz bandwidth and one for 40 MHz bandwidth. The standard allows 40 MHz bandwidth in jurisdictions which allow it. Naturally, using larger bandwidth will lead to more throughput. In fact, it should lead to proportionately more throughput. But examining any of the tables shows that the 40 MHz column is not exactly double the corresponding 20 MHz column. The reason lies in disparate overheads that will be explored below.

There are two further sub-columns for each column. Each sub-column relates to a choice of guard interval (GI) duration. A short GI is 400 ns long and a long GI is

800 ns long. The long GI is usually considered the default because it is also the value of legacy GI from 802.11a. There is a different value for throughput under each of the GI sub-columns.

We will now calculate the throughput figures in the tables above. The best way to do this is to trace back from the transmitter antenna. If there is a single transmitter antenna sending order M QAM symbols over a bandwidth of B, we have already concluded that throughput out of the transmitter antenna in Mbps is:

$$\text{Throughput out of antenna} = B \log_2(M)$$

If we use multiple antennas in true MIMO (Chap. 9), we are basically multiplying this throughput by the number of antennas we have available. Thus, for N_{rx} receive antennas and assuming N_{rx} is less than or equal to N_{tx}:

$$\text{Throughput out of antennas} = B \log_2((M) \times N_{rx}$$

Tracing back, there is a guard interval inserted into the time-domain OFDM symbols before they are sent. In Sect. 7.10, we discussed that guard intervals contain cyclic prefixes. Cyclic prefixes contain repeated information, which boils down to no information. Thus, the time we are sending the GI is "wasted" from the point of view of carrying data. GI is not truly a waste because it provides protection against time offsets and ISI, but it is going to eat into our throughput. Thus, the data rate at the output of GI insertion at the transmitter is higher than the data rate at its input. The fraction between the two is the ratio of OFDM symbol time to the total of OFDM symbol time and GI time:

$$\text{Data rate before GI insertion} = B \times \log_2\left((M) \times Nrx \times \frac{t_{\text{OFDM}}}{t_{\text{OFDM}} + t_{\text{GI}}}\right)$$

where t_{OFDM} is the duration of the OFDM symbol and t_{GI} is the guard interval duration. In 802.11, the duration of the OFDM symbol is 3.2 μsec, while the guard interval could be either 400 or 800 ns. Thus, the fraction is either 3.2/4 or 3.2/3.6. Using a shorter GI allows for more efficient communication with more data bits being pushed through, but using a longer GI allows for more ISI protection.

Going further back in the transmitter chain, we are faced with the IFFT. This will take 64 input samples and produce 64 output samples (in 20 MHz mode). The rate calculated above is the rate at the input of GI insertion and thus the rate at the output of the IFFT. This would also be the rate at the input of the IFFT, *if* all 64 subcarriers carried data.

We have already seen in Sect. 7.13 that not all subcarriers are data-bearing. A significant minority are either nulls or pilots. In Chap. 11, we will see that most CSI is obtained from the header of the packet. However, if there is high Doppler, or if we are dealing with something critical, we must continually improve our estimates. This is the case, for example, with frequency offset estimation. This is where pilots come in. Pilots carry known information and are thus not data carrying.

There are also null subcarriers. We send no information on null subcarriers because we know for certain that data on these subcarriers will be destroyed. You will find null subcarriers in three locations: at the two peripheries of the channel and in the middle of the band. The peripheral null subcarriers are inserted because any OFDM symbol passes through a filter before it is transmitted. The filter ensures the transmitted signal fits the mask described by the standard and is thus standard-compliant. Filters have roll-off. And roll-off means that peripheral frequency components will be severely compromised; thus, they are best left at nulls. The central subcarriers are also nulled because they correspond to DC in baseband. All systems have AC coupling at one point in the analog chain. One should never count on DC components to carry data.

Combining all the above, we deduce that data throughput before the FFT is a large fraction of that at the output of the FFT. This is because of the 64 or 128 subcarriers of the OFDM symbol; some are either null or pilot subcarriers. Thus, the data rate at the input of FFT:

$$\text{Data rate at input of FFT} = B\log 2(M) \times Nrx \times \frac{t_{\text{OFDM}}}{t_{\text{OFDM}} + t_{\text{GI}}}$$

$$\times \frac{N_{\text{data}}}{N_{\text{data}} + N_{\text{pilot}} + N_{\text{null}}}$$

where N_{data} is the number of data subcarriers, N_{pilot} is the number of pilot subcarriers, and N_{null} is the number of null subcarriers. We define populated subcarriers as:

$$N_{\text{populated}} = N_{\text{data}} + N_{\text{pilot}}$$

So, we can restate data rate as:

$$\text{Data rate at input of FFT} = B\log 2(M) \times Nrx \times \frac{t_{\text{OFDM}}}{t_{\text{OFDM}} + t_{\text{GI}}} \times \frac{N_{\text{data}}}{N_{\text{populated}} + N_{\text{null}}}$$

It is important to notice that accounting for the guard interval overhead and accounting for the non-data subcarriers are completely different things. It is sometimes tempting to think of the two as the same overhead in the time and frequency domains, but they are independent. Pilots and nulls are inserted before IFFT is performed at the transmitter. Once the IFFT is performed, it produces 64 time-domain samples. Some of these samples are then taken and repeated as a cyclic prefix that is inserted in the GI.

Going back further through the transmitter chain, we account for the baseband modulator, interleaver, deparser, puncturer, and channel encoder. The modulator, interleaver, and deparser have no impact on the data rate obtained above. None of them introduce any physical layer overheads. The data bitrate at their inputs is the same as the data bitrate at their outputs. They all reshape the data in important ways, but without introducing overheads.

The story is very different for the encoder and puncturer. When considering data rates, it is useful to consider the encoder and puncturer as a single unit. This unit has a coding rate r, picked from one of the allowed coding rates according to the list of MCS in the physical layer standard. Coding rate is the ratio of bits carrying unique data. Thus, the actual data rate that the physical layer receives from the MAC layer, which is also the input of the encoder, can finally be derived as:

$$\text{Data rate from MAC} = B \log 2(M) \times Nrx \times \frac{t_{\text{OFDM}}}{t_{\text{OFDM}} + t_{\text{GI}}} \times \frac{N_{\text{data}}}{N_{\text{populated}} + N_{\text{null}}} \times r$$

This is the data that the MAC layer experiences whenever there is a successful transmission. There are many reasons this data rate will not be identical or even close to the data rate observed by the user. However, this *is* the data rate that the standard lists because it *is* the data rate that the physical layer provides to upper layers. This is also often the rate used by ethical manufacturers to advertise their products, although many would use the raw data rate available at the output of the transmitter antenna.

So, what sets the physical layer data rate apart from the rate observed by the user? Some reasons include:

- We are not transmitting all the time. In fact, we are not even transmitting most of the time. If, over a window of 1 h, we transmit for only 1 min, should we use the peak rate available in the minute as throughput, or average this over the entire hour? The answer is that we should penalize ourselves whenever we try to transmit but fail to do so, but we should never average data out over times we were not even trying to transmit.
- One reason we might try to transmit and fail to do so is if we find the channel busy. This is part of the MAC layer and is handled by the multiple access technique.
- Another reason is that we tried to transmit and ended up with a collision. This can also be expanded to the fact that there will be erroneous packets that will need retransmission. So perhaps the user throughput should be the MAC throughput multiplied by $1 - PER$.
- The header of the packet is not considered in the above calculation. We take pilots and GI into account, but the header is not calculated as an overhead. The PHY header should be seen as an overhead to the MAC layer. MAC headers should be seen as an overhead to higher layers, and so on.
- Sometimes we are transmitting successfully, but our payload carries no data. Wi-Fi allows special packets to be sent to establish CTS and RTS or to acknowledge receipt, manage the network, or sound the channel.
- Ultimately, there is a finite non-zero BER and PER even when packets manage to go through, which means that some bits are thrown away. In fact, you can deduce the good bitrate as $(1 - BER) \times throughput$.

So why does the standard allow you to use multiple MCS? A simplistic way to view this is that the standard allows different designers to pick the MCS they want for their transceiver. For example, my transceiver can use MCS 3, while another uses MCS 8 and both would be standard-compliant. The two can detect each other and will avoid each other, but they cannot communicate because they are using different modulation and coding rates.

In practice, the different MCS are expected to be supported by *all* platforms, or at least those MCS that have the same number of spatial streams. Thus, a transceiver must be designed in such a way that it can switch between any of the MCS in one of the tables above. This means that it can use any of the QAM constellations and any of the coding rates supported by the standard.

This allows any two standard-compliant terminals to communicate with each other. They will use initial handshaking and the high throughput signal field of the preamble (Chap. 11) to agree on an MCS and then use this to push the payload through.

But why do they *need* multiple MCS? This allows them to adapt to different channel conditions. For example, if the channel is "good," meaning that fading has a high amplitude and noise is low, then the link can extract a very decent goodput from 64-QAM with little coding. If channel conditions worsen, sticking to this same MCS will lead to a lot of dropped packets and a lot of retransmissions. So, in this case, it is better to switch to a lower constellation with a lot of coding, for example, BPSK and coding of ½. Such an MCS will have a low throughput according to the tables above, but most of this low throughput will get through correctly.

The above discussion has implications on the processing units and the control of our transceivers. All the blocks we have discussed so far must be designed with support for all the modes in the tables. So, you cannot design a mapper/demapper that supports only one constellation. It must support BPSK, QPSK, 16-QAM, and 64-QAM. And this cannot be by using four independent mappers/demappers for the four constellations. Instead, the same hardware must be reused for all of them.

Sometimes the above reconfigurability is rather simple. For example, supporting all the coding rates in the standard means we include the puncturer/depuncturer for all the rates, which is very simple in hardware. Sometimes things are a little more challenging. For example, if we want to support both 20 MHz and 40 MHz modes, the FFT must be able to work on 64 points or 128 points. This is much harder to do and requires a deep dive into throughput, power, and latency requirements (Chap. 8).

The other aspect of the reconfigurability is who makes the decision to switch MCS and based on what. There must be a central "intelligence" that looks at metrics and picks an appropriate MCS. The metrics will include the following:

- Channel state information. Better channels promote bigger constellations and less coding.
- PER. If we are using an inappropriate MCS, PER will shoot up.
- Throughput. If we are using an inappropriate MCS, throughput will plummet.

In fact, all three factors are telling the "intelligence" the same thing. Channel state information combined with MCS will determine PER. PER will require retransmissions which will waste throughput. But if PER is low, then there are few retransmissions and using an MCS with a low throughput is a waste. Perhaps an appropriate metric is goodput. The intelligence should pick the MCS that maximizes goodput by picking the optimal PER and throughput.

This "intelligence" will be one of many optimization or machine learning algorithms. The design and tradeoffs involved in optimization are beyond the scope of this book. But the combination of reconfigurable hardware with an intelligence that optimizes the mode leads to the vast field of cognitive radios which we will revisit in Chap. 10.

7.15 We Can Finally Look at Everything in the Receiver

We have systematically looked at the transmitter chain several times so far. There is good reason we could easily do this: standards normally concern themselves only with the transmitter. The wireless standard describes a viable way in which multiple users can communicate in a shared channel. This requires balancing conflicting requirements to maximize the instantaneous throughput that a certain user observes while allowing as many users as possible to simultaneously access the channel.

Thus, the standard will describe ways to regulate how and when users manage to put data on the channel. And when they do put data on the channel, how this data should look like. This is all a result of what the transmitter does. Ultimately, the standard wants to describe a spectral mask that puts limitations on how the signal radiated out of standard-compliant platforms should look like.

Figure 7.28 shows the mask for 20 MHz mode in 802.11n. The x-axis is frequency offset from the main carrier. Since we are in 20 MHz mode, the bandwidth of the channel extends from −10 MHz to 10 MHz. Recall that the 20 MHz is passband bandwidth. The y-axis is power level relative to the carrier. Thus, we can have carrier-level power extending from −9 to 9 MHz, meaning only 18 MHz are useful in the channel. In Sect. 7.14, we defined null subcarriers at the periphery. These nulls absorb the roll-off we see in Fig. 7.28 between 9 and 11 MHz and −9 and −11 MHz. The mask also dictates that to be standard-compliant, we must be 20 dB below the carrier at an offset of 11 MHz, 28 dB below at 20 MHz, and 45 dB at any frequency above or below 30 MHz. Figure 7.29 shows a similar mask for 40 MHz mode.

We have not had a similar comprehensive discussion of the receiver chain. Standards do not normally talk about the receiver. This is intentional to give designers the freedom to implement the receiver the best way they can. Any design decisions at the receiver will not affect other users, and therefore you should feel free to innovate.

Of course, this is not set in stone. The standard *will* often talk about the receiver, making recommendations, or even demands, about it. For example, in Chap. 6, the standard is explicitly stating that the channel decoder must use the Viterbi algorithm because the channel encoder is a non-recursive non-systematic convolutional

Fig. 7.28 Transmitter
spectral density mask for
20 MHz mode

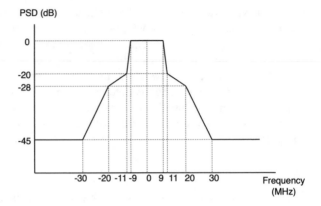

Fig. 7.29 Transmitter
spectral density mask for
40 MHz mode

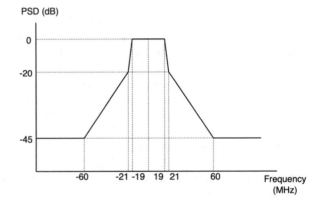

encoder. The standard will also demand a certain latency budget from the receiver, requiring it to acknowledge within a certain timeout window.

At this point, we are ready to describe the receiver chain. Because it is rather long, we divide it between two figures, the first of which is Fig. 7.31. The receiver is significantly larger than the transmitter. This statement is true for every part of the physical layer. Receiver RF is more complex than transmitter RF. The ADC is more complicated than a DAC. And the receiver baseband radio is much more complicated than the transmitter baseband radio.

The receiver baseband contains many more blocks than the transmitter, also many of the blocks that have counterparts in the transmitter are more complicated in the receiver. The baseband receiver has two types of blocks:

- Blocks with a counterpart in the transmitter. These are used to invert the impact of the corresponding transmitter block to extract data bits for the MAC layer. Sometimes the blocks are exact mirror images of the transmitter counterpart. For example, the FFT and the IFFT are almost identical in terms of operation and are identical in terms of complexity. The puncturer and the depuncturer are both very simple. The deinterleaver and the interleaver do the same thing. Other times

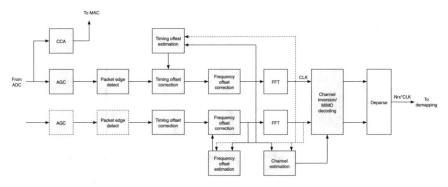

Fig. 7.30 Receiver chain part 1

the blocks are asymmetric, for example, the channel decoder is nowhere near as simple as the channel encoder.

- Blocks with no counterpart in the transmitter. These blocks are concerned with estimating and inverting the effects of the channel and receiver non-idealities. These blocks are to a large extent responsible for the relative complexity of the receiver. This includes channel estimation, channel inversion, frequency offset estimation and correction, packet edge detection, and timing offset estimation and correction.

Most of the blocks in Figs. 7.30 and 7.31 were discussed in earlier chapters or will be discussed in later chapters. Some of the bocks do not fit neatly under any chapter and thus will be exclusively discussed here.

Clear Channel Assessment (CCA)
CCA is essentially a MAC block. However, it must measure physical signals and is thus often combined with the physical layer. CSMA-CA, used to allow multiple access in Wi-Fi, requires terminals to measure power in the channel before attempting to transmit. If the channel is seen to be "clear," the station can send after a random wait period.

CCA thus tells the MAC layer whether the channel is clear or not. It does so by measuring energy in a window containing a number of received samples. Notice that CCA does not and should not care about the nature of the energy on the channel. Whether the energy is someone talking to us, someone talking to someone else, or a microwave oven, the channel is not clear.

CCA calculates energy by adding up the autocorrelation of the detected signal over a window L samples long. The longer the window, the more resistant CCA is to being triggered by noise. To decide if the channel is clear or unclear, this energy is compared to a threshold. The value of the threshold will depend on the standard, noise, experience, and length of window (if energy is not normalized):

$$\text{Channel is clear if} : \sum_{L} x(n)x^\times(n) < \text{Threshold}$$

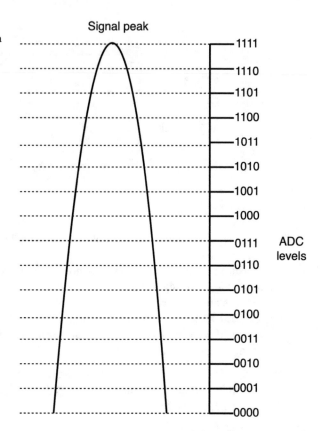

Fig. 7.31 Dynamic range of a 16-level ADC. This is a simple 4-bit unsigned ADC meaning that the signal is always positive

Automatic Gain Control (AGC)

AGC is tightly related to RF, but because it requires significant arithmetic, it is performed in baseband. Automatic gain control is integrated with the ADC, LNA, and VGA in the RF and mixed signal sections of the chain. So, to understand AGC, we must examine its interaction with the analog and mixed-signal parts of the physical layer.

Figure 7.31 shows a very simple 16-level ADC. It digitizes an analog signal by dividing it between 16 levels, producing the result in four bits. The results we derive for this oversimplified case can be extended to more realistic cases. ADCs in practical systems are signed and have 12 output bits or more, digitizing into 2048 or more levels.

The ADC should digitize the analog signal so that the maximum analog signal level corresponds to the all 1's digital word. The zero analog signal level should correspond to the all 0's digital word. But what if the transceiver is tuned this way and then the transmitter and the receiver move relative to each other? In such a case, the signal level at the receiver will increase or decrease.

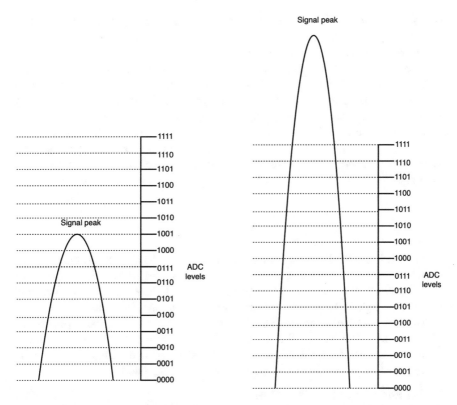

Fig. 7.32 Dynamic range with the analog signal not corresponding to the ADC range. Left, attenuated signal, right, clipped signal

If the receiver and transmitter move farther away from each other, the analog signal will attenuate because of increased path loss. This will lead to the maximum swing of the analog signal not covering the entire range of the ADC as shown in Fig. 7.32 (left). Headroom at the top and bottom of the signal will be wasted without being digitized. But more critically, distinctions between signal levels within the range of the signal will be less pronounced. This will translate into increased errors and worse BER performance.

If the receiver and the transmitter move closer together, this causes the analog signal to magnify as path loss decreases. As shown in Fig. 7.32 (right), this causes some of the peaks and troughs of the signal to be clipped. If we are dealing with positive and unsigned numbers, clipping only occurs on the high end. With signed signals, it should happen both ways. This clipping is an extreme form of nonlinearity and will lead to loss of information. This translates directly into increased BER.

We must manage the whole setup so that the range of the swing of the analog signal fully occupies but does not exceed the range of the ADC input. This is done by managing the gains of a variable gain amplifier (VGA) and the low noise amplifier

(LNA). In some RF chains, these two entities are separate; in some they are both identified as the LNA.

The AGC is the unit that tells the VGA what value of gain to use. It does this by measuring the energy in a window and comparing it to certain thresholds to decide what gain level should be used to compensate for path loss. Power measurement is usually done using preambles in the header because we are aware of the energy levels that these preambles *should* have if gain were properly tuned.

In 802.11n, the field most often used for this is the HT-STF (high throughput short training field). The HT-STF has a periodicity in time domain of 16 samples. We calculate energy in a window of multiples of 16 samples. The more the multiples, the better the result in the presence of noise. The energy metric is:

$$E = \sum_{L} x(n) x^{\times}(n)$$

The VGA and the LNA do not have a knob that allows us to adjust the gain on a continuous scale; instead they have a finite number of discrete gain settings. The AGC compares E from above against several threshold levels and decides which of the gain settings in the amplifiers to choose based on the range where E lies.

The energy of the header of an 802.11n packet is different from energy in the rest of the packet. This allows the header to be more resistant to noise and makes estimates obtained from the header (Chap. 11) more accurate. This discrepancy in energy must be taken into consideration when adjusting gain levels.

Packet Edge Detection (Sect. 11.8)
This detects whether a packet is being sent over the air and gets the receiver ready to start decoding. Packet edge detection is a coarse operation that will "trigger" within a very wide window in time once the packet has started. Packet edge detection is a little like CCA, but it is distinct. For example, packet edge detection cannot use energy as a metric because this would cause us to trigger due to many non-Wi-Fi terminals that operate in the ISM band.

Channel Estimation
This is where we determine the channel coefficient, h. This is always done by comparing known data at the transmitter and the receiver. In Wi-Fi this is often done using preambles in the header (Sect. 11.1). In cellular systems pilots and interpolation are used (Sect. 12.4).

MIMO channel estimation is significantly more complicated than SISO and requires pilots or preambles to be multiplexed in time. Channel estimation can happen using samples before they pass through the FFT, in which case it is called time-domain channel estimation. If it uses samples after they pass the FFT, it is called frequency-domain channel estimation.

Frequency Offset Estimation (Sect. 11.6)
This is a particularly important step in OFDM. In Sect. 7.3, we talked about the impact of frequency offsets on intercarrier interference. In Sect. 5.5 we saw how it

can completely ruin the constellation. In Sect. 11.6, we will see how preambles or pilots can be used to estimate how much offset there is between the transmitter and the receiver. Like channel estimation, this could happen in frequency domain or in time domain. Frequency offset estimation is so critical that it is often divided into two steps: coarse estimation for an initial estimate and fine estimation for a continuous improvement of these initial coarse estimates.

Frequency Offset Correction (Sect. 11.6)
This will also be discussed in more detail in Sect. 11.6. It follows hand in hand with frequency offset estimation. After the frequency offset is known, frequency offset correction is responsible for removing this offset from the baseband signal.

Timing Offset Estimation and Correction (Sect. 11.7)
In Sect. 11.8, we will see that packet edge detection produces a very coarse estimate of when the packet begins. We need a more accurate decision on when OFDM symbols begin and end so that we can do FFT properly. This is done by timing offset estimation and correction. This block is much more important in single-carrier systems. In OFDM, there is a large tolerance of timing offsets. In fact, anywhere in the cyclic prefix is fine, though it leads to loss of protection against ISI.

Deparsing
This is the inverse of parsing. Deparsing is a parallel to serial conversion; it gathers multiple streams of bits into a single stream. This happens in the bit-domain since it comes before the demapper. In 802.11n deparsing is not necessarily done on a bit-by-bit basis. Parsing and deparsing are done in units of s bits, where $s = \max \{N_{BPSC}/2, 1\}$. Thus, it is on a bit-by-bit basis for BPSK and QPSK, but two bits by two bits for 16-QAM.

Baseband Demodulation (Sect. 5.6)
This is also known as demapping (Fig. 7.33). In baseband demodulation we convert QAM symbols to bits. This is done by comparing the equalized received symbol to the reference constellation and observing the closest symbol to find that maximum likelihood solution. Alternatively, we could produce soft metrics instead of making a hard decision about the received bits. Baseband demodulation is deeply entangled with channel decoding when soft metrics are used. It is also deeply intertwined with channel inversion, especially in MIMO systems.

FFT
This was the topic of this chapter. It corresponds to IFFT in the transmitter. FFT performs OFDM demodulation by removing the entanglement between subcarriers. It moves symbols from time domain to frequency domain where they can be demapped into bits and decoded.

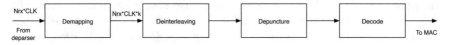

Fig. 7.33   Receiver chain part 2 (bit-domain)

MIMO Decoding/Channel Inversion (Chap. 9)

Channel inversion is the process of exposing the observation to the inverse of the channel. This was discussed in the SISO context in Sect. 5.6. In MIMO, inverting the channel is significantly more complicated and often involves baseband demodulation along the way. MIMO channel inversion, sometimes alone and sometimes in combination with baseband demodulation, is known as MIMO decoding. Algorithms to perform channel decoding are covered in Chap. 9, and their implementation in hardware is the topic of Chap. 10.

Depuncturing (Sect. 6.4)

This is the counter to the puncturer in the transmitter. It is a bit-domain frequency-domain block that inserts dummy bits in place of stolen bits before channel decoding. Depuncturing allows us to use the same channel decoder for multiple coding rates.

Deinterleaving (Sect. 6.13)

This is the inverse of interleaving. It was discussed in Sect. 6.13. It is implemented in the same way as interleaving, specifically by using two port memories with one disordered address generator.

Channel Decoding (Chap. 6)

This is the process of determining the data bits from the coded bits. Coded bits introduce redundancy to allow error detection and correction at the receiver (Chap. 6).

Table 7.6 lists the receiver blocks. The first column lists corresponding transmitter blocks. Receiver blocks without a transmitter counterpart usually counter some sort of channel effect. Sometimes it is a little misleading to say we are trying to overcome channel effects, because we are really inverting receiver effects. This is the case, for example, with frequency offset estimation. And is packet edge detection really inverting the channel? In fact, receiver blocks should more accurately be seen as reversing transmitter processing or not.

Frequency Domain and Time Domain

All the blocks in Table 7.6 are classified as either time domain or frequency domain. This classification defines whether they act on samples in the time domain or the frequency domain. In the transmitter, everything that comes before the IFFT is in frequency domain. Blocks after the IFFT are in time domain. In the receiver, it is the opposite. Thus, blocks like the decoder or the demapper are in frequency domain. Blocks like packet edge detection are in time domain. Note that blocks and their counterparts in the transmitter and the receiver must be in the same domain. Thus, the interleaver and the deinterleaver are both frequency-domain blocks.

The FFT is defined as N/A in the table because it is neither a time-domain nor a frequency-domain block. It transforms between the two. Some blocks, specifically those involved in estimating channel effects, are defined as E in the table which stands for either. These blocks can be in the frequency domain or in the time domain. For example, algorithms exist that allow estimation of frequency offset using

Table 7.6 Receiver subsystems. The second column indicates the transmitter counterpart, if any. The third column marks subsystems as time domain or frequency domain. The fourth column classifies them into bit and symbol domain

Receiver block	Transmitter counterpart	T/F	S/B
CCA	N/A	F	S
AGC	N/A	F	S
Packet edge detection	N/A	F	S
Timing offset estimation	N/A	E	S
Timing offset correction	N/A	F	S
Frequency offset estimation	N/A	E	S
Frequency offset correction	N/A	F	S
FFT	IFFT	N/A	S
Channel estimation	N/A	E	S
Chanel inversion/MIMO decoding	N/A	T	S
Deparsing	Parsing	T	S
Demapping	Baseband modulation	T	N/A
Deinterleaver	Interleaver	T	B
Depuncture	Puncture	T	B
Decoder	Channel encoder	T	B

symbols before the FFT or symbols after. Thus, there are time-domain and frequency-domain possibilities for timing offset, frequency offset, and channel estimation.

Bit Domain and Symbol Domain

The table also classifies blocks as either symbol domain or bit domain. This defines whether they act on QAM symbols or on bits. The dividing line is obviously the demapper. All blocks after the demapper act on bits, and all blocks before act on QAM symbols. Notice that depending on the MIMO decoding algorithm, the demapper might be part of channel inversion.

The bit domain is entirely consumed with channel decoding and its accessories. Designers often address the bit domain as a single entity to optimize it and use synergies. In Figs. 7.30 and 7.33, clock rates are marked on some arrows. These rates will be explored in detail in Sect. 10.12. But if you assume that samples are coming in from the ADC at a rate of CLK, the rate will increase after the deparser as spatial streams are combined and will increase further after the demapper after symbols are broken down into bits.

Chapter 8
Algorithm to Hardware: FFT as a Case Study

8.1 Design Adaptation Flow

Figure 8.1 shows a flowchart for the adaptation of algorithms to hardware. This is *very* different from the design flow described in most hardware design books. In fact, the output of the design flow in Fig. 8.1 is the input to the traditional hardware design flow. An ASIC design flow, for example, starts from a block diagram and converts it to a netlist which is then placed and routed into a layout that can be taped out to a fabrication facility. The FPGA design flow is very similar but has a different final objective. The FPGA design flow should also ideally start from a block diagram, it also transforms it into a netlist, but the result is a programming file rather than a layout. Neither the ASIC nor the FPGA design flows are what we talk about here.

In the preface we discussed how hardware designers must focus on hardware design and how algorithm designers must focus on the algorithm. There is an area of knowledge missing in the middle; this is where an algorithm is transformed into a representation ready for the hardware flow. This is what we have been doing for most of this book. However, we have been doing so implicitly through examples. In this chapter, we will introduce a systematic and explicit approach to this process. But we will still use a specific case study.

In Chap. 7, we discussed OFDM and the DIT FFT algorithm in detail. The FFT has a regular implementation. It has a single processing unit, namely, the butterfly. The butterfly itself is a rather simple multiply-accumulate. The signal flow graph of the FFT is one-to-one with its direct hardware implementation, which is exactly why we will use the 16-point DIT FFT as a case study for this chapter.

Using a specific case allows us to make solid conclusions and to describe the concepts of this chapter in a concrete way. The mundane nature of processing and signal flow in FFT means that the conclusions we make are not constrained by any special case, and thus we can generalize them to most other systems and subsystems. It also means we can focus on more important concepts, especially as they relate to time-sharing and memory partitioning.

© The Author(s), under exclusive license to Springer Nature Switzerland AG 2023
K. Abbas, *From Algorithms to Hardware Architectures*,
https://doi.org/10.1007/978-3-031-08693-9_8

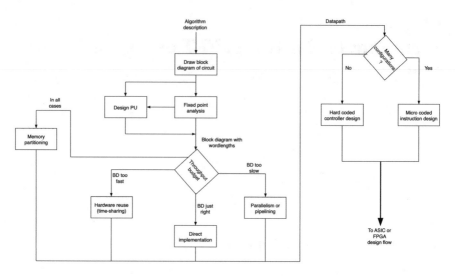

Fig. 8.1 Hardware adaptation flow

The flowchart in Fig. 8.1 describes what happens along the way from an algorithm to the start of hardware design. It is also an outline for this chapter and thus will be expanded upon for the remainder of the chapter. But we will describe it in broad strokes for the rest of this section.

The flow diagram starts with a description of the algorithm in a high-level floating-point convention. This could be from a simulation environment, from pseudocode, or even from an informal but solid description of the algorithm. This high-level description is the output of the algorithm design phase and should very firmly describe how the algorithm works.

We transform this high-level description into a block diagram. The block diagram isolates operations and describes how signals flow between them. These operations are translated into processing units, where each *type* of operation is a separate type of processing unit. This gets the algorithm closer to hardware by doing two things: isolating operations and describing how signals flow between these operations.

Operations in the block diagram are still floating-point; hardware will typically use fixed-point arithmetic. Thus, we perform a fixed-point simulation to determine the size of busses, registers, and arithmetic processing units. The block diagram can be directly transformed into a circuit. The PUs are designed in isolation and then as many units as there are in the BD are instantiated. Connections in the block diagram become wires in the circuit. This one-to-one implementation is called the direct implementation.

This direct implementation is rarely, if ever, an acceptable hardware implementation. Instead, we should carry out a throughput analysis where we calculate the throughput required by the application and the throughput that the PUs can provide. There are three possible outcomes from this analysis:

- The throughput required is exactly or slightly lower than that provided by the direct implementation. In such case, the ease of the direct implementation should promote it even if it provides slightly higher throughput than required.
- The throughput required is much higher than that the direct implementation can provide. Either parallelism or pipelining is required to cover the gap.
- The throughput required is much lower than that provided by direct implementation. Hardware reuse should be employed.

Whatever the decision, we now have a strategic decision on architecture. This allows us to determine storage requirements. We might need registers, memories, or a combination of the two. These will be needed to provide pipelining and to store intermediate values. Memory requirements and partitioning can become particularly challenging for hardware reuse architectures. The combination of processing units and associated storage forms what is called the datapath—the part of the circuit that performs processing.

The datapath needs an "intelligence" to manage it from cycle to cycle. This intelligence is the controller. Depending on the degree of flexibility required from the circuit, the controller could either be hardwired, or it could be programmable. The last part of the adaptation design flow in Fig. 8.1 is thus designing the controller; this involves all the details of the signals it exchanges with the datapath on a cycle-by-cycle basis.

The combination of the block diagram, fixed-point analysis, architectural decision, memory design, and controller design is the output of our design flow. These documents can be used to directly inform how the hardware description language of the circuit should be written and can thus kick off the hardware design flow.

8.2   Components of a Digital Architecture

In this section, we will discuss the building blocks of a typical digital system. These are macro-categories under which we classify our hardware units. For the rest of this chapter, we will use these building blocks to describe the hardware implementation of the 16-point FFT, our design example.

8.2.1   The Datapath

This is where processing takes place. The datapath is part of a paradigm where the design is split into a controller and a datapath according to Fig. 8.2. The datapath is processing-intensive and does most of what the circuit needs to do, but it is not intelligent. The controller does not do any processing, but it needs to be aware of what we are doing now and what we need to do next; thus, it is intelligent.

Fig. 8.2 Datapath and
controller exchanging
control and status signals

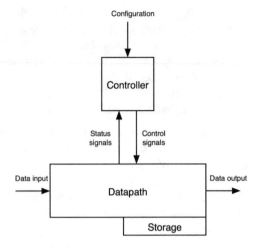

The datapath and the controller communicate as shown in Fig. 8.2. Selected datapath signals are given as inputs to the controller. We call these status signals. Status signals provide a complete picture of the current state of the datapath to the controller. This picture, combined with the awareness of the past state of the circuit, will allow the controller to determine what to do next.

The controller imposes this decision about what to do next by providing inputs to the datapath. We will call these inputs control signals. Control signals will either set the modes of the processing units (if they have multiple modes), reroute signals between them, or manage data storage.

8.2.2 Processing Units (PUs)

These are the bulk of the datapath where all the digital signal processing takes place. PUs are usually arithmetic or logic circuits. Arithmetic processing units tend to be the critical path of the circuit. When processing involves multiplication, the largest multiplier is often the critical path or part thereof. When multipliers are absent, the largest adder will dominate the critical path.

Other parts of the system (control and memory) are often simpler and faster than arithmetic. Thus, the most interesting design decisions focus on processing units. In some complicated algorithms (Chap. 10), figuring out how to implement the PU in hardware is a challenge. But even with simple PUs, as is the case in this chapter, there are many interesting decisions that need to be made. Examples of said decisions include the number of PUs, their connection, and scheduling their use.

8.2.3 Storage

This is the totality of registers and memories used in the design. If the design is pipelined, and most designs do contain some degree of pipelining, then pipeline registers account for some of the storage needs. If time-sharing is used, then intermediate storage is almost always necessary (Sect. 8.8). Most on-chip storage is in embedded SRAM blocks. The amount of memory used and its distribution among different banks are tasks that are very resistant to automation and thus require a lot of designer involvement. Pipeline registers will be very hard to separate from the datapath and are thus usually included in it. Memories on the other hand can and should be treated separately because they need special consideration and their layout is often different from that of random logic.

8.2.4 Flow Control

Processing units and storage form the datapath. The other part of the circuit is the controller. Flow control units are hardware pieces that bridge the controller to the datapath. They are the actuators of the controller, or its means to affect change into the datapath. Flow control units are usually multiplexers or registers with enable signals. Multiplexers help reroute signals so that the inputs of PUs or storage are programmable. Enabled registers allow the outputs of PUs to be stored in any of several locations. Multiplexers need someone to provide their select lines every cycle. Enabled registers also need someone to provide their enables. This someone is the controller. In very simple datapaths without the need for reconfigurability, flow control reduces to a trivial case of hard wiring.

8.2.5 Controller

The controller sets the pace of the circuit. It determines what we are doing now and what we should do next. The controller reads the situation by observing status signals from the datapath. It then issues control signals that reshape the datapath based on the current situation. Control signals will go to the PUs to reconfigure them, to flow control circuits to determine routing, and to memories to provide addresses and write enables.

There are two broad strategies for implementing controllers: either as hard-coded state machines or as programmable instruction memories. Hard-coded controllers are rigid but easy to implement, understand, and debug (Sect. 8.11). Programmable controllers blur the line between application specific datapaths and processors and can be challenging to implement (Sect. 8.12).

8.2.6   Address Generators

Address generation is part of the controller, but because it has a very specific nature, we need to discuss it separately. When memories are used in a design, particularly a time-shared design, the addresses to read from and write to are supplied every cycle. This is very design-specific and normally needs to be hand-designed based on the application.

The designer might decide to generate these memory controls using counters if the addresses follow a regular pattern, or they might want to fold the addresses into the main controller as one of the controller output (control) signals. Again, the decision is design-specific and must be made on a case-by-case basis.

8.3   16-Point FFT: The SFG as "the Algorithm"

Figure 8.3 shows the butterfly (signal flow) diagram of a 16-point decimation in time FFT. This will be our case study for the rest of this chapter, so we will briefly review what is going on in the diagram. For more details review Sect. 7.7.

The diagram is regular. It goes through four "stages" from left to right. The number four is $\log_2 16$. In every stage, we perform eight complex multiplications by twiddle factors. The twiddle factors are written on the arrows. Although the result of multiplication is added for half the samples and subtracted for the remaining half, we only need to perform the multiplication eight times every stage. Thus, there are

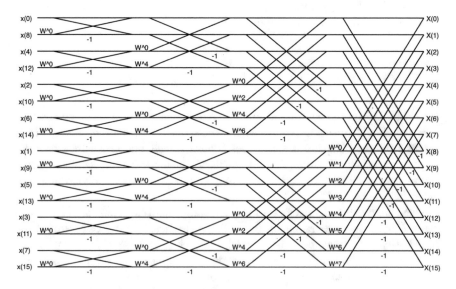

Fig. 8.3 Signal flow graph of 16-point DIT FFT

$8 * \log_2 16$ complex multipliers. At every intersection of branches, there is a complex addition. Thus, there are 16 additions for every stage for a total of $16*\log_2 16$ complex adders.

Inputs come in from the left-hand side in time domain and exit from the right-hand side in frequency domain. Samples are ordered in the frequency domain, but they are disordered in the time domain. The disordering comes from splitting indices into even and odd while going back at every stage. Bit reversal (Sect. 7.12) can be used to deduce the order of input indices.

8.4   Prepare Your Block Diagram and Isolate Processing Units

The first step in moving an algorithm to hardware is to draw its block diagram. This helps us determine the types and numbers of processing units and how routing will work in the circuit. It might even give some insight into the controller. As we will see in Sects. 8.5 through 8.14, we rarely rely on the block diagram to directly realize hardware, but it is an important reference point.

Algorithms are usually described in some high-level description method written by a systems designer. The systems designer is concerned with finding algorithms that perform well and have predictable complexity. They do not concern themselves with how these algorithms can be moved to hardware. For the FFT, we can think of the signal flow graph in Fig. 8.3 as a succinct description of the algorithm. In fact, for FFT the system description is unusually close to hardware because it shows how the signal is processed in parallel as it moves from left to right.

Figure 8.4 shows the signal flow graph in Fig. 8.3 translated into a block diagram. Each block in Fig. 8.4 is a butterfly from the SFG in Fig. 8.3. The internal construction of the butterfly is shown in Fig. 8.5. The butterfly accepts three inputs and produces two outputs. The three inputs are the two data points and a twiddle factor. The outputs are two data points consisting of the summation or subtraction of the first input and the second input multiplied by the twiddle. Specifically:

$$Output1 = Input1 + Input2 * Twiddle\_factor$$

$$Output2 = Input1 - Input2 * Twiddle\_factor$$

Figure 8.5 is thus the processing unit (PU) we will be using in our design. Determining the processing units you need is a very important design step. Most systems require several different types of PUs. For example, in Chap. 6 we needed PMUs, BMUs, and a traceback unit. In Chap. 10, we will use two very complicated processing units. For the FFT, there is only one PU.

One step that should take some focus is optimizing the hardware design of the PU. Because most PUs are repeated many times, savings in speed or area will add up in the final circuit. For example, in the PU in Fig. 8.5, we decided to use two adders to add and subtract the two outputs. But because adders are often much faster than

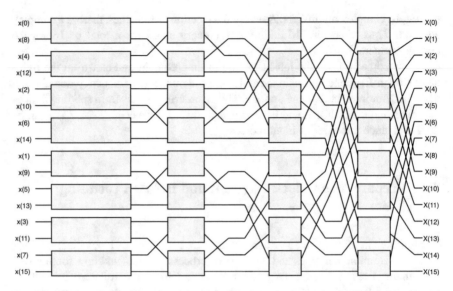

Fig. 8.4 Block diagram of 16-point FFT. Each block is a butterfly from Fig. 8.3. The internal construction of the block is shown in Fig. 8.5

Fig. 8.5 Internal construction of the butterfly, the PU of our circuit

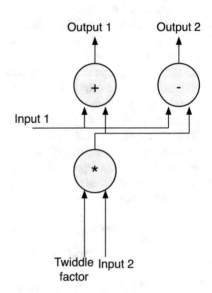

multipliers, perhaps we could have reused the same adder to perform the two. On the other hand, should we internally pipeline between the multiplier and adders in Fig. 8.5? These questions are not trivial and require a full understanding of the complexity of processing and control in the overall circuit. We will discuss these system-level issues in detail, but for now be aware that the implementation of the PU is a first-tier priority because it propagates through the rest of the design.

The block diagram in Fig. 8.4 shows the PUs connected to form a 16-point FFT. The block diagram can be used as a guide to wire PUs when writing the hardware description language of the circuit. The diagram is missing the twiddle inputs, which can be deduced from the signal flow graph in Fig. 8.3.

Figure 8.4 is directly related to the SFG even though it might be a little hard to trace at first. You can see the pattern after observing a few PUs. The top left butterfly, for example, acts on input 0 and input 8 to produce output 0 and output 8. Thus, in Fig. 8.4, we connect the top-right PU to these inputs and outputs. In the second column of butterflies, they act on inputs and outputs that are more separated than the first column. This is reflected in the block diagram, leading to more skewed wires the deeper into the SFG we go. The effort invested into drawing the diagram in Fig. 8.4 will save a lot of time in wiring decisions later.

8.5   Perform Fixed-Point Analysis and Determine Bus Sizes

The block diagram in Fig. 8.4 and the PU in Fig. 8.4 are missing one thing: width of busses. The signal flow graph assumes a floating-point implementation of the FFT. This is generally the case for any algorithm when you see it the first time. Systems designers need an easy-to-use programming environment for prototyping; this allows them to quickly make changes and test their impact. These are usually environments that run on a general-purpose processor and use floating-point registers.

Hardware registers are of limited size and arithmetic done in hardware is always fixed-point. Fixed-point arithmetic is fast and power efficient. Thus, the multiplier and adders in Fig. 8.5 are fixed-point and assume fixed-point input and output registers.

So, we must move the design from floating-point to fixed-point. This means choosing the sizes of all registers and in turn determining the size of adders and multipliers. We cannot choose adders and multipliers to be arbitrarily large. In Sect. 3.9, we saw that the delay of both increases linearly with wordlength. The area of adders increases linearly, while the area of multipliers increases quadratically. Thus, arbitrarily large arithmetic is inefficient. On the other hand, choosing adders and multipliers that are too small will lead to very high quantization noise in the output. This shows up as a shift in the SNR-BER curve of the transceiver. Thus, the choice of the size of busses is important.

Finishing the fixed-point design involves three things:

- Determining the wordlength of inputs to the circuit. For the FFT, these would be the time-domain samples.
- Determining the size of arithmetic units in the circuit.
- Using these two to determine the wordlength of the outputs from the circuit. For the FFT, these would be the frequency-domain samples.

Usually, the wordlength of the inputs is already a given design constraint. It will be dictated from the output wordlength of the previous circuit in the overall system. For example, the inputs of the FFT in 802.11n come from the offset correction blocks according to Fig. 7.30, and thus that block's output wordlength is our input wordlength. Ultimately, the receiver PHY inputs come from the ADC, which dictates the global wordlength of the system as a whole.

Less often, the design is also constrained by the width of its outputs. These are cases where the input to the next block needs to be a certain width; thus, the current block must somehow maintain this resolution. This is not the case for the FFT as used in a transceiver. For the FFT, we try to maintain the best performance while using the smallest possible PUs, and whatever the output wordlength we end up with would be the input wordlength to the next block.

Chapter 3 discusses how fixed-point analysis can be performed in detail. But the main question is how to determine that a fixed-point performance is "good." The only fair metric is the SNR-BER curve or rather the degradation in its performance.

For example, assume we simulate a transceiver with a floating-point FFT and plot its BER-SNR curve. We then model the FFT using fixed-point arithmetic and choose 16-bit butterflies. When we draw the BER-SNR curve, we see a hit of 0.5 dB. This means that to achieve the same BER as the floating-point curve, the fixed-point curve needs 0.5 dB of extra SNR. Is this "good"? There is no easy way to tell. It depends on your quality-of-service requirements, how you budgeted loss of SNR in the first place, and how major blocks in the rest of the system impact BER through quantization noise.

In short, deciding the size of PUs requires experience and trial and error. But it also requires a global view of the system. Determining the fixed-point performance of a block in isolation is always hampered by the lack of meaningful metrics and by the complexity of its interaction with other subsystems. There is no way around having a flexible and efficient simulation that we can rerun at will.

But even if we assume the block is designed in isolation, making fixed-point decisions can be challenging. For example, by examining Figs. 8.3 and 8.4, we notice that the MACs (butterflies) are arranged in stages. If we assume that the MACs in the first stage contain 12-bit multipliers, their noiseless outputs will be 24-bit (Sect. 3.4). This requires 24-bit multipliers in the second stage, 48-bit multipliers in the third stage, and 96-bit multipliers in the final stage. This is obviously untenable, the outputs of each stage of MACs will thus have to be concatenated to allow subsequent multipliers to be of reasonable size. For example, if the output of the first stage is truncated down to 16 bits, then the second stage MACs will be 16 bit instead of 24. The decision about the amount of truncation will be made based on the fixed-point simulation and how much performance deviates from floating point.

One impact that this has on hardware design is that you must make the wordlength of your PUs a parameter. Thus, even though the block diagram in Fig. 8.4 suggests that all PUs are identical throughout the FFT, we know that PUs in different stages will have different-sized multipliers and adders. To allow hierarchical design without using more PU variants than necessary, the PU should have the wordlength as a parameter (also called a generic in VHDL) until they are instantiated in the actual design.

8.6   Try Out (and Immediately Hate) the Direct Implementation

The easiest way to implement a design is to take its block diagram and realize it so that every block in the diagram has a corresponding PU in hardware. The result is a fully parallel expanded design that we will call a direct implementation. The direct implementation is easy to implement, interpret, and debug. Its controller is usually trivial. The direct design is also very efficient, with PUs doing something useful most of the time. On the negative side, direct implementations often produce much higher throughput than needed, especially when pipelined. This means it uses too many resources and dissipates more power than needed.

The block diagram in Fig. 8.4 is itself the direct hardware implementation. This is especially true when combined with the fixed-point analysis from Sect. 8.5. You can use a hardware description language or a schematic capture method to directly implement the block diagram. The PU will be described only once and instantiated as many times as necessary. Connections between PUs can be seen in Fig. 8.4, which is why the block diagram is a good reference to have around. Twiddle factors are visible on the signal flow graph in Fig. 8.3.

From Fig. 8.4, we see there are 8 PUs for every stage. There are four stages for a total of 32 PUs. From Fig. 8.5, each PU is a single complex multiplier and two complex adders. Thus, there are $8*4 = 32$ complex multipliers and 64 complex adders. This is consistent with the complexity of the DIT FFT algorithm in Sect. 7.7. However, we will shortly see that this consistency is not always necessary or efficient.

For starters, Fig. 8.3 shows that many of the twiddles are trivial, and thus many of the multipliers are unnecessary. This is particularly true for null twiddles, where:

$$W_N^0 = e^0 = 1$$

This is not a small saving. In Fig. 8.4, 15 of the 32 butterflies are "special." They do not perform a multiplication, or rather they multiply by unity. This makes them trivial in hardware. In general, the last stage of any DIT SFG will contain one such twiddle, increasing by a factor of 2 moving back till the first stage is all null twiddles. Thus, there are $1 + 2 + 4 + \ldots + N/2$ null twiddles, which in this case is $1 + 2 + 4 + 8 = 15$.

The above discussion can be generalized to hardware implementations beyond FFT, where certain optimizations can be made to the design once it has been fully realized. This usually involves looking for constants in the block diagram and realizing which of these constants introduce a special case to the arithmetic in the PUs.

Because the twiddle factor is a sinusoid, we should investigate if any other twiddles in Fig. 8.3 also introduce special cases like the null twiddles. For example, consider the $N/2$ twiddle:

$$W_N^{N/2} = e^{-2\pi jN/2N} = e^{-j\pi} = -1$$

Unfortunately, Fig. 8.3 shows that no $N/2$ twiddles are ever used. This is also true for any size FFT. On the other hand, $W_N^{N/4}$ twiddles do exist, which can also simplify to something useful:

$$W_N^{N/4} = e^{-2\pi jN/4N} = e^{-j\pi/2} = -j$$

There are seven such twiddles in Fig. 8.3; thus, 22 total butterflies have now been reduced into trivial multiplications, leaving only 10 butterflies with actual (parallel) multipliers.

But if we take this a step further, even these 10 butterflies do not contain actual multipliers. The discussion in Sect. 3.4 distinguished between full multipliers with two variable inputs and constant multipliers where only one input is a variable. All the butterflies in Figs. 8.3 and 8.4 have constant twiddles and thus have one fixed input to the multiplier. These constant multipliers are often much cheaper to implement than general multipliers.

The above simplifications are all a result of us going forward with a direct unrolled implementation of the block diagram. This allows every PU to perform only one task and thus allows us to define some of its inputs as trivial or constant. When we start to reuse PUs in Sect. 8.8, we will lose this advantage.

Once we determine what the hardware looks like, we can estimate its performance. Table 8.1 lists performance parameters for the three arithmetic components used to build the FFT. These figures are arbitrary but representative and do not reflect any technology. Size (area) is measured in the number of logic gates used. Delay is measured in ns, and power is measured in μW. The table lists the variable multiplier even though we will not use it in this section, but we will need it once we start to do time-sharing.

Registers are also included to be used in Sect. 8.6. We list the delay of registers as null. This is obviously a simplification. It is true that register delay is much lower than that of arithmetic, but it is not null. However, as we will find out in Sect. 8.6, register delay is added only a single time to any critical path calculation that we make. Thus, it represents a common floor for all our delay calculations and should not have a significant impact on apple-to-apple comparisons in upcoming sections. We do have to take register area and power into consideration when we look at pipelining.

Table 8.1 Size, delay, and power of adder, multiplier, and register

Block	Size (gates)	Delay (ns)	Power (μW)
Variable multiplier	3000	3	10
Adder	500	1	2
Constant multiplier	2000	2	4
Register	100	0	1

Power figures are for a cycle of 10 ns

We can try to find resource and performance estimates for the circuit in Fig. 8.4 by counting the number of PUs and determining performance per PU. But not all PUs in Fig. 8.4 are equal. Twenty-two of the 32 PUs do not contain a multiplier, or to be more accurate they do a trivial multiplication. The remaining 10 will contain constant multipliers. Each of the 32 PUs will contain 2 complex adders. Thus, our total budget is 10 constant multipliers and 64 adders. This is an area of $2000 * 10 + 500 * 64 = 52$ kgates. If we assume that power figures in Table 8.1 are averages, then power dissipation in the FFT will be $4 * 10 + 2 * 64 = 168$ μW.

Determining how fast the FFT can operate is a bit more challenging. We must look at Fig. 8.4 and imagine that all the time-domain samples become available at the same time then figure out when the last of the outputs in the frequency domain becomes available. The path from the inputs to this latest output is called the critical path, and it determines the frequency of operation of the overall circuit.

In the FFT, in every stage each path must pass through an adder. In every stage, there is also a constant multiplier, except for the first stage where all multipliers are trivial. Note that in all stages, there are trivial multipliers, but there are also non-trivial multipliers which determine the delay of that stage. Thus, there are multiple critical paths. Every critical path goes through an adder in the first stage and a multiplier and adder in every stage thereafter.

Thus, any critical path will have four adders and three complex multipliers between the time-domain and the frequency-domain outputs. This is a total delay of $2*3 + 4*1 = 10$ ns, which means the FFT can be operated at a 100 MHz clock and will produce outputs at such a throughput.

This throughput is too high. In 802.11n, inputs to the FFT are provided at a rate of 20 MHz or 40 MHz. The fully unfurled design we get from directly implementing the block diagram in Fig. 8.4 expects a full frame of 16 samples to be available every 100 ns; this means samples must be coming in at a rate of 1.6 GHz.

This is a major problem with direct implementations. The rate we get from the direct circuit is always either too high or too low. We must devise ways to deal with this by trading off area and speed. In Sect. 8.7, we will deal with the case where the rate is lower than we need. In the rest of the chapter, we will deal with the more common case where the direct circuit is faster and bigger than we need.

8.7 What Can You Do If the Direct Implementation Has a Throughput Deficit: Parallelism and Pipelining

What if the direct implementation does not give us a sufficient throughput? In Fig. 8.4, the implementation produces a frame at a rate of 100 MSps. If the sample is a single frequency-domain output rather than a full frame of 16 samples, then the implementation provides outputs at a rate of 1.6 GSps. We will obviously not need outputs at a higher rate in any realistic radio. In some other systems, the direct implementation provides a lower throughput than needed. This is often the case for example with video processing hardware. We will assume that the 1.6 GSps provided by the FFT is not enough, just to probe the options available, knowing well that we are doing this only for the sake of generalizing to other applications.

Our first option to obtain higher throughput is parallelism. This was considered in detail in Sect. 3.8. Hardware implementations get their efficiency relative to software from parallelism. In fact, the very direct implementation in Fig. 8.4 is highly parallel. But by parallelism in this section, we do not mean using parallel PUs to implement the algorithm as in Sect. 8.5. We mean parallel instances of the block diagram in Fig. 8.4 as a whole. In other words, if we want to double the throughput of Fig. 8.4, the easiest solution is to have two copies of Fig. 8.4 operating in parallel as in Sect. 3.8.

Parallelism increases throughput by as many times as we repeat the subsystem. It also increases the area and power dissipation by the same ratio. It is a very direct and logical trade-off without much additional overhead. In Sect. 8.14, we will see some neat tricks we can play on parallelism to break this tight correlation of throughput, area, and power.

The main challenge in parallelism is being able to provide the inputs at a fast enough rate. Notice that the implementation in Sect. 8.5 already required inputs at an astounding rate of 1.6 GHz. Any degree of parallelism will increase the rate at which the subsystem consumes data by the same order. Thus, parallelism of order 2 requires inputs at a rate of 3.2 GHz which will be deparsed on a cycle-by-cycle basis between the two FFT circuit instances, giving each its inputs at an effective rate of 1.6 GHz. The two instances produce outputs at a rate of 1.6 GHz which are parsed into a single stream at 3.2 GHz or two parallel streams at a rate of 1.6 GHz.

The other option to increase throughput above that of the direct implementation is pipelining. Pipelining is performed by inserting registers in the critical path to break it. In Sect. 8.6 we traced the critical path of the direct implementation. This tracing assumed the implicit presence of registers at all the inputs and outputs (x's and X's). In fact, this is how we define a path: a path is any combinational logic between two consecutive registers. This is a neat little trick you need to use when calculating the critical path of a circuit: always assume the presence registers at all the inputs and outputs.

In Fig. 8.6, we internally pipeline the direct implementation from Fig. 8.4. registers marked as circles. The input/output registers (circles) at $x(0)$ to $x(15)$ and $X(0)$ to $X(15)$ were also implicitly present in the direct implementation. The difference in the "pipelined" design in Fig. 8.6 is that registers are inserted in internal nodes. Figure 8.6 shows "full pipelining" where pipeline registers are inserted at all possible internal node, which in this case is at the output of every stage (column) of PUs. These registers break the critical path into shorter paths. We must introduce registers even in non-critical paths to ensure that data arrives at different PUs in the right cycle.

The critical path in Fig. 8.6 is now a single PU. Thus, the length of the critical path is $2 + 1 = 3$ ns. This is roughly a threefold improvement over the direct implementation. However, we have introduced additional area and power to account for the internal registers. Table 8.2 compares the unpipelined (direct) and pipelined designs.

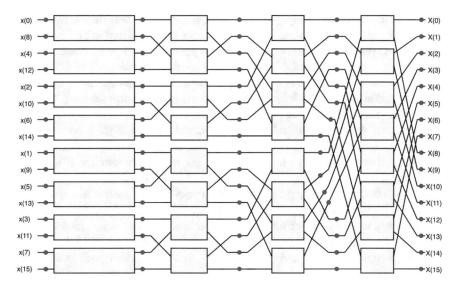

Fig. 8.6 Block diagram with inter-stage pipelining

Table 8.2 Comparison of performance metrics for a direct unpipelined and pipelined implementation of 16-point DIT FFT

Metric	Unpipelined	Pipelined
Critical path delay (ns)	10	3
Combinational area (kgates)	52	52
Register area (kgates)	3.2	8
Total area (kgates)	55.2	60
Combinational power	168	560
Register power	32	267
Total power	200	827

Pipelining obviously improves throughput. It has a minor negative impact on area. Notice that curiously there is a register area for the unpipelined design. This takes into consideration the fact that even the unpipelined design has input and output registers. Thus, the first and final column of registers in Fig. 8.6 should also be counted for Fig. 8.4.

Power needs a little bit of explaining. In Sect. 8.14, we will find that power dissipation increases proportionate to operating frequency. Thus, the power dissipation figures for all units in Table 8.1 must be scaled up by the ratio of improvement in clock frequency when calculating power dissipation for Fig. 8.6. The figures listed in Table 8.1 are for a cycle of 10 ns, so they need to be scaled up by 10/3 for the pipelined design. For example, register power dissipation is 10/3 µW instead of 1 µW.

Pipelining improves throughput by raising the clock frequency. It leads to an increase in latency when counted in number of cycles. However, latency in absolute time units might not change because the clock period drops relative to the unpipelined case. Area increases only slightly to account for the additional registers, but power dissipation shoots up mainly due to the improvement in clock frequency.

The main challenge in pipelining is to align signals so that they arrive at the proper point in the proper cycle. This is not a major challenge for the FFT because it has balanced paths. But subsystems which have paths with different number of units will require careful insertion of shift registers to align signals properly. This is a general theme we will find in upcoming sections. Any departure from the direct implementation in Sect. 8.5 leads to complications in the control scheme.

Another question regarding pipelining is how far to take it. In other words, how deeply should we cut the critical path. Why should we stop at the level we did in Fig. 8.6? Why not go into the PU in Fig. 8.5 and break it down? PU delay is 3 ns consisting of one adder delay in cascade with one constant multiplier. If we break down inside the PU by inserting registers between the adders and the multiplier, the critical path delay will become 2 ns, accounting only for the constant multiplier. The number of registers will roughly double, and latency (in cycles) will also double.

So, should we or should we not do this? There is no one-size-fits-all answer to this question. It will always depend on the requirements of the design and tolerance for latency, throughput, power, and area. However, when PUs are as small as the MAC unit in the FFT, it is rare for designers to need to pipeline inside the PU. The gain in throughput is often unnecessary, and the cost in control, power, and latency is often prohibitive.

One might wonder if it is possible to go inside the arithmetic units and pipeline them. In Sects. 3.3 and 3.4, we saw the internal construction of adders and multipliers. Should we consider pipelining between the half adders and full adders that make up the N-bit adders and multipliers of the PUs? This is never done. You should *not* try to pipeline *inside* adders and multipliers. Pipelining at this level makes latency intolerable and control impossible. But also, N-bit adders and multipliers are implemented in ASICs and FPGAs as standard cells from standard cell libraries. These implementations are optimized and efficient, and you should not play around with them.

8.8 Calculating Needed Throughput: Application and Implementation

In Fig. 8.1 we first did something called a "throughput budget" before deciding the architecture of the circuit. In this step, we use the required throughput and the speed of processing units to determine the number of needed PUs. This will determine the architecture of the circuit and will propagate into memory partitioning and controller design. This step requires information about the position of the design in the overall system; this will determine the required throughput. It also requires familiarity with the implementation platform, which will determine the operating speed of the PUs.

To keep going forward, it helps to have a specific example. We will use the 20 MHz mode of 802.11n. In this mode, the FFT produces outputs at a rate of 20 MSps. We will assume a 16-point FFT. We know that this mode does not exist in

802.11n (20 MHz uses 64 subcarriers), but it allows us to keep using the hardware from Sect. 8.5. We will also use the delay values obtained from table. Our processing unit is a MAC with a delay of 2 + 1 ns = 3 ns. Thus, our givens are:

- System requirements: FFT at the rate of 20 MSps.
- PU speed: MACs operating at 3 ns.

To clarify, the sample in the 20 MSps is a single frequency-domain sample, not an entire OFDM symbol. The throughput budget thus proceeds as follows:

- The time available to produce one frequency-domain sample is 1/20 = 50 ns.
- When the 16-point FFT in the direct implementation finishes calculation, 16 frequency-domain samples are available.
- We can take 16*50 ns = 800 ns to finish the 16 samples or in other words to finish the entire processing done by the direct implementation.

The hardware budget proceeds as follows:

- To finish 16 frequency-domain samples, we need 4*8 = 32 MAC operations.
- The MAC PU has a period of 3 ns.
- If we use 32 MACs, we might be able to get 16 samples in 3 ns (as in full pipelining).

But do we need to? The answer is obviously not. It is obvious from the above that our circuit is much faster than it needs to be, but how much faster exactly?

- To finish 16 samples, we need 32 MAC operations.
- To finish 16 samples, we have 800 ns available.
- Thus, each MAC operation could take 800/32 = 25 ns to finish.
- But each MAC operation only takes 3 ns to finish!

The above discussion assumes that all 32 MAC operations are performed in sequence. The SFG clearly shows many of them happen in parallel; thus, this discussion is conservative. But even with this conservative caveat, we can clearly see that our PUs are 25/3 times faster than they need to be. This leads to the concept of time-sharing or hardware reuse where processing units are reused to perform multiple operations. This theory of hardware reuse was discussed in Sect. 3.8. In upcoming sections, we will see how this can be applied to the above case study and how hardware reuse impacts storage and control.

8.9   Time-Sharing Processing Units

From Sect. 8.8, we conclude that we have plenty of reserve throughput that allows us to time-share our PUs. The level of time-sharing available is 25/3. Thus, we can use one eighth of the 32 PUs. However, we will shortly find out that the deeper the order of time-sharing, the more complicated control becomes. So, it is often more useful to back off from the maximum possible order of reuse to an order that makes sense in terms of the function of the circuit.

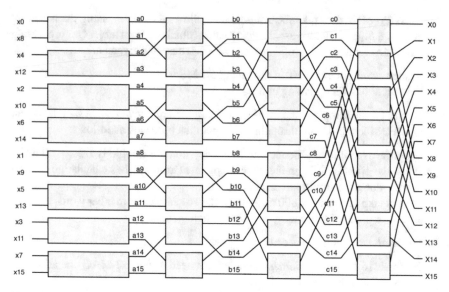

Fig. 8.7 Block diagram with internal signals marked

With block diagrams like Fig. 8.4, our first impulse should be to use one stage (column of butterflies) and reuse it four times. Figure 8.4 is replicated in Fig. 8.7 with internal signals marked. This labeling will be very important when we discuss memory and controller design. Figure 8.8 shows a single-stage implementation of the FFT. We will use eight PUs (shown as shaded blocks) and a single memory bank to keep intermediate results. Figure 8.8 has a hardware reuse order of 4. In the previous section, we saw that a reuse order of 8 was safe; thus, an order of 4 should be more than adequate. But let us check.

The column of PUs in Fig. 8.8 will be used over four cycles to produce a single frame of 16 frequency-domain samples. This means that every 4 cycles, we get 16 outputs. But what is the cycle time? The critical path in Fig. 8.8 is a single PU, which according to Sect. 8.6 is 3 ns. However, this assumes that the PUs used in Fig. 8.8 are identical to those used in Fig. 8.4.

In the direct implementation in Sect. 8.6, we found that none of the PUs have full multipliers. Many of the PUs contained trivial multiplications, and those that did not had constant multipliers. In Fig. 8.8, the PUs are reused multiple times. This means that each cycle, they multiply by a different twiddle factor. Thus, none of the multipliers in Fig. 8.8 are trivial, and none of them are constant multipliers. So, the PU in Fig. 8.8 contains a full multiplier and two complex adders.

Our PU now has a critical path of 4 ns instead of 3 ns. Over 4 cycles, it takes $4*4 = 16$ ns to produce 16 output samples. We are required to produce the 16 samples in 50 ns (Sect. 8.8); thus, we are safe. The main challenge in Fig. 8.8 is to ensure that the memory can produce and accommodate data at the rate required by the bank of PUs.

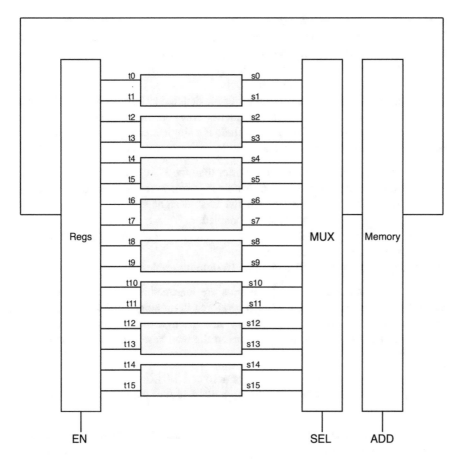

Fig. 8.8 Single-stage time-sharing. Shaded blocks are MAC PUs

Figure 8.8 shows PUs with two inputs and two outputs. But according to Fig. 8.5, each PU actually has three inputs. The third input is the twiddle factor. In the direct and pipelined implementations, these inputs could be assumed hardwired because each PU had only one input. In the time-shared implementation, the twiddle factor will vary on a per-cycle basis depending on which column of PUs the bank in Fig. 8.8 is currently implementing. We will ignore twiddle inputs for the rest of this discussion. This will facilitate the discussion because it allows an even number of inputs and outputs to be needed from each PU. Just be aware that if the controller is not somehow toggling the twiddle factor inputs, then they too must be read every cycle from the memory.

Figure 8.8 contains a single memory bank. The memory is assumed to be two-port, although the conclusions we make can be easily extended to the single-port case. The ADD bus of the memory is actually two address busses, one for each of the ports. The port on the right-hand side of the memory is dedicated to reading. The bus on its left-hand side is dedicated to writing. The two ports can be accessed independently and simultaneously, meaning we can perform one read operation and one write operation per memory cycle. In a single-port memory, we would be able to perform *either* a read or a write every cycle.

The read port of the memory is a register file denoted as Regs in Fig. 8.8. This is a set of 16 registers, each with an independent (write) enable bit. Each register feeds 1 of the 16 inputs of the 8 PUs. Because there is a single memory but there are 16 PU inputs to be read every cycle, demand on the memory must be calculated:

- The memory cycle must be much shorter than the PU cycle.
- We clock the memory 16 times in a single PU cycle to read the required 16 inputs.
- Simultaneously, we cycle through the enable signals of the Regs. For each *memory cycle*, a different register is enabled.
- After 16 memory cycles, the registers contain a full time-domain frame of 16 inputs. We can then clock the Pus.
- One PU cycle later, the PUs produce 16 output samples.

After the first PU cycle, the 16 outputs are intermediate signals. These are a0 through a15 from Fig. 8.7. These must be stored in memory to be read as inputs in the next PU cycle. Again, the memory has a single write port, but there are 16 samples to write, which means that as in the read case, we must overclock the memory:

- The 16 outputs are fed to the inputs of a 16×1 MUX.
- For 16 memory cycles, we cycle through the values of the SEL bus and write the value each time.

The above sequences are repeated three more times. Thus, we cycle through the memory reading a0 through a15 for 16 cycles. Then one PU cycle later, we get b0 through b15. These must then be stored in memory to be processed in the following cycle. At the end of the fourth PU cycle, the outputs of the PUs are the 16 frequency-domain samples.

In the above sequence, we repeatedly used the terms "memory cycle" and "PU cycle." It will take 16 "cycles" to read the 16 inputs and 16 "cycles" to write the 16 outputs. But what do we mean by "cycle" here? We cannot possibly mean the same thing when we talk about PU cycles as when we talk about memory cycles. It would be prohibitive for the PU to wait 32 of its own cycles for inputs and outputs to be read and written. This would mean that even if we are clocking the PU at a period of 4 ns, it is effectively working at 1/33 this frequency.

Instead, we operate the memory faster than the PU, allowing the memory to be read and/or written multiple times for every single PU cycle. This is possible because memories can normally be operated at multiples the clock frequency of PUs. Note that while using multiple clocks in a single design can be very challenging, using multiple clocks that have integer multiple frequencies is a special and easy case.

In the example in Fig. 8.8, we need the memory to operate 32 times for every PU cycle. This is alleviated by the fact that the memory is two-port because we can read and write simultaneously from memory. Thus, memory needs to be clocked 16 times per PU cycle. The PU operates at 250 MHz, which requires the memory to operate at 16 times the rate, specifically 4 GHz. While memories can usually be clocked fast, 4 GHz is not achievable and there will be contention at the memory ports. This is not a special problem with the FFT; it is a general issue that plagues most hardware reuse architectures.

Another problem with the setup in Fig. 8.8 is where the control signals come from. The control signals are, namely, the SEL lines for the 16×1 MUX and the EN signals for the 16 input registers as well as the addresses for the two ports of the memory. This is a very large set of signals that will change every memory cycle.

In Sect. 8.10, we will address the memory contention problem. In Sects. 8.12 and 8.13, we will address how the controller generates control signals for time-sharing. But first let us assess the time-shared approach in terms of area, power, and throughput.

The PUs occupy $(3000 + 1000) * 8 = 32$ kgates. Assuming the memory and register files occupy identical areas and the MUX is negligible, memory/register area is $100 * 16 * 2 = 3.2$ kgates. The total area is thus 35.2 kgates. This is as compared to 55.2 kgates for the direct design. But wait, Fig. 8.8 has one fourth of the PUs in Fig. 8.7. We should have seen the area drop to nearly 25%; it barely dropped to half. The reason is the PUs in Fig. 8.8 use full multipliers, while those in Fig. 8.7 used constant multipliers. The general lesson here is that one of the few things the direct implementation has going for it is the ability to optimize individual PUs.

Power dissipation is $(8 * (10 + 4) + 16 * 2 * 1) * 10/4 = 360$ µW. We are scaling the power to the clock frequency of the single PU column. This power dissipation is much higher than direct design despite the smaller area, mainly due to the higher clock. This power figure is optimistic because it assumes the memory is clocked at the same rate as the PU. Memory power will be much higher than above.

To estimate delay, we first assume full pipelining between the PUs and the memory-MUX-Regs combo. This allows us to read the input vector for the next processing cycle while the current processing cycle is proceeding. This is different from the row of Regs in Fig. 8.8, because Regs are responsible for holding the values constant at the input of the PU column for the duration of the PU cycle. The pipeline registers mean that after an initial latency, the PU is never idle. Thus, we need to pass through only 4 PU cycles to produce 16 outputs. This means that throughput is $16/(4 * 4)$ GSps = 1 GSps. This is as opposed to a throughput of 5.3 Gsps in the pipelined design and 1.6 Gsps in the direct implementation.

So, which is better? The pipelined design provides higher throughput at a higher power. The direct design provides a slightly higher throughput at a somehow larger area and a lower power. Obviously, there must be a unified metric to allow us to figure out which is "better." We will discuss this further in Sect. 8.14. But first, we must address the elephants of memory contention and controller design.

8.10   Resolving Memory Bandwidth Limitations

The 4 GHz bandwidth required from the memory in Sect. 8.9 is too high. Memory contention is a very common problem with hardware reuse. We will always end up using multiple PUs. These PUs could potentially have multiple inputs and outputs. Because the inputs to the PUs are their outputs in a previous cycle, the data must all go to and come from a single memory. This will always require the memory to be clocked at a significantly higher rate than the PU.

A common way to address this problem is to split the memory into multiple banks. The individual banks have independent ports and address busses and thus can be accessed in parallel. But memory bank splitting comes with its own set of headaches. There are two prices we pay for memory bank splitting:

- Control is more complicated than using a single bank. We provide multiple addresses and write enables every cycle instead of a single set of memory controls.
- Partitioning cannot be done automatically. To split memory into two banks, you do not randomly assign data to either bank. You must keep track of which results go to which bank every cycle, or you will still have contention.

Assume in our example that memories can only be clocked at 1 GHz. Thus, we cannot use a single memory and clock it at 4 GHz. Instead, if we use four memory banks in parallel, we can effectively read from/write to them at a rate of 4 GHz. So, the contention problem in Sect. 8.9 can be solved by partitioning the memory into four banks. Each bank has two ports, which means that 16 reads and 16 writes can be performed over only four memory cycles.

In Fig. 8.8, we used a 16 × 1 multiplexer to connect the PU outputs to the memory. Figure 8.9 shows this approach adapted to multiple memory banks. There are four 16 × 1 multiplexers, one each for the four memory banks. The advantage of this approach is that it is very flexible. It allows every PU to write into every memory bank. This avoids the contention that could happen with hardwiring. The disadvantage is that the area of the multiplexers can start to become significant. But more importantly, this kicks complexity down the road to the controller. The controller will now need to provide four sets of select lines for each of 4 bits instead of one set in Sect. 8.9.

The other option is to restrict which PUs can write to which banks. This will reduce the hardware requirements of the multiplexers and will relieve the controller. However, it requires very careful assignment of data to avoid contention. To understand why, let us take a particular example. Assume the assignment in Fig. 8.10 is true, meaning that PU0 can only accept inputs from the topmost memory bank. This means that in the first step, inputs x0, x8, x4, and x12 must all be in bank 0. But where do we write the outputs a0 through a3? By inspecting Fig. 8.9, we conclude that we should write them to bank 0 because the topmost PUs will still act on them in the second step. b0 through b3 start to become more challenging; b0 and

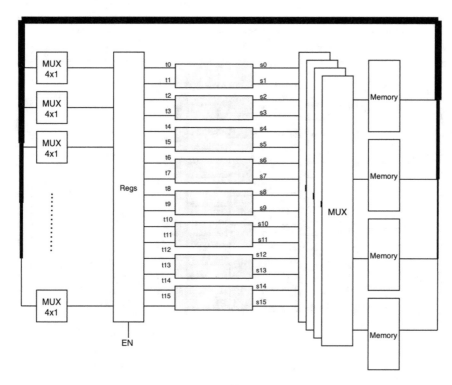

Fig. 8.9 Fully flexible assignment using 16×1 MUXes

b1 can safely be written to bank 0 because they are still read in PU0 and PU1 in the third step. But b2 and b3 are processed in PU2 and PU3, so they must be written to bank 1. Can we write them to bank 1?

By inspecting Figs. 8.9 and 8.10, we find that there are two strategies to reading as was the case for writing. In Fig. 8.9, each PU may accept an input from one of four registers. These registers are fed from the four memory banks. Because the PU input needs to be single, these four inputs are then multiplexed. Thus, we will need 16 4 × 1 MUXes with their associated select lines as well as 16 registers with their associated enables. Again, we are kicking all this complexity down the road to the controller.

The other option according to Fig. 8.10 is to insist that every PU can only read from a single memory bank. In this case, we will not need multiplexers and will only need 16 enabled registers, reducing to the original case in Fig. 8.9. On the other hand, we run the risk of contention. In fact, Fig. 8.10 is incapable of handling the FFT in Fig. 8.3. This is because once it reaches the third step, it will need to read b2 and b3 as inputs to PU3, but PU3 is only capable of reading from bank 1, while b2 and b3 are outputs from PU1, which was only capable of saving in bank 0. So again, we run into contention and must rethink our partitioning and assignment.

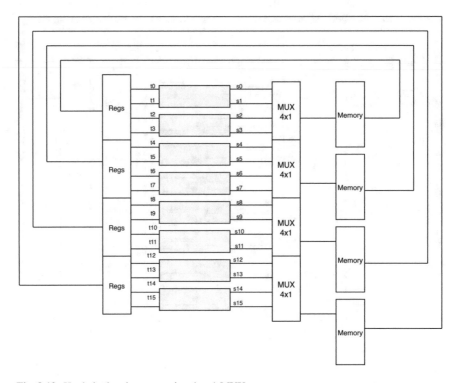

Fig. 8.10 Hardwired assignment using 4×1 MUXes

The answer does not have to be either of the extremes of Figs. 8.9 and 8.10. We can introduce some degree of flexibility, allowing some PUs to read from or write to multiple banks. Or perhaps we can get away with introducing flexibility only to the inputs or only to the outputs. But the common theme is that once time-sharing is used, designer involvement and good understanding of both the underlying algorithm and the hardware are a must.

8.11   Think About Your Controller

We have been postponing the discussion of the controller for a bit now. In Sect. 8.9 we talked a lot about things the controller should handle. The controller determines the flow of data through the datapath. It observes what we are doing now and decides what we should do next. According to Sect. 8.2, the controller and datapath communicate only through status and control signals. The datapath tells the controller what is going on using status signals. The controller reconfigures the datapath and tells it what to do next using control signals.

In Sect. 8.6, the fully expanded design barely had any control signals; thus, its controller can be trivial. On the other extreme, the time-shared design in Sect. 8.9 might reach into dozens of control signals, requiring a very aware controller. We will consider the fully flexible time reuse case in Fig. 8.9 as an example for controller design because it is the most interesting.

Figure 8.9 shows the datapath only. We consider memories used to store intermediate values as part of the datapath. MUXes are actuators that reconfigure the hardware based on what the controller decides. At this point, we should name the control and status signals that the FFT datapath in Fig. 8.9 exchanges with its controller.

It is easier to start by tallying control signals. The column of 16 4×1 MUXes will need two select lines each. This is 16×2 select lines. We will label them sel $\_ i \_ k \_ m$, where k is the MUX index and m is the select line index. The index k thus runs between 0 and 15; m always goes between 0 and 1. We will also use the shorthand sel $\_ i \_ k$ for two-bit select bus for the k'th input MUX. In other words, sel $\_ i \_ k =$ sel $\_ i \_ k \_ m$ for $m = 0$ through 1. For the registers, we need 16 enable bits indexed $EN0$ through $EN15$.

At the output, there are four 16×1 MUXes, each needing four select lines. As with the input multiplexers, these select lines will be indexed sel $\_ o \_ k \_ m$, where k runs between 0 and 3 and m between 0 and 3. Again sel $\_ o \_ k$ is the select bus for the k'th output multiplexer.

There are four memory banks; we will assume each is a very small memory with only eight locations. Thus, each memory has a three-bit address. However, we need two address busses per memory since they are using two-port memories. The address busses will be indexed add $\_ k \_ p \_ m$. Index k is for the bank, running between 0 and 3. Index p is for the port, with 0 representing the read port and 1 the write port. Index m is for the bits in the address bus, thus running between 0 and 2. Again add $\_ k \_ p$ is shorthand for one address bus in its entirety. Table 8.3 lists all the control signals that the controller needs to provide the datapath.

WE $\_ k$ are write enables to the write port of the memory banks. There are four of them, one for each bank. The address provided to the write port will not necessarily result in a write. We want to plan for the very common possibility that we will not write to a certain bank in a certain cycle.

The status signals are harder to figure out just by looking at the block diagram. To deduce them, we must know the sequence of steps we go through to finish the FFT. We have been postponing any notion of sequencing or timing so far, but the status signals will become more evident once we discuss specific controller implementations in the coming two sections.

Table 8.3 Control signals provided by the controller to time-shared FFT

Name	Type	Size	Number
sel _ i _ 0 through sel _ i _ 15	MUX sel (input)	2	16
EN0 through EN15	Register enables	1	16
sel _ o _ 0 through sel _ o _ 3	MUX sel (output)	4	4
add _ 0 _ 0 through add _ 3 _ 0	Read port add	3	4
add _ 0 _ 1 through add _ 3 _ 1	Write port add	3	4
WE _ 0 through WE _ 3	Write enables	1	4

There are 92 control bits

8.12   Hard-Coded Controllers: Finite State Machine

Our first option to implement the controller is to create a custom solution. This will lead to an efficient though inflexible controller. The sequence of operations is fixed, and options in flow are limited and predetermined. There are multiple approaches to hardwired control, but finite state machines are perhaps the most used. Finite state machines can be coded systematically and are reliable. In a finite state machine, the status of the circuit is divided into several states. Each state produces specific control signals to the datapath, and each state will predictably move to another state under certain conditions which can be conclusively deduced from the status signals. Thus, an FSM is characterized by the following:

- The number of states of the circuit is finite.
- The control signals produced by the controller are a function of solely the current state.
- The next state to move to can be concluded from the current state and the status signals exclusively.

We will continue to use the fully flexible hardware reuse example to illustrate FSM design. This example has a particularly challenging controller because it uses two different clocks for the memory and the PUs. This allows us to address this common issue. The example also has a large number of control signals and a complex but regular flow.

The FSM (finite state machine) can be defined using a state transition diagram as seen in Fig. 8.11. The state diagram makes it easy to visualize how states move into other states. The FSM can also be defined using a state transition table as in Table 8.4. The state table makes it easy to denote the control signals in every state. Combined, the table and the diagram allow easy coding of the FSM.

Every FSM starts from a reset state. Reset is a safe point from which the circuit can start operation. It is also an escape state that we can go to in case something goes wrong. Normally, control signals in the reset state are defined to be passive, meaning that registers are disabled, write enables are de-asserted, counters are reset, etc.

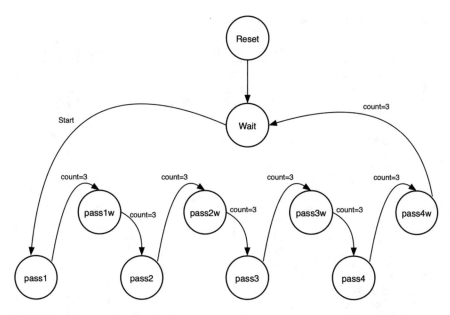

Fig. 8.11 FSM for the time-shared architecture in Fig. 8.9

All states in an FSM are reachable from other states under certain conditions that can be conclusively deduced from the status signals. The reset state is an exception, where it is also reachable through an external interrupt signal, which we also call "reset." This reset signal is an input to both the datapath, where it clears registers, and the controller, where it forces it to go to reset state. Reset *should always* be an external signal that allows the user to salvage operation if something goes horribly wrong. However, we will define the reset signal as a status signal in Table 8.5. Notice that this breaks the definition of status signals from Fig. 8.2, where status signals are outputs from the datapath rather than external inputs to the circuit.

Any state machine should leave the reset state in the immediate next cycle. It is important to understand which clock the FSM uses. The FSM will need to exert full control over both the PU and the memories. Since the memories clock multiple times for every PU cycle, the FSM needs to be able to keep up with the faster of the two. Thus, the FSM will use the memory clock.

Once the FSM leaves the reset state, we cannot immediately start performing the FFT. For starters, the transceiver does not always want to perform FFT; most of the time it is just sitting there doing nothing. But even when we want to perform FFT, we must wait before the entire time-domain symbol (16 samples in this case) are stored in the memory.

Thus, after leaving the reset state, we move into a "wait" state. In this state, we observe an external signal from the global system. We will call this signal "start." We only leave the "wait" state when the "start" signal arrives. When the "start" signal asserts, we know that there are 16 valid samples in the memory banks.

Table 8.4 State transition table for the time-shared FFT

State	Sel_i_k	En_k	Sel_o_k	Add_k_0	Add_k_1	WE_k	State count reset
Reset	–	0	–	–	–	0	1
Wait	–	0	–	–	–	0	1
Pass1	0xxx 1xxx 2xxx 3xxx x0xx x1xx x2xx x3xx xx0x xx1x xx2x xx3x xxx0 xxx1 xxx2 xxx3	8888 4444 2222 1111	–	State_count	–	0	0
Pass1w	–	0	0000 2222 1111 3333	–	State_count	1's	0
Pass2	0xxx 1xxx 2xxx 3xxx x0xx x1xx x2xx x3xx xx0x xx1x xx2x xx3x xxx0 xxx1 xxx2 xxx3	8888 4444 2222 1111	–	State_count	–	0	0
Pass2w	–	0	028A 46 CE 139B 57DF	–	State_count	1's	0
Pass3	0xxx 1xxx 2xxx 3xxx x0xx x1xx x2xx x3xx xx0x xx1x xx2x xx3x xxx0 xxx1 xxx2 xxx3	8888 4444 2222 1111	–	State_count	–	0	0
Pass3w	–	0	0246 8ACE 1357 57DF	–	State_count	1's	0
Pass4	0xxx 1xxx 2xxx 3xxx x0xx x1xx x2xx x3xx xx0x xx1x xx2x xx3x xxx0 xxx1 xxx2 xxx3	8888 4444 2222 1111	–	State_count	–	0	0
Pass4w	–	0	0819 2A3B 4C5D 6E7F	–	State_count	1's	0

Table 8.5 Status signals provided by the time-shared FFT
to the FSM

Signal	Size	Type
Reset	1	Global interrupt
Start	1	Global control
State_count	2	Counter

At this point we can kick off the FFT. We will need four PU cycles to transform
the 16 time-domain samples to frequency domain. This factor of 4 comes from the
fact that the single column of PUs in Fig. 8.9 must be reused four times to realize the
FFT in Fig. 8.4.

However, these four PU cycles will involve four memory cycles each. Every
memory cycle reads a single sample from each memory bank; we need to read four
samples from each memory (Fig. 8.9), which is where *this* factor of 4 comes in. We
can think of states as cycles of the memory or as cycles of the PU. Because there is a
rather regular pattern to control signals in the four cycles, we will define states based
on PU cycles although we know the FSM uses the memory clock.

As soon as we move out of the wait state, we go to a state called "pass1". In
"pass1", the column of processing units in Fig. 8.9 implements the leftmost column
of PUs in Fig. 8.4. We stay in this state for one PU cycle, which is four FSM
(or memory) cycles. To allow the FSM to know when it can exit "pass1", we
introduce a counter called "state_count". This counter has a reset called
"state_count_reset", which is added to the list of control signals. The counter output
"state_count" is a status signal (Table 8.5).

This suggests that the counter will be folded in with the datapath even though
practically it is usually built in the controller. The counter is always reset unless we
are in states "pass1" through "pass4w", where it is enabled. The counter counts from
0 to 3 and then rolls over. This means as we exit the wait state and go into "pass1",
the counter is ready to start at 0. It will count up to four times; when the counter
reaches 3, the state machine knows to move to the next state.

From the state transition diagram in Fig. 8.11, the next state is "pass1w", where
we spend four (memory) cycles to write the results of the first pass. We then move to
"pass2", where the counter is once again enabled, counting four cycles, then exiting
to "pass2w", and so on.

Defining the rest of the control signals for the "pass" states is where most of the
challenge of time-sharing comes in. Most of the control signals involve the memory
and its associated multiplexers. We list these control signals in Table 8.3. This table
will be invaluable when coding the FSM; thus, the time spent on constructing it is
well worth it. We will come back to this table and how it is filled shortly.

Addressing memory can be aided by having a table tracking the contents of
memory after every write. This is summarized in Table 8.6. Table 8.6 shows the
control signals up to and including state "pass2w". The specifics are derived from the
connections in Fig. 8.9 and the banks and multiplexers in which data is found. Once
"pass4w" is concluded, the memory contains frequency-domain samples, and we
can go back to the "wait" state to await a new start.

Table 8.6 Memory contents
after each stage

Bank	Add	P1	P1w	P2w	P3w	P4w
0	0	x0	a0	b0	c0	X0
	1	x8	a2	b4	c8	X1
	2	x4	a1	b1	c1	X2
	3	x12	a3	b5	c9	X3
1	0	x2	a4	b2	c2	X4
	1	x10	a6	b6	c10	X5
	2	x6	a5	b3	c3	X6
	3	x14	a7	b7	c11	X7
2	0	x1	a8	b8	c4	X8
	1	x9	a10	b12	c12	X9
	2	x5	a9	b9	c5	X10
	3	x13	a11	b13	c13	X11
3	0	x3	a12	b10	c6	X12
	1	x11	a14	b14	c14	X13
	2	x7	a13	b11	c7	X14
	3	x15	a15	b15	c15	X15

This arrangement allows data to be read in order in every cycle.
The contents can be easily deduced from Fig. 8.7

Back to Table 8.3, states pass1 through pass4w last 4 FSM (memory) cycles each. Thus, for these states, control signals come in four copies corresponding to each cycle. For example, En_K is 16 bits, but 4×16 values are listed in four rows for each cell to cover the four cycles used to read from the memory. Filling the rest of the table is contingent on an understanding of which inputs we should arrange for the PUs and where they come from.

In the first cycle of pass1, we read $\times 0$ from bank 0, x2 from bank 1, x1 from bank 2, and x3 from bank 3. These are all stored in addresses 0 in the banks according to Table 8.6. Since State_count is also 0 in this cycle, we can use this status signal as Add_k_0. The first input of PU1 should select bank 0 to read x0; the rest of its inputs are going to be don't cares because they will not store in this cycle. The first input of PU2 should select bank 1, and the rest of its inputs are not storing. It follows that the select lines for the input multiplexers of the first PU are 0xxx, the second PU 1xxx, the third PU 2xxx, and the final PU 3xxx. This can be seen in the first row of Pass1 in Table 8.3 under the Sel_i_k column. Note that x stands for don't care while each nibble is a 2-bit select bus for each of the input multiplexers.

Since pass1 is a read state, the write multiplexers are irrelevant, and all write enables should be disabled. Thus WE_k is all 0's and Sel_o_k is all don't cares. The Regs are very relevant though. In the first row of the En_k signal, the values are 8888, where each nibble is a four-bit word, leading to 16 enables for 16 registers. The 8888 values enable the registers for the first inputs of every other PU.

In the second cycle of Pass1, State_count increments to 1. We read $\times 8$, $\times 10$, $\times 9$, and $\times 11$ out of the four memories using State_count as an address. This feeds the second input of every other PU, requiring En_k of 4444. The significant nibbles of

Sel_i_k will shift by one relative to the first cycle so that the second set of inputs select the four memory banks, respectively. The same logic applies to the remaining two cycles of Pass1. At the end of Pass1, the column under P1 in Table 8.6 is registered in the register file in the order shown in the table, which is also the order in Fig. 8.7.

When State_count reaches 3, the state machine moves to the pass1w state. The counter will start counting again from 0, and again we stay in this state four cycles, leading to four entries per cell in Table 8.6. We can guess a lot of the control signals of pass1w using the inverse logic of that used for pass1. Since pass1w is a write state, then En_k is all 0's, and Sel_i_k is all don't cares. We only need to figure out WE_k and Sel_o_k. For the four cycles of pass1w, we need to write 16 outputs from the PUs to 16 locations in the four memory banks. By design, this means each memory bank will write each cycle. Thus We_k will be all 1's for the duration of Pass1w since none of the banks will ever not be writing. Also, we will be writing into address 0 in all banks in cycle 1, moving to address 1 in cycle 2, and so forth. Thus, as in the pass1 state, Add_k_1 will be State_count, while Add_k_0 is don't care.

This leaves us with Sel_o_k, the select lines of the output multiplexers. From Fig. 8.7, the outputs are produced in order as a0, a1, a2, ... According to Table 8.6, they are stored in the order a0, a2, a1, a3, a4, a6, a5, a7, ... This is the order in which they should be read in the following PU cycle according to Fig. 8.7. In the first cycle of pass1w, outputs a0, a4, a8, and a12 are written. There are 16 bits in Sel_o_k divided into four groups of 4 bits each for the four multiplexers. The select lines are 0000 to choose the first output of every other PU. In the second cycle, we need to skip the second output of every other PU and store the first output of the second, fourth, and so on. This is to achieve the memory structure of pass1w in Table 8.6. This leads to Sel_o_k of 2222. In the third cycle, we go back to storing the second outputs of off-numbered PUs; thus, Sel_o_k is 1111, and finally 3333.

In state pass2, data is stored in the same order as in state pass1; this means that the control signals will be identical to those of pass1. We can thus move to pass2w. The output b0 through b15 are produced in order from the PUs. They are stored disordered as shown in Table 8.6. This allows them to be read in order in state pass3. As with state pass1w, there are no challenging signals in the pass2w state except Sel_o_k. For the first cycle, we store b0, b2, b8, and b10. This means the select lines for the four multiplexers are 028A. Similarly, we can deduce the signal for the remaining three cycles. What helps here is that the indices of b are also the select lines of the MUXes. Thus, in the fourth cycle, we write b5, b7, b13, and b15. Sel_o_k will thus be 57DF.

Because data is written in pass2w in the order that it should be read in pass3, we can copy the control signals of pass1 or Pass2 into pass3. Thus, we move to pass3w. As in pass2w, filling Sel_o_k for Pass3w is straightforward by copying the indices of the entries for pass3w in Table 8.6. In the first cycle, we copy indices in the 0 locations of all four banks; in the second, indices in locations 1; and so forth.

Pass4 is a copy of pass3, pass2, and pass1. This is all contingent on the write states arranging data in the banks in the order in which they should be read in the next state. State pass4w is a little different. Data is not arranged in an order that suits a following read state, because there is no following read state. Instead, we arrange the outputs so that they are the properly indexed frequency-domain samples. You can deduce the structure of the Sel_o_k signals by assuming the outputs of the PUs in order are d0 through d15, and then map the d's to the corresponding frequency domain samples X0 through X15. Sel_k_o will then be the indices of the d signals along the same lines of how we deduced it for pass1w through pass3w.

This exercise shows that a lot of effort must be invested in figuring out control signals for the FSM. However, documenting the signals the way we have shown is invaluable once coding starts. It is also a great help if we need to make changes to the state machine. The next step should always be trying to simplify the state machine by reducing the number of states. As with most issues related to control, this requires familiarity with the circuit at hand.

In our example, you might have noticed that we did not make any use of the fact that the memory is two-port. In fact, this state machine could very well work with a single-port memory. The port is used for reading the pass states and for writing in the passw states. With two ports, however, we can fold some of the states into each other. Specifically, after an initial latency, each passw state can be merged with the following pass state. Thus, once we finish the first frame of data, we can start reading in the new pass1 state while writing the final pass4w state. This assumes that we go to a following OFDM frame immediately after finishing the first one. This requires us to distinguish between a "wait" state and a condition where we have a continuous flow of data into the FFT. The bottom line is, once you finish the design of your FSM, it is only the beginning of the end, because a lot of fundamental changes can spring from the design exercise.

8.13   Instruction-Driven Controllers: Flexibility and Programmability

FSMs are the prototypical hardwired controller. They are efficient, easy to characterize, and easy to debug. They are also inflexible. An FSM designed for an FFT will only drive an FFT. You might be able to introduce "modes" to a controller designed using a state machine, but once it is fabricated, this controller can only do the set of things it was designed to do.

This might not be suitable for datapaths with a very large number of modes and is especially not suitable for circuits where we want to keep open the possibility that the user might want to exercise the circuit in novel ways that the designer did not envision. The alternative to hardwired controllers is to think of your circuit as a processor. In this case, the datapath acts as an ALU, and the controller reconfigures this ALU and routes data to, from, and within it.

This approach leads to the controller shown in Fig. 8.12. We will consider four cases for this setup from the simplest to the most complex:

Fig. 8.12 Instruction
memory-based controller

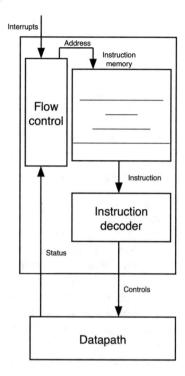

- A fully expanded instruction with forward-going counter
- A coded instruction set with a forward counter
- A controller with unconditional flow control (looping)
- A controller with conditional flow control (looping and branching)

8.13.1   Fully Expanded Instruction

The simplest instruction-driven controller is one where the instruction is a concatenation of all the control signals that the controller provides to the datapath. For example, in Table 8.3, there are 92 control bits produced by the FSM. The concatenation of all these bit forms a single instruction, and one instruction is saved per memory (controller) cycle.

In this case, the controller is just the instruction memory and a counter. The counter provides the address to read from the instruction memory. In every controller cycle, the counter is incremented, and a new instruction is read. This instruction is then deparsed according to Fig. 8.13 into control words that are routed to the datapath.

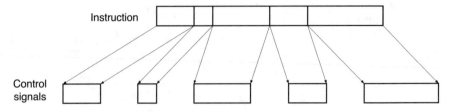

Fig. 8.13 Controller deparsing an instruction into control signals

There are two major problems with this approach:

- The length of the instruction is rather large. Here each entry in the instruction memory is 92 bits. While memory is relatively cheap, our design example is also simple. In more realistic designs, the controller could produce hundreds of control bits, many of which follow regular patterns or are very sparse. This could lead to very unusual memories with very wide entries but very few locations. Notice that the number of locations in the memory is related to the length of the program, while the width of the memory is related to the size of the instruction, which in this case is a blind concatenation of all control bits including memory addresses.
- The program counter only counts up. This means the program can only flow forward. Any state machine is going to have loops and conditional branches; things cannot just flow in a single stream, otherwise the controller is not really observing the datapath.

8.13.2 Coded Instructions

We can solve the instruction size problem by first discussing what we mean by an instruction set. An instruction set is the total library of instructions that the designer can use to write their program. An instruction with n bits has 2^n variations and thus has an instruction set with 2^n instructions.

In our design example, we used a 92-bit instruction. This has 2^{92} possible permutations. We are not going to need that many instructions. You can guess the number of instructions you will need if you look at the FSM. But as a rule, your instruction set will not include more instructions than your longest program. For FFT, 256 instructions are more than enough to encode everything we want to do. This would be an 8-bit instruction.

This necessarily makes the controller more complex. Let us take a simpler example for illustration. Assume we need only as many instructions as there are states in the FFT FSM. We have ten states and thus need a 4-bit instruction. Assume that instruction 0000 then represents the reset state, while instruction 0010 represents pass1. There must be a block that translates these four bits into the corresponding control bits for the two states in Table 8.3. This block is the instruction decoder. It takes 4 bits and expands them into 92 bits. In the fully expanded instruction, the instruction decoder was 92–92 and was thus a simple deparser.

Things can also be complicated by how much flexibility you want to leave the programmer. In other words, do you really want to hard-code memory addressing as we suggested above? After all, having a small instruction for pass1 means we are tightly associating the addressing pattern in Table 8.3 with this instruction. Or perhaps it is better to allow the programmer to freely read and write to and from the memory addresses. In this case, perhaps the instruction should have portions dedicated to the operand, memory addressing, data routing, and so on. Which of these should be encoded and which should be left expanded is up to the designer.

8.13.3   Introducing Absolute Looping

The instruction counter cannot simply just keep counting up. For starters, most FSMs contain loops. Loops can either cause the FSM to stay in the same state for several cycles or cause it to loop across different states. Looping on the same state a fixed number of times can be achieved by repeating the instruction as many times as we need. This is only viable if the number of iterations is limited.

All iterations can also be managed by causing the counter to count in a non-BCD sequence. For example, 0, 1, 2, 0, 1, 2, 0, 1, 2, 3, 4.. is a sequence that loops over instructions 0–3 three times. The instruction counter thus starts to become more of an address generator than a counter. This approach will only work with absolute loops, meaning that we loop a fixed number of times. This is viable, for example, in the four cycle loops we must make in each of states pass1 through pass4w in the FSM.

8.13.4   Introducing Conditions

One thing has been missing from this whole discussion: datapath status signals. We have not allowed (nor required) the controller to observe the datapath in any way. This has limited us by making us unable to create any conditional branchings or loops. For example, when does the FSM exit the wait state? It does so only when the start signal is asserted. There is no way the controller as discussed so far can make this decision.

To allow this, the controller must contain a set of registers to store status signals from the datapath. The counter (or rather address generator) from the previous section is upgraded to a flow control unit. This unit can read these status registers, carry out comparisons on them, and determine the address of the next instruction based on the result.

The controller will also need to allow external hardware interrupts to change the operation. The simplest and most necessary interrupt is perhaps the rest signal. Reset, according to our discussion of FSM design, should push the state machine into a well-defined reset state regardless of anything else. So, it should set the address of the instruction memory regardless of flow control.

8.13.5 The Ultimate Result

Instructions are very hard to write. The expanded instruction is long and requires familiarity with the hardware. The coded instruction requires as much familiarity while adding a layer of requiring the programmer to remember which instruction does what. Perhaps there could be a piece of software that stands between the programmer and the instruction. This software will take a high-level instruction from the programmer, like "process this vector," and translate it into as many instructions as necessary to implement this on the hardware. This is starting to sound a lot like a compiler.

We also need to define which status signals are registered for the flow control unit and what processing it can do on them to make flow control decisions. It should definitely be able to compare each of them to an absolute number provided by the instruction and should perhaps be able to compare some of them to each other.

This whole setup is starting to approach a microprocessor, albeit one with a specialized ALU. The border between a microprocessor and such an "accelerator" is sometimes blurry and can be philosophical; but we will take a jab at it in the coming section.

8.14 Time-Sharing Revisited: Reconfigurable Hardware and Accelerators

Most of the time PUs are too fast for a direct implementation (Sect. 8.6). What makes sense is to use fewer PUs and reuse them to obtain only the throughput we need. The degree of time-sharing we can do is deduced from a throughput budget (Sect. 8.8). The main cost of time-sharing hardware is complicated control (Sect. 8.12) and the need to manage and partition a lot of memory (Sect. 8.10). But we also see that power dissipation is a concern because of the higher clocking rate.

Combined with the instruction-based controllers in Sect. 8.13, we can use time-shared architectures in very flexible ways. For example, there is no reason that we must use exactly eight PUs as in Fig. 8.9. We could have used fewer than eight or more than eight PUs. The more PUs we use, the fewer passes we must make before we finish a set of 16 FFT symbols.

This means that we can use any number of PUs to perform any size FFT. The eight PUs in Fig. 8.8 can be used to perform a 32-point FFT, but every pass will take two cycles instead of 1, and the frame is finished in eight cycles instead of four. We will also need more memory and more complicated control to store intermediate values.

If the controller is instruction-driven, we can change the size of the FFT on the fly. This means that we can trade off speed and power for more complicated algorithms. This approach to design fits well with cognitive radios discussed in Sect. 7.14 and accelerators in Sect. 10.9. The preponderance of modes and choices in modern radios means that hardware must be able to switch on the fly between multiple modes.

When designing PUs and their interconnections, it is helpful to write hardware descriptions in a scalable fashion. To be specific, it is very helpful to make the number of PUs a parameter of the design. Combined with the instruction-based controller, this gives users the ultimate level of flexibility. Specifically, there are two degrees of flexibility available:

- Before the hardware is implemented, we can decide the number of PUs. This allows trading off performance and cost at the implementation stage.
- After the hardware is realized, we can further trade off on the fly by changing the program.

8.15   Which Architecture Is "Best"?

Our discussion of area, speed, and power in Sect. 3.9 was lacking an understanding of how power and speed trade off. The main form of power dissipation we consider in this book is dynamic power. This is an active form of power that is burned off in MOSFET channel resistances when values switch at the outputs of logic gates. The equation for dynamic power dissipation is:

$$P = \alpha C f V_{DD}^2$$

where C is the total capacitance switched in the circuit. Because not all capacitances are switched every cycle, we include an activity factor α which is the average percentage of cycles that capacitive nodes are switched in the circuit. Thus, αC is the average switched capacitance every cycle. f is the frequency at which the circuit is switched; the higher the operating frequency, the proportionately more the circuit dissipates power. And finally, V_{DD} is the supply voltage to the circuit. This is where the strongest dependence of power comes in. Power dissipation is thus a function of area (through capacitance), operating frequency, and above all supply voltage.

The operating frequency of the circuit is inversely proportional to the delay in the critical path. Delay in the critical path is inversely proportional to supply voltage. Thus, operating frequency is directly proportional to supply voltage:

$$f \propto \frac{1}{t_{pd}}$$

$$t_{pd} \propto \frac{1}{V_{DD}}$$

$$f \propto V_{DD}$$

This direct dependence of frequency on supply is a little misleading. It used to be true for long channel devices, but for reasons beyond the scope of this book, the dependence is sublinear for modern devices. In such devices current available to charge and discharge nodes grows weakly with supply, and thus delay scales down slowly.

We also need to qualify the fact that we ignored power dissipation mechanisms other than dynamic power. Other forms of power dissipation include leakage and short circuit current. While it is not acceptable to ignore leakage in submicron technologies, it is fine to consider it negligible in deep submicron. In such devices, multiple gates are used to effectively suppress leakage, causing active power to dominate again. But these devices also suffer from significant short circuit power.

The fact that both power and frequency depend on supply voltage can be used to play interesting tricks to reduce overall energy consumption. Consider Fig. 8.14, where two parallel circuits are used instead of one. Assume that a single block can provide us with the throughput we need, S. Thus, two blocks will provide double the needed throughput $2S$. But if we operate each block at half the clock frequency, it will produce a throughput of only $S/2$; thus, the two together will provide a throughput of S.

Fig. 8.14 Parallel implementation

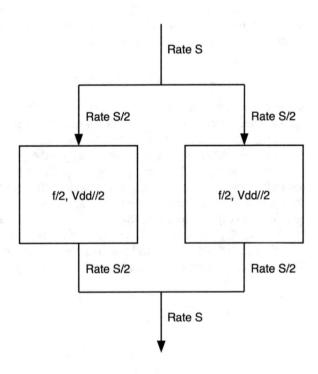

If we are operating the blocks at only half their maximum frequency, we can pull the supply down to half. This leads to an overall reduction in power dissipation to one quarter of the original value:

Unreduced supply, single block :

$$P = \alpha C f V_{DD}^2$$

$$\text{Frequency} = f$$

$$\text{Throughput} = S$$

Unreduced supply, two blocks :

$$P = 2\alpha C f V_{DD}^2$$

$$\text{Frequency} = f$$

$$\text{Throughput} = 2S$$

Reduced supply, two blocks :

$$P = 2\frac{\alpha C f}{2}\left(\frac{V_{DD}}{2}\right)^2 = \frac{\alpha C f V_{DD}^2}{4}$$

$$\text{Frequency} = f/2$$

$$\text{Throughput} = \frac{S}{2} + \frac{S}{2} = S$$

So, what is the limitation on this? In other words, if we use ten parallel units, can we reduce power to one hundredth the value? There are two limits, the first and more obvious is area. The second is noise margins. The lower the supply voltage, the less noise the logic values can tolerate, and the more errors start to arise from bits randomly switching at the outputs of logic gates.

Pipelining can also be used to create the same effect. Assume we have an unpipelined design clocked at a rate f, producing the needed throughput at a rate S. We pipeline this design so that the critical path is now one third as long; in other words, we can clock it at a rate of $3f$, to produce a throughput of $3S$. But if we only need a throughput of S, we can clock the pipeline at only one third of the $3f$ clock. This means we can reduce the supply to one third of the original values, which will reduce the active power to almost one ninth. Pipelining costs a lot less area than parallelism and gives the same impact on power. It does, however, introduce extra latency.

We should talk a bit more about what we mean by "power" and "speed." We will talk in detail about the difference between throughput and clock frequency in Sect. 10.12. But here, we want to expand on how to compare two hardware implementations using a single figure of merit. In other words, which is better, the time-shared architecture in Sect. 8.8 or the direct architecture in Fig. 8.5?

Power or throughput on their own is not enough for a fair comparison. Consider two blocks A and B that do the same thing. Block A produces a throughput at a rate S and consumes power P. Block B produces a throughput of $2S$ at power $2P$. The two blocks are equally good. In fact, block B is just two blocks A in parallel. There is no difference in the quality of implementation of the two.

A good comparison of two circuits that do the same thing is their power for the same throughput, or throughput for the same power. Alternatively, we can calculate the energy required to produce a single output sample. The time to produce a sample is the reciprocal of throughput. Thus, the energy to produce a single output is the product of power and the reciprocal of throughput.

But energy per output is only good when comparing two circuits that implement the exact same algorithm. How about a metric that can compare two circuits that do completely different things? In other words, we want to compare the cleverness of the hardware and its efficiency while normalizing everything that has to do with the algorithm.

For example, if Anne implements a Viterbi decoder that dissipates power P_a and produces throughput S_a and John implements an FFT that dissipates power P_b and produces throughput S_b, who is the better hardware designer?

At first this might seem like a silly question because we are comparing Viterbi and FFT, which is worse than comparing apples to oranges. But if we can normalize the complexity of the two algorithms, we can compare them. To normalize complexity, we first quantify it. The best way to do this is to measure the rate at which operations are done in each algorithm. But first, we must agree what an "operation" is. Normally, a real addition is a good enough unit. Multiplications can be easily broken down into and counted as additions. Operations less complicated than additions are usually simple enough to be ignored.

Thus, let us assume that Viterbi needs X_a operations (additions) to finish one sample. The rate at which operations are performed to produce a Viterbi throughput of S_a is $X_a S_a$. This rate is measured in MOPS (mega operations per second). If we divide this by the power, we get MOPS/mW, which measures the rate at which this hardware can perform operations for each mW of power we give it. Thus, if we compare $X_a S_a / P_a$ and $X_b S_b / P_b$, we can tell who is more clever between John and Anna. The higher the MOPS/mW, the better the design regardless of the complexity of the algorithm.

Let us take a numerical example. Laila designed a circuit to perform data compression. The circuit consumes 500 µW and produces a throughput of 100 MSps. To produce one sample of compressed data, it performs 10,000 adds. Vivec designed a circuit to perform MIMO decoding. The design consumes 1000 µW and has a throughput of 20MSps. To finish one sample, we perform 200,000 adds. To the best of our knowledge, who designed "better hardware"?

Laila's design does additions at a rate 100*10,000 MOPS = 1000 GOPS. Thus, for every µW, she performs 1000/500 = 2 GOPS/µW. The same metric for Vivec is 20*200,000/1000 = 4 GOPS/µW. This means that for every µW of power we give each designer, Vivec can produce double the operation rate that Laila can, which means that regardless of the algorithm, Vivec is doing a better job.

Chapter 9
MIMO

9.1 What Can You Do with Multiple Antennas?

Increasing goodput can be achieved through larger baseband modulation constellations, better channel coding, or wider bandwidth. Usually, a new standard will include a combination of all the above. However, channel capacity is always an upper threshold that shackles the goodput you can achieve to SNR.

With advancement in technology, and especially as communication in higher bands becomes more widely adopted, multiple antennas can be economically integrated on mobile platforms. This allows communication schemes with arrays of small antennas at the transmitter, the receiver, or both. These antennas can be used to raise the channel capacity and thus increase goodput in more fundamental ways than any of the techniques we have discussed so far.

Based on the multiplicity of antennas at the transmitter and the receiver, we divide multi-antenna systems into three categories. Multiple antennas at the transmitter with a single antenna at the receiver is called a MISO (multiple input single output) system and will be explored in detail in Sect. 9.3. Multiple antennas at the receiver only will result in a scheme called SIMO (single input multiple output) and is explored in detail in Sect. 9.2. Multiple antennas at both ends will result in a situation called MIMO (multiple input multiple output) and is explored for the rest of the chapter. Sometimes MISO and SIMO systems are also called MIMO systems, although it is far less confusing to reserve the term for systems with multiple antennas at both ends.

K. Abbas, *From Algorithms to Hardware Architectures*,
https://doi.org/10.1007/978-3-031-08693-9_9

9.2   Receiver Diversity

In a system with multiple antennas at the receiver only, we can use the multiple antennas to add receiver diversity to the system. The paths from the transmitter to the multiple receiver antennas can potentially be independent. This improves the ability of the receiver to fight fading because if one antenna experiences a deep fade or shadow, the other antennas may not, allowing us to receive acceptable SNR (Fig. 9.1).

In systems with $N_{tx} \times N_{rx}$ antennas, there can still be some degree of receiver diversity if $N_{rx} > N_{tx}$. The balance of antennas will be used to introduce diversity. In all cases, receiver diversity is contingent on the placement of antennas on the receiving platform. The antennas should be far enough from each other that the streams arriving from the transmitter are independent.

To develop a quantitative model for receiver diversity, we must reimagine it as a way to improve effective SNR observed by the receiver. On each of the antennas, we have both signal power and noise power. The noise processes on the antennas are decorrelated. The signal on the multiple antennas is correlated. Thus, there must be ways to combine the total signal on the three antennas so that the overall SNR is better than that of a single antenna.

There are three schemes to combine the multiple received signals, namely:

- Selection diversity
- Equal ratio combining (ERC)
- Maximal ratio combining (MRC)

In selection diversity we pick the antenna with the highest estimated SNR and discard the signals from all other antennas. While we perform channel estimation, we also produce estimates of SNR. In Fig. 9.2, there are three SNR estimates for the three antennas. This is not because the noise power on each antenna is different (it is not) but rather because signal power is different. The selector block is a decoder that

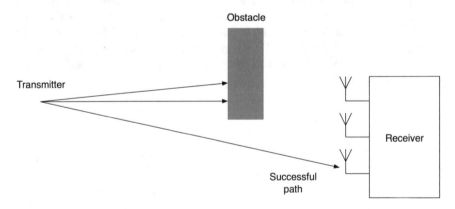

Fig. 9.1 Conceptual view of receiver diversity

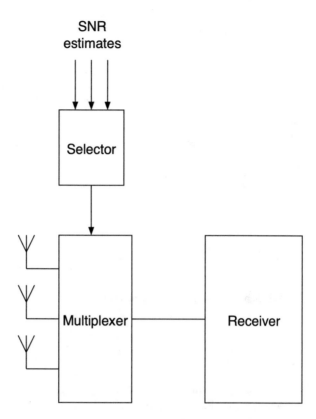

Fig. 9.2 Selection receiver diversity

determines the antenna with the highest SNR and translates this into select lines for the multiplexer. The multiplexer will route the antenna with the highest SNR to the rest of the receiver.

As a side note, the multiplexer in Fig. 9.2 is an analog multiplexer. Its input from the antennas is analog, although the select line provided by the selector may be digital. But how frequently are we supposed to re-estimate the best antenna? The ideal solution is to make the decision continuously. This means that the selector in Fig. 9.2 is always looking at the SNR estimates and always updating the select lines to the multiplexer. This allows us to get the best performance possible.

Having the selector be a perpetual monitor on SNRs is not practical. For instance, in packet-based communication, we calculate channel estimates, and thus SNR, during the header of the packet. While pilots can be used to refine these values during reception, the values are roughly constant till the packet ends. In such a case, it does not make sense to update the selector decision in Fig. 9.2 except when new channel estimates are calculated, which in packet-based communication will only happen once per packet. In cellular systems (Sect. 12.4), SNR estimates are updated more often due to high Doppler, and thus selection must keep pace.

We can show that using N_{rx} receiver antennas and using selection diversity, the effective SNR is improved by the ratio:

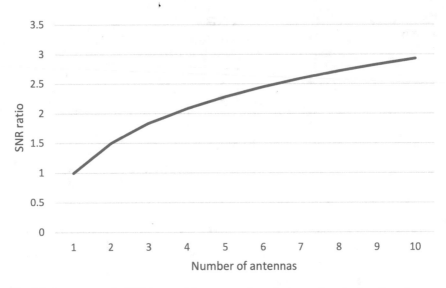

Fig. 9.3 Improvement in SNR (as a ratio) versus number of antennas for selection diversity

$$\frac{SNR_{N_{rx}\text{antenna}}}{SNR_{1\text{antenna}}} = \sum_{i=1}^{N_{rx}} \frac{1}{i}$$

Figure 9.3 plots this ratio of SNRs on the y-axis versus the number of antennas on the x-axis. The observed SNR more than doubles if four antennas are used at the receiver. The curve looks like it has a saturating behavior, but this is not true. Even for an arbitrarily large value of N_{rx}, any ΔN_{rx} will introduce a measurable improvement in perceived SNR.

However, there are diminishing returns, and the largest percentage improvement in SNR is obtained from the first few added antennas. There are other practical limitations that impose a cap on N_{rx}. There is the logistics of attaching the multiple antennas, the opportunity cost of not using spatial multiplexing (Sect. 9.5), and the limits on MIMO capacity imposed by antenna correlation. In practice, there is no realistic situation where adding more than ten antennas would make sense from the point of view of receiver diversity.

But also, selection is the worst performing receiver diversity scheme. Its mistake is that it assumes that if an antenna is not receiving the highest signal power, then it is worth nothing. This is not true, a path that is not carrying the maximum power is still carrying signal power. There must be a way to use this signal power to improve the overall SNR, especially given the fact that noise on the different antennas is decorrelated.

The simplest way to combine the signals on the multiple antennas is to simply add them. This is called equal ratio combining. Before signals are added, they must be co-phased. This means that the phase components of their channel coefficients must be inverted. If we do not do this, the signals will certainly be out of phase with each other some of the time and the perceived SNR will be worse than using a single antenna as they destructively interfere.

The received signal with equal ratio combining is:

$$y = \sum_{i=1}^{N_{rx}} y_i = \sum_{i=1}^{N_{rx}} h_i x = x \sum_{i=1}^{N_{rx}} h_i$$

The channel coefficients we are using here are amplitude only because of co-phasing. This does not mean the channel is in-phase only, but it does mean we have already inverted channel phases. Received signal power is thus:

$$|y|^2 = |x|^2 \left(\sum_{i=1}^{N_{rx}} h_i \right)^2$$

We can square the summation of the channel coefficients instead of squaring their absolute values because they are already amplitude-only.

Noise power in each antenna is the same because all the antennas are at the same temperature. This does not mean that all antennas have the same SNR because they have different signal powers. If a single antenna has a noise power of N_0, then the total signal is suffering from noise N:

$$N = N_{rx} N_0$$

We can add the noise powers because all the noise components are decorrelated and independent from each other.

Signal to noise ratio for the combined signal is thus:

$$\frac{|x|^2 \left(\sum_{i=1}^{N_{rx}} h_i \right)^2}{N_{rx} N_0} = \frac{|x|^2}{N_0} \frac{\left(\sum_{i=1}^{N_{rx}} h_i \right)^2}{N_{rx}}$$

To obtain an expression for the average SNR, we apply expectation to find the average value of the Rayleigh fading channel coefficient:

$$\mathrm{SNR_M} = E\left\{ \frac{|x|^2}{N_0} \frac{\left(\sum_{i=1}^{N_{rx}} h_i \right)^2}{N_{rx}} \right\} = \frac{|x|^2}{N_{rx} N_0} E\left\{ \left(\sum_{i=1}^{N_{rx}} h_i \right)^2 \right\}$$

$$= \left\{ 1 + \frac{(N_{rx} - 1)\pi}{4} \right\} \frac{|x|^2}{N_0}$$

This is a better performance than selection diversity. However, it is still not optimal. The problem with ERC is that we assume we must combine all the signals with equal weights. In fact, we can multiply any of the signals by any gain we want.

If all the gains we multiply by are pure real, we can still add all the signals in-phase. Thus, we can restate the received signal as:

$$y = \sum_{i=1}^{N_{rx}} G_i y_i = \sum_{i=1}^{N_{rx}} G_i h_i x$$

$$|y|^2 = |x|^2 \left(\sum_{i=1}^{N_{rx}} G_i h_i \right)^2$$

These gains also multiply noise power, leading to total noise power:

$$N = N_0 \sum_{i=1}^{N_{rx}} G_i^2$$

Again, we are using the decorrelation of noise processes to spread the square on individual noise signals. Let us use a limited case with three antennas to expand the expression of SNR:

$$\text{SNR} = \frac{|x|^2}{N_0} \frac{(G_1 h_1 + G_2 h_2 + G_3 h_3)^2}{G_1^2 + G_2^2 + G_3^2}$$

What values of gain G would maximize this perceived SNR? To find out, partially differentiate with respect to any of the gains:

$$\frac{\partial \text{SNR}}{\partial G_1} = 0$$

$$2h_1 (G_1^2 + G_2^2 + G_3^2)(G_1 h_1 + G_2 h_2 + G_3 h_3) = 2G_1 (G_1 h_1 + G_2 h_2 + G_3 h_3)^2$$

$$h_1 (G_1^2 + G_2^2 + G_3^2) = G_1 (G_1 h_1 + G_2 h_2 + G_3 h_3)$$

We find two similar equations by differentiating with respect to the other two gains. The three results of differentiation are:

$$h_1 (G_1^2 + G_2^2 + G_3^2) = G_1 (G_1 h_1 + G_2 h_2 + G_3 h_3)$$

$$h_2 (G_1^2 + G_2^2 + G_3^2) = G_2 (G_1 h_1 + G_2 h_2 + G_3 h_3)$$

$$h_3 (G_1^2 + G_2^2 + G_3^2) = G_3 (G_1 h_1 + G_2 h_2 + G_3 h_3)$$

These three equations are only solved by:

$$(G_1^2 + G_2^2 + G_3^2) = (G_1 h_1 + G_2 h_2 + G_3 h_3)$$

And thus:

$$h_1 = G_1$$

$$h_2 = G_2$$

$$h_3 = G_3$$

This is called maximal ratio combining (MRC) and is the optimal way to perform receiver diversity. In MRC, each antenna signal is multiplied by its own channel coefficient before combining. This means that a stronger path is given more gain while a weaker path is given less attention. This allows all signal powers to be used while giving cleaner signals more weight.

Using the optimal gain values and taking expectations of the Rayleigh variables, we find that the SNR improvement ratio of MRC is N_{rx}. MRC improves the perceived SNR by as many times as there are antennas. This might sound too good to be true, and in a way it is. There are fundamental limits on how any multiple antenna scheme can improve the system. To understand these limits, consult Sect. 9.5 on the limits of true MIMO's impact on channel capacity. The same limits on antenna correlation also affect receiver and transmitter diversity.

9.3   Transmitter Diversity

Transmit diversity schemes work when there are multiple antennas at the transmitter and a single antenna at the receiver. However, transmit diversity can also be applied to systems where there are multiple antennas at the receiver if the number of transmit antennas is larger. In fact, multiple receive antennas improve transmit diversity by adding some measure of receive diversity.

Transmit diversity relies on a scheme called space-time coding (STC). STC at the transmitter creates redundancy of sent symbols to introduce error correction. The earliest STCs were in fact based on trellis codes. More practical STCs are based on block coding and are thus called STBCs (space time block codes). STBCs send the same symbol multiple times on multiple antennas and in multiple time slots. This creates a matrix of data where columns represent time slots and rows represent antennas. Because there are repeated symbols in the matrix, there will be redundancy, fractional coding rate, and error correction capacity.

The simplest STBC is the Alamouti code. Despite being the simplest, it is also the most practical and most desirable because of a very special property. The Alamouti code uses only two antennas at the transmitter. The matrix of symbols is:

$$\begin{bmatrix} x_1 & x_2 \\ -x_2^* & x_1^* \end{bmatrix}$$

To reiterate, rows are antennas, columns are time slots. Thus, in the first time slot, the first antenna transmits x_1 while the second antenna transmits $-x_2^*$. In the second time slot, the first antenna transmits x_2 and the second antenna transmits x_1^*.

What is unique about the Alamouti code is that it is the only STBC that is rate 1. In two time slots, we transmit two unique symbols. There is no redundancy. The diversity still exists both in the spatial dimension and in time. Alamouti code provides SNR improvement equal to an MRC scheme with two antennas.

9.4   Beamforming

Beamforming is an umbrella term of techniques that use large antenna arrays at either or both the transmitter and the receiver to improve performance. In beamforming, the antenna arrays are used to create radiation patterns around the transmitter or the receiver. This can be achieved even if individual antennas are isotropic by pre-distorting the signal in each antenna in a specific way.

Beamforming allows the transmitter to pour most of its data in the spatial direction of the receiver, thus reducing its waste on omnidirectional radiation toward terminals that are not interested in the signal. The receiver can also use the array to raise its effective antenna gain (Sect. 4.5) toward the direction of the transmitter instead of looking for the signal in all directions.

There are two benefits to the directionality introduced above. First, it improves the perceived SNR at the receiver by making sure more useful transmit power reaches the receiver. But it also reduces interference between multiple users. The transmitter is focusing most of its power toward the receiver, thus not affecting other platforms in the environment. The receiver is only looking toward the transmitter, thus avoiding most of the other interfering transmitters in the environment. This reduces collisions, thus improving the performance, especially in multiple access techniques like CSMA-CA. This will translate into less retransmissions and thus less power being expended to communicate the same amount of data.

Beamforming benefits from increasing the number of antennas. More antennas mean the ability to shape an even sharper beam. However, there are always the technical problems of mounting the antennas on a small platform, as well as the computational power required to pre-process and/or post-process the beamformed data. Better technology means the ability to mount more antennas and to process more efficiently, and thus the trend is toward more beamforming.

Beamforming has historically been associated with radar and sonar systems. But 802.11n brought multiple antennas to Wi-Fi and LTE brought them to cellular systems. This makes beamforming a possibility for both. The 802.11n standard acknowledges beamforming as a tacit possibility but does not really provide a standard way to implement it. Using beamforming in 802.11n was thus very rare, with multiple antennas being used either for diversity or spatial multiplexing.

802.11ac and particularly 802.11ax offer more antennas. Both recognize the possibility of using beamforming more systematically and thus explicitly discuss it. Beamforming on 802.11ac access points is thus common.

While beamforming in 4G cellular systems is virtually nonexistent, this will change with the evolution of 5G. 5G cellular systems use high-frequency carriers that offer a larger bandwidth, smaller antennas, and significant challenges to propagation. High-frequency signals have a hard time propagating through barriers and are susceptible to interference. The fact that antennas are small though allows us to throw a lot of antennas at the problem, specifically in the base stations. These antennas can be used to beamform, reducing some propagation problems, particularly interference in the downlink.

Notice that all beamforming techniques are contingent on knowledge of the channel state information. The transmitter needs to modify its coefficients so that the beam is directed toward the receiver. However, to do this it must know the channel state information at the receiver (CSIR), which allows it to determine the position of the receiver. Since only the receiver knows this CSIR, we need a feedback channel to communicate this information back to the transmitter.

A special form of beamforming in multi-user systems that can find its way into 5G systems is multi-user MIMO (MU-MIMO) (Sect. 12.7). In MU-MIMO, platforms may or may not have multiple antennas, but the base station has at least as many antennas as there are multiple users in the environment. The base station will then use channel state information from all the terminals to create a channel matrix that allows it to decouple their transmissions. This is a form of MAC that we have not seen before. Multiple users are not dividing time or frequency; instead they are using spatial diversity to use the exact same channel at the exact same time. Whether MU-MIMO is beamforming or spatial multiplexing is a matter of semantics and philosophy.

9.5 Spatial Multiplexing and True MIMO

Single input (SIMO) and single output (MISO) systems are sometimes classified as MIMO. However, a colloquial use of the acronym preserves MIMO for systems where there are multiple antennas at both the transmitter and the receiver. In such case the system will be able to perform "spatial multiplexing." This means the system will send different data on the different antennas at the same time and in the same frequency band but still be able to decode and receive it due to spatial diversity. This improves the performance of the system by improving its throughput, as opposed to the improvement in PER seen with diversity schemes.

Let us consider a system where the number of receiver antennas and transmitter antennas is the same:

$$N_{tx} = N_{rx}$$

There is no loss of generality in this assumption. When the number of antennas at either end is smaller, that number will limit the number of independent spatial streams, with the balance of antennas at the side with more antennas providing antenna diversity or the ability to beamform.

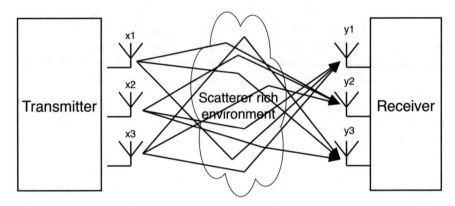

Fig. 9.4 System with three antennas at both the transmitter and the receiver. Each "path" shown here is flat band only in narrowband transmission

Consider the system in Fig. 9.4. There are three transmitter and three receiver antennas. The observation at the first receiver antenna y_1 is:

$$y_1 = h_{11}x_1 + h_{12}x_2 + h_{13}x_3$$

If the system only had antenna 1 at the transmitter and the receiver, then the first term in the above equation would be the "signal" term. But independent data is sent from antennas 2 and 3 at the same time and in the same band. This creates the two last terms of the above equation. These are "interference" terms that destroy the received data on antenna 1 irreversibly.

But we do not have only one receiver antenna; we have three. Thus, we have three observations, with y_2 and y_3 being:

$$y_2 = h_{21}x_1 + h_{22}x_2 + h_{23}x_3$$
$$y_3 = h_{31}x_1 + h_{32}x_2 + h_{33}x_3$$

We have three equations for the three observations. So, what are the knowns and the unknowns in the equations? The three observations y_1, y_2, and y_3 are known. The channel coefficients h_{ij} are all known from channel estimation. Obviously, performing channel estimation for a MIMO system is going to be a little more challenging than in SISO, but as we will see in Chap. 11, it is doable with the right header or pilot structure.

Thus, the only unknowns at the receiver are the sent symbols x_1, x_2, and x_3. In short, we have three equations with three unknowns, which can be solved to obtain the values of the x's. Notice that each channel coefficient h_{ij} is a complex scalar Rayleigh faded coefficient. This is the case in flat-faded channels and also on a per subcarrier basis in OFDM. We will shortly see that MIMO works only in scatterer-rich environments, which are by definition frequency-selective. Thus, MIMO is always associated with the use of OFDM.

In MIMO, it is often helpful to restate the problem in matrix form. The three equations above can be restated as:

$$\begin{bmatrix} y_1 \\ y_2 \\ y_3 \end{bmatrix} = \begin{bmatrix} h_{11} & h_{12} & h_{13} \\ h_{21} & h_{22} & h_{23} \\ h_{31} & h_{32} & h_{33} \end{bmatrix} \begin{bmatrix} x_1 \\ x_2 \\ x_3 \end{bmatrix}$$

And in matrix-vector form:

$$y = H.x$$

which allows us to find the solution for x just by inverting the channel matrix:

$$x = H^{-1}y$$

The above discussion is missing an important factor: noise. We must add an AWGN process to each of the observation equations:

$$y_1 = h_{11}x_1 + h_{12}x_2 + h_{13}x_3 + n_1$$
$$y_2 = h_{21}x_1 + h_{22}x_2 + h_{23}x_3 + n_2$$
$$y_3 = h_{31}x_1 + h_{32}x_2 + h_{33}x_3 + n_3$$

The three noise signals are independent from each other. This is by the very nature of AWGN, which is auto-decorrelated, making it only logical that different processes on different antennas are also independent. This means that at any point in time the three noise signals are different and independent. The noise signals are complex because we are dealing with complex sent baseband signals and observations. However, AWGN power is a function of temperature only. Since the three antennas are usually in the same operating conditions, this means that the variance of the three noise processes is the same.

In matrix form, the MIMO equations become:

$$y = H.x + n$$

Thus, inverting the channel does not produce the sent signals, rather our best estimate of the sent vector:

$$\hat{x} = H^{-1}y$$

The quality of our estimate depends on the amount of noise. We will always assume perfect channel estimates, meaning that we have a true view of channel state information. In a realistic situation, the channel estimation process itself is affected by the fact that preambles and pilots used in channel estimation will be loaded with noise. In other words, we should have said that:

$$\widehat{x} = \widehat{H}^{-1} y$$

But by assuming perfect CSIR, we will always make the following simplification:

$$H \approx \widehat{H}$$

We made an assertion above that is not true. We said that since we have three equations in three unknowns, the equations must have a solution. This is true if and only if the three equations are independent. If one or more equations are dependent, we will not be able to estimate any of the unknowns. Strictly speaking, such a deficient set of equations does not have no solutions; it has an infinite number of solutions. Whether the equations are independent is not always immediately obvious. The equations can be derived from each other through complicated relationships and still be dependent.

The best way to check if a system of equations is independent is to check if the matrix they form is invertible. A matrix that has an inverse must be square and must have a non-zero determinant. In cases where the determinant is zero, the inverse shoots up to infinity; in other words it ceases to exist.

We spent most of Chap. 3 discussing how practical hardware systems use fixed-point registers and perform fixed-point arithmetic. Thus, we must develop a view of the channel matrix where it too is fixed-point. In a fixed-point system, the channel matrix does not have to be truly singular (zero determinant) for us to fail to invert it. Instead, it is enough for the determinant to be so small that it appears to be null in the wordlength of the registers we use.

We can call this "numerical non-invertibility." In reality, and probably on a floating-point processor, the channel is not actually singular. However, because the inverse is very large, it will quickly saturate all the fixed-point registers of the matrix, so that all elements seem to be identical at the maximum value that the register can hold. This is as good as infinity for the fixed-point registers, and the matrix is thus non-invertible on our implementation platform.

This is easier to understand through an example. Consider the following channel matrix, which is 3×3 and consists of three perfectly independent row vectors. The vectors are marked as $H1$, $H2$, and $H3$ respectively:

$$H = \begin{bmatrix} H1 \\ H2 \\ H3 \end{bmatrix} = \begin{bmatrix} 1 & 2 & 3 \\ 3 & 2 & 1 \\ 5 & 15 & 5 \end{bmatrix}$$

The inverse of this matrix is:

$$H^{-1} = \begin{bmatrix} -0.0625 & 0.4375 & -0.05 \\ -0.125 & -0.125 & 0.1 \\ 0.4375 & -0.0625 & -0.05 \end{bmatrix}$$

Now assume that the third vector starts to lose some of its independence so that only 90% of its contents come from the original independent vector, and the remaining 10% is the summation of H1 and H2:

$$H = \begin{bmatrix} H1 \\ H2 \\ 0.9H3 + 0.1(H1 + H2) \end{bmatrix} = \begin{bmatrix} 1 & 2 & 3 \\ 3 & 2 & 1 \\ 4.9 & 13.9 & 4.9 \end{bmatrix}$$

The inverse then becomes:

$$H^{-1} = \begin{bmatrix} -0.0569 & 0.4431 & -0.0556 \\ -0.1361 & -0.1361 & 0.1111 \\ 0.4431 & -0.0569 & -0.0556 \end{bmatrix}$$

The inverse increases ever so slightly. Let us try to introduce more clear dependence by making the third row only 10% dependent on the original third row:

$$H = \begin{bmatrix} H1 \\ H2 \\ 0.1H3 + 0.9(H1 + H2) \end{bmatrix} = \begin{bmatrix} 1 & 2 & 3 \\ 3 & 2 & 1 \\ 4.1 & 5.1 & 4.1 \end{bmatrix}$$

$$H^{-1} = \begin{bmatrix} 0.3875 & 0.8875 & -0.5 \\ -1.025 & -1.025 & 1 \\ 0.8875 & 0.3875 & -0.5 \end{bmatrix}$$

The inverse is clearly increasing.
If the third row of the channel matrix is only 1 part per 10,000 dependent on H3:

$$H = \begin{bmatrix} H1 \\ H2 \\ 0.0001H3 + 0.9999(H1 + H2) \end{bmatrix} = \begin{bmatrix} 1 & 2 & 3 \\ 3 & 2 & 1 \\ 4.0001 & 4.0011 & 4.0001 \end{bmatrix}$$

And the inverse is:

$$H^{-1} = \begin{bmatrix} 0.4999 & 0.5004 & -0.5 \\ -1 & -1.025 & 1 \\ 0.5004 & 0.4999 & -0.5 \end{bmatrix} \times 10^3$$

We can see that the more dependent the vectors, the closer to singularity we get. In a fixed-point system, we do not have to wait until the matrix becomes truly singular. It is enough for numbers in the inverse to become large enough that they saturate the maximum capacity of the fixed-point registers. For example, if we use 8-bit.registers, then the last example will already be singular for all intents and purposes even though the third row is still strictly independent.

Now it is time to consider how much MIMO improves communication. To do this, we look at the improvement to channel capacity as we increase the number of antennas. The advantage of channel capacity as a metric is that it normalizes all other aspects of communication, including bandwidth. The capacity of a faded multi-antenna channel is:

$$c = \log_2 \left\{ \det \left(I_N + \frac{HH^*\text{SNR}}{N} \right) \right\}$$

where N is the number of antennas, which we will assume is the same on both sides of the channel. *SNR* is the SNR at the receiver and H is the channel coefficients matrix. The SNR needs a bit of explanation. This is the SNR of the entire system, meaning that it is the total power on all the antennas divided by the noise on a single antenna. Thus, SNR when divided by N in the expression above is the SNR of a single antenna. To be clear if BN_0 is the noise power on a single antenna, then:

$$\text{SNR} = \frac{\sum_{\text{antennas}} \text{Signal power}}{BN_0}$$

$$\text{SNR}\big|_{\text{single antenna}} = \frac{\sum_{\text{antennas}} \text{Signal power}}{NBN_0}$$

This is a good definition of SNR because it allows us to compare systems with different numbers of antennas fairly. If we had measured SNR on an antenna-by-antenna basis, then a 2×2 system with an SNR of 10 dB would consume double the power of a 1×1 system at 10 dB. According to our definition above, total power at the receiver is the same for the same SNR regardless of the number of antennas.

Another issue we have here is the presence of the channel matrix. This is a stochastic process, and thus the capacity as defined above is not a single number but rather a probability. In Sect. 5.8, this caused us to consider outage rather than measure capacity in bps/Hz. To avoid this, we obtain the expectation of the capacity above:

$$E[c] = E\left[\log_2 \left\{ \det \left(I_N + \frac{HH^*\text{SNR}}{N} \right) \right\} \right]$$

Because the channel is a unity gain process, its expectation and the expectation of its second norm are unity. Thus, the channel capacity is:

$$E[c] = \log_2 \left\{ \det \left(I_N + \frac{\text{SNR}}{N} \right) \right\}$$

We will expand the determinant for a 3×3 case and by induction extend this to larger numbers:

$$E[c] = \log_2 \left\{ \det \left(\begin{bmatrix} 1 + \dfrac{\text{SNR}}{N} & 0 & 0 \\ 0 & 1 + \dfrac{\text{SNR}}{N} & 0 \\ 0 & 0 & 1 + \dfrac{\text{SNR}}{N} \end{bmatrix} \right) \right\}$$

$$E[c] = \log_2 \left\{ \left(1 + \frac{\text{SNR}}{N} \right)^3 \right\}$$

$$E[c] = 3 \log_2 \left\{ 1 + \frac{\text{SNR}}{N} \right\}$$

And by induction, we reach the final and very important expression of capacity under spatial multiplexing:

$$E[c] = N \log_2 \left\{ 1 + \frac{\text{SNR}}{N} \right\}$$

$$E[c] = N \log_2 \left\{ 1 + \text{SNR}|_{\text{single antenna}} \right\}$$

But how do we interpret this result? From the very basic discussion of spatial multiplexing we went through, we expect MIMO to improve our capacity by as many times as we add antennas. So, does this pan out? And if not, how far off are we?

Figure 9.5 shows capacity per Hz on the y-axis versus the number of antennas. There are three curves for 20 dB, 30 dB, and 40 dB. Note that these SNR values are total values as defined above. To reiterate, as the number of antennas drops, the available power is concentrated in the remaining antennas so that the SNR per antenna in an $N \times N$ system is double that in a $2N \times 2N$ system.

The graph shows clearly that capacity increases as we increase the number of antennas. At lower values of N, the increase is close to linear, but as we go further, there are diminishing returns. The linearity keeps going further the higher the SNR. This tells us a few things:

- MIMO is a miracle. It has increased channel capacity without increasing power, bandwidth, coding, or any other aspect of the system.
- MIMO increases capacity, which means that it not only increases the performance of the system but also increases the theoretical upper bound on the system.
- Increasing the number of antennas gives diminishing returns, but we can see that larger antenna arrays are more justified in systems that observe better SNR.

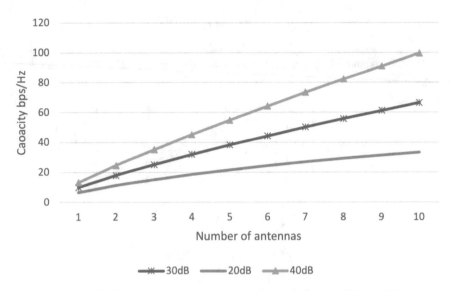

Fig. 9.5 Scaling of MIMO capacity with the number of antennas for three different SNR values

The diminishing returns shown above are fundamental. But there is another limit on MIMO capacity. The derivation above assumed the norm of the channel is unity and that the spatial streams in the MIMO channel are completely independent from each other. If spatial streams are correlated, the channel loses invertibility and the determinant drops, approaching null. The capacity will start to deteriorate as dependent streams start to fail.

Commercial platforms have a certain form factor limited by economics and practical considerations. Increasing the number of antennas while keeping the same form factor can only mean squeezing antennas closer together. This increases correlation between antennas and leads to deterioration of capacity below that predicted in the above discussion.

Thus, the capacity curves in Fig. 9.5 are theoretical upper limits. In practical systems antenna arrays above eight are impractical. In small form factors, even four antennas can be hard to support. The choice of the number of antennas is further compounded by the available transmit power. As shown in the curves of Fig. 9.5 to extract practical improvements out of large antenna arrays, we need high SNR.

9.6 Zero Forcing and SIC

Zero-forcing equalization is a method through which frequency-selective channels in wideband single-carrier systems can be inverted. In the frequency domain, zero forcing simply means multiplying the observation by the inverse of the channel.

But in the context of MIMO-OFDM, zero forcing means nullifying the effect of each antenna on all other antennas given knowledge of the channel matrix. Zero forcing is the simplest and least performing of all the MIMO decoding algorithms. Describing ZF as interference cancellation makes it sound complicated. In reality, all we do is solve the simultaneous observation equations, or in other words we multiply the observation vector by the inverse of the channel matrix:

$$\widehat{x} = H^{-1}y = H^{-1}(Hx + n) = x + H^{-1}n$$

In Sect. 9.9 and onward, we will see algorithms to implement matrix inversion in parallel. In Sect. 9.10, we will see an alternative algorithm for matrix inversion called the singular value decomposition. In Chap. 10, we will discover hardware architectures to efficiently implement all the algorithms we discuss in this chapter. But the bottom line is that zero forcing is simply matrix inversion.

To avoid numerical problems that arise from singular channels and systems with asymmetric number of antennas, we use the pseudoinverse of the channel instead of its inverse:

$$\widehat{x} = H^{+}y$$

where:

$$H^{+} = (H^{*}H)^{-1}H^{*}$$

For non-singular square channel matrices, the pseudoinverse reduces to the inverse. Given the fact that matrix inversion is extremely computationally intensive, the hardware cost of calculating the pseudo-inverse is not significantly more than that of the true inverse, with the added advantage that the pseudoinverse always exists.

While ZF is extremely simple, we can extend it into ZF-SIC (zero forcing, successive interference cancellation). This is an algorithm that uses calculated signal values to successively calculate improved versions of other streams. If you order the sequence so that better SNR signals come later, this can improve performance significantly. This is best explained by an example. In a 2 × 2 system:

$$\begin{bmatrix} y_1 \\ y_2 \end{bmatrix} = \begin{bmatrix} h_{11} & h_{12} \\ h_{21} & h_{22} \end{bmatrix} \begin{bmatrix} x_1 \\ x_2 \end{bmatrix} + \begin{bmatrix} n_1 \\ n_2 \end{bmatrix}$$

$$\begin{bmatrix} \widehat{x}_1 \\ \widehat{x}_2 \end{bmatrix} = H^{+} \begin{bmatrix} y_1 \\ y_2 \end{bmatrix}$$

But:

$$y_1 = h_{11}x_1 + h_{12}x_2 + n_1$$

$$y_2 = h_{21}x_1 + h_{22}x_2 + n_2$$

Now, using the estimate we obtained for x_2:

$$r_1 = y_1 - h_{12}\widehat{x}_2 = h_{11}x_1 + h_{12}(x_2 - \widehat{x}_2) + n_1$$

$$r_2 = y_2 - h_{22}\widehat{x}_2 = h_{21}x_1 + h_{22}(x_2 - \widehat{x}_2) + n_2$$

And ignoring the error in estimating x_2:

$$r_1 = h_{11}x_1 + n_1$$

$$r_2 = h_{21}x_1 + n_2$$

This has now reduced to a receiver diversity situation like Sect. 9.2. The two receiver antennas are providing diversity to the first transmitter antenna. The two samples r_1 and r_2 can be used to maximally combine and calculate an optimal value for x_1:

$$\widehat{x}_1 = \left(h_{11}^* r_1 + h_{21}^* r_2\right) / \left(h_{11}^* h_{11} + h_{21}^* h_{21}\right)$$

While the simplicity of ZF allows such successive algorithms to be well understood, there is nothing stopping us from using them with more complicated algorithms such as MMSE.

9.7   MMSE

Zero forcing has dismal performance. It assumes we know nothing about noise in the channel and thus produces a suboptimal solution. Specifically, ZF will multiply the observations by a very large gain when the channel has a deep fade (H is small). However, it also ends up multiplying noise by the same gain, which leads to a deterioration of the observed SNR for all streams.

Assume a 2×2 system:

$$\begin{bmatrix} y_1 \\ y_2 \end{bmatrix} = \begin{bmatrix} h_{11} & h_{12} \\ h_{21} & h_{22} \end{bmatrix} \begin{bmatrix} x_1 \\ x_2 \end{bmatrix} + \begin{bmatrix} n_1 \\ n_2 \end{bmatrix}$$

The inverse of the 2×2 channel can be calculated heuristically as:

$$\begin{bmatrix} \widehat{x}_1 \\ \widehat{x}_2 \end{bmatrix} = \left\{ \begin{bmatrix} h_{22} & -h_{12} \\ -h_{21} & h_{11} \end{bmatrix} \begin{bmatrix} y_1 \\ y_2 \end{bmatrix} + \begin{bmatrix} h_{22} & -h_{12} \\ -h_{21} & h_{11} \end{bmatrix} \begin{bmatrix} n_1 \\ n_2 \end{bmatrix} \right\} * 1/\det$$

Now assume that spatial stream 1 is weak. This means that signals going from transmitter antenna 1 to receiver antenna 1 are highly attenuated, at least relative to antenna 2. This means that $h_{11} \ll h_{22}$. To compensate for this, the inverse multiplies y_1 by the large value of h_{22} to retrieve $\hat{x}_1$. Unfortunately, it also ends up multiplying the noise on antenna 1 by this same large value. Thus, ZF leads to magnification of noise on the weak stream because it is trying to retrieve the weak signal.

The problem here is that ZF is not trying to find a good solution; it is just trying to find *a* solution. We need an estimator that can distinguish two cases: a case where the path is weak where it needs to multiply by a smaller gain to avoid maximizing noise and a case where the path is strong and it can retrieve a strong signal.

Specifically, we need to find the estimate $\hat{x}$ that minimizes the error. The error is the difference between the estimator and the true signal. Because errors are zero mean, we find the minimum of the square error. And because error is stochastic, we should minimize the mean of the square error rather than the square of the error; thus, we find $\hat{x}$ that satisfies:

$$\min E\left\{ (x - \hat{x})^2 \right\}$$

This is called the minimum mean square error (MMSE) estimator, and we can show that the estimate that achieves this is:

$$\hat{x} = H^*(HH^* + I/\text{SNR})^{-1}y$$

When all SNRs are very high, the MMSE solution reduces to ZF as the SNR matrix tends toward null. This makes sense because high SNR means that signals are strong, and noise is very weak on all spatial streams. In such case, the ZF solution cannot mess up the solution and will be optimal.

The SNR matrix is a diagonal matrix containing the reciprocals of the SNR values of each antenna in order. So, for a 3×3 antenna, the matrix is:

$$I/\text{SNR} = \begin{bmatrix} 1/\text{SNR}_1 & 0 & 0 \\ 0 & 1/\text{SNR}_2 & 0 \\ 0 & 0 & 1/\text{SNR}_3 \end{bmatrix}$$

For spatial streams with low SNR, the SNR matrix adds a large number to h_{xx}; this prevents the inverse from being too magnified. But why do we specifically choose to do this? In other words, why not add double the reciprocal of the SNR, or half? The answer is clear from the derivation above; this is the estimator that minimizes the mean square error.

If all the streams have the same SNR, then the SNR matrix will be a diagonal matrix with identical values. It makes sense that noise power on all antennas will be the same because they have the same temperature. Notice that this does not mean that the noise processes on the three antennas are identical or even correlated; they just have the same power.

SNR values on the three antennas will thus depend solely on the signal powers on the antennas which will depend on how power was distributed at the transmitter and how we define the channel matrix. For example, we normally do not include fast fading variations in signal power in the calculation of the SNR; these are rolled into the channel coefficients. However, long-term fading and shadowing should affect the values of SNR. Thus, it makes sense to not assume that the SNR values on the different antennas are identical.

In fact, the MMSE estimator does not assume anything about the shape of the SNR matrix. It does not need to be diagonal, and the estimator works for a general matrix G:

$$\widehat{x} = H^*(HH^* + G)^{-1}y$$

This matrix can be non-diagonal if the noise is spatially colored, meaning that there is some correlation between the noise on the different antennas. This also works in cases where there is interference instead of or in addition to noise.

9.8 Maximum Likelihood Decoding

In Sect. 5.6, we measured the distance between received observations and the possible sent signals to determine the most likely sent symbol. This was a simple and effective way to decode the sent symbols. In fact, in Sect. 5.6 we said that the solution is optimal.

So why not do the same for MIMO? In fact, we did something very similar with both ZF and MMSE in Sects. 9.6 and 9.7. In both sections we talked about how to obtain estimates $\widehat{x}$ of sent symbols. But these estimators will not coincide with any of the symbols on the constellation. Thus, after obtaining $\widehat{x}$, we take its elements one by one and do hard decision to obtain the closest constellation symbol for each antenna.

In short, we inverted the channel, multiplied by the observation, then did hard decision decoding on each stream individually as if it were a SISO channel. However, neither ZF nor MMSE will lead to maximum likelihood performance as described for SISO in Sect. 5.6, and we want to find the ML solution because it gives the optimal BER-SNR performance.

Inverting the channel then multiplying by this inverse to obtain $\widehat{x}$ leads to loss of information. Every observation y equation includes information from all the transmitted signals through the channel coefficients. When we invert, we are separating these dependencies and causing each estimator to be dependent only on the corresponding observation.

In other words, to reach the optimal solution we need to measure distances between vectors instead of scalars. We should measure and minimize distances the same way we did in Sect. 5.6, but thinking in terms of matrices and vectors:

$$\hat{x} = \min_{x} \; |\,|x - H^{-1}y|\,|$$

y is the entire observation vector. x is a vector possibility of the symbols sent from the transmitter. $\hat{x}$ is a vector of the optimal estimators.

The channel is still inverted to reverse its effect, but distance measurement is done on the whole vector, rather on an element-by-element basis. The product $H^{-1}y$ is a column vector. The possible sent x is also a vector, and thus its difference from $H^{-1}y$ is also a vector. We measure the distance as the second norm of this vector; thus, the distance between vectors a and b is:

$$\text{distance} = ||a - b||$$

$$= \sqrt{(a_0 - b_0)^2 + (a_1 - b_1)^2 + (a_2 - b_2)^2 + \ldots (a_{N-1} - b_{N-1})^2}$$

The square root is very annoying to calculate in hardware. But as in the SISO case (Sect. 5.6), the $\hat{x}$ that minimizes the distance is also the $\hat{x}$ that minimizes the square of the distance, and thus:

$$\hat{x} = \min_{x} ||x - H^{-1}y||^2$$

The maximum likelihood solution is nearly impossible to calculate in MIMO systems with many antennas and a high modulation order. To understand why, we need an example. Consider a 3×3 system that uses QPSK modulation. How many vectors x do we need to measure the distance of y from? The first element of the vector x could take four values corresponding to the four QPSK symbols. But for each of the four possibilities for the first element, the second element can also have four possibilities. And for each possibility of the second element, the third element can have four of its own.

Thus, in the 3×3 QPSK system, we need to compare the distance to $4 \times 4 \times 4 = 4^3$ possible sent vectors. And in general, in a system with k-QAM and N_{rx} spatial streams, we need to perform the following number of distance calculations:

$$\text{number of nodes visited} = k^{N_{rx}}$$

In 802.11n we could go up to 64QAM with four spatial streams, requiring 64^4 nodes to be visited. By node visits we mean distances calculated from y. This is over 16 million comparisons. This is untenable. The situation is much worse in 802.11 ac where we can use 256 QAM with 8 spatial streams, requiring $1.8 * 10^{19}$ calculations. But even for modest systems, the ML problem in MIMO is still computationally prohibitive. Even with three antennas and 16QAM, we need 4096 calculations, which is extremely difficult to perform with a good throughput.

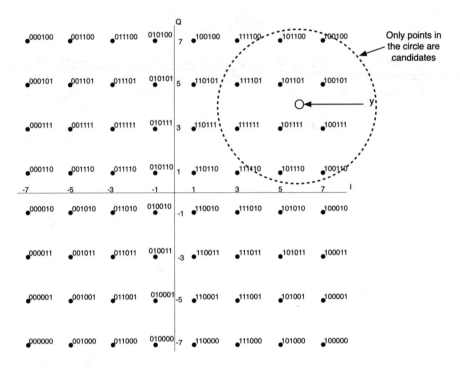

Fig. 9.6 Visiting fewer nodes in a SISO system

To be clear, the problem with the ML solution in MIMO is not that the distance calculation itself is complicated. It is nearly identical in complexity to the calculation in the SISO case. The main issue is that the problem scales horribly. And yet, the performance of ML in terms of BER is something we cannot ignore. It is so tempting that we must find a viable way to do it.

Figure 9.6 shows a 64 QAM single antenna constellation with a single observation point y. To find the ML solution for this observation, we need to calculate the distance to 64 constellation points and find the point that minimizes the distance. But we showed in Sect. 5.6 that successive comparisons can quickly cast aside many possible candidates. For example, if we notice that both the real and imaginary parts of y are positive, we can eliminate ¾ of all constellation points as candidates, and we need to perform only 16 comparisons.

In Fig. 9.6, the observation point y must be already multiplied by the channel inverse before the distance can be calculated to the constellation points. A systematic way to perform the eliminations in the previous paragraph is to draw a circle around the observation point. Constellation points that lie within this circle should be visited. Points outside it can safely be ignored because the point with minimum distance cannot be outside the circle if the circle is not empty.

There is a need to intelligently pick the radius of the circle in Fig. 9.6. If the circle is too small, we run the risk of having no constellation points within. If the circle is too large, then we visit way more points than we need to. In the extreme, if the circle is huge, it will include all the points of the constellation and we are back to measuring k distances. The radius of the circle is evidently related to SNR. The less noise power in the system, the tighter the circle can be, because the closer the observation is to its closest constellation point.

The above discussion can be extended to MIMO. However, if we first invert the channel and multiply this by the observation vector, we end up decoupling the observations from each other, reducing to the zero-forcing solution. We must measure vector norms as discussed above. The comparison takes place in N_{rx} dimensions, and around the observation vector we draw a hypersphere in that many dimensions. Candidate vectors within the hypersphere are visited. Those outside it are not. This is a class of MIMO decoding called sphere decoding. We need a thorough understanding of QR decomposition before we further discuss sphere decoding in Sect. 9.13.

The ML solution is the maximum theoretical performance that a system can provide. But in Sect. 9.7, we also deduced that the MMSE solution is optimal. Is there a contradiction here? No, the ML solution is optimal; it gives us the best estimates of sent symbols that give the lowest possible BER. MMSE simply reduces the mean square error between the estimate and the sent symbol.

Another way of stating this is that MMSE assumes the possible symbols that could have been sent are any point in the entire plane. After the MMSE estimator is calculated, we need to perform hard or soft decisions on each of the estimates to obtain the demapped bits. The ML solution recognizes that only certain points in the plane are possible and calculates the most likely among them; thus, it produces the true optimal solution.

9.9   QR Decomposition for Matrix Inversion

This chapter is coming to a clear conclusion: whatever you do, MIMO decoding seems to involve matrix inversion. So, we must find ways to invert matrices in hardware. To do that, we first describe algorithms for matrix inversion and then translate said algorithms to hardware. We could find the inverse by calculating the determinant and the adjoints the same way we do in a hand calculation. This is viable only for very small matrices, usually falling apart in hardware for anything larger than 2×2. So, we must find more scalable algorithms.

Matrices with special properties are easier to invert than general matrices. By special properties we usually mean a sort of sparsity. Special matrices include diagonal matrices, upper-triangular matrices, and lower-triangular matrices. The channel matrix is none of the above. It is a general complex matrix and is thus very hard to invert. Therefore, we will first decompose the channel matrix into matrices with better properties before doing the inversion.

A matrix decomposition is the process of breaking down a matrix into the product of several constituent matrices. Some matrix decompositions exist for all matrices; some exist only for non-singular matrices, some exist only for square matrices, and others exist for matrices of arbitrary dimensions.

One of the most common decompositions we come across in linear algebra is the LU decomposition which breaks down the matrix into an upper triangular matrix and a lower triangular matrix. Both triangular matrices are much easier to invert than a general matrix. However, the LU decomposition is hard to adapt to hardware and is thus more commonly seen as part of software libraries.

In hardware, the QR decomposition is more common. The QR decomposition breaks down a general matrix H into two matrices: Q and R. R is an upper triangular matrix. Q is a unitary matrix. Unitary matrices are unity norm matrices whose inverse is also their Hermitian (complex transpose), thus:

$$QQ^* = Q^*Q = I$$

The QR decomposition exists for square and rectangular matrices where the number of rows is greater than the number of columns. In square matrices both R and Q have the same dimension as the original matrix. In rectangular matrices with dimensions $m \times n$ with $m > n$, the Q matrix is still square at $m \times m$ and R is rectangular at $m \times n$. The R matrix is triangular only in the top nxn section, with the bottom $m - n$ rows being null.

So, why is the QR decomposition useful? Because it finally means we can calculate the inverse of a square channel matrix:

$$H^{-1} = (QR)^{-1} = R^{-1}Q^{-1} = R^{-1}Q^*$$

The calculation of Q^* in hardware could not be simpler. Transposition simply means we read the matrix column by column rather than row by row. This can be done by adapting the address generator, essentially a zero-cost operation. Conjugation can be done using a single adder per element to calculate its negative. R^{-1} is also easy to calculate because R is upper triangular. Its inverse will also be upper triangular. This allows us to use successive simplifications to find R^{-1}.

Taking the 3×3 case as an example, and assuming the inverse of the upper triangular matrix is V:

$$RV = VR = I = \begin{bmatrix} R11 & R12 & R13 \\ 0 & R22 & R23 \\ 0 & 0 & R33 \end{bmatrix} \begin{bmatrix} V11 & V12 & V13 \\ 0 & V22 & V23 \\ 0 & 0 & V33 \end{bmatrix} = \begin{bmatrix} 1 & 0 & 0 \\ 0 & 1 & 0 \\ 0 & 0 & 1 \end{bmatrix}$$

This allows us to readily calculate $V33$:

$$V33 = \frac{1}{R33}$$

And $V22$:

$$V22 = \frac{1}{R22}$$

And $V11$:

$$V11 = \frac{1}{R11}$$

The non-diagonal elements are slightly more complicated to calculate. But only slightly so. For example, looking at element $V12$:

$$R11.V12 + R12.V22 = 0$$

But since we already know $V22$:

$$V12 = -\frac{R12}{R22.R11}$$

We can use similar successive derivations to find heuristic equations for all elements of the inverse matrix that do not require the extensive calculations of the adjoints of a general matrix. Specifically:

$$R22.V23 + R23.V33 = 0$$

$$V23 = -\frac{R23}{R22.R33}$$

And:

$$R11.V13 + R12.V23 + R13.V33 = 0$$

$$V13 = \frac{R12.R23}{R11.R22.R33} - \frac{R13}{R33.R11}$$

This is the first time we find a way to implement matrix inversion that can be translated to hardware.

There are still two huge gaps on the quest for MIMO decoding:

- How do we perform QR decomposition? The algorithm is discussed in detail in Sect. 9.11 for real matrices and Sect. 9.12 for complex matrices
- How do we implement any of this in hardware? This is the topic of Chap. 10.

9.10 Decoding by Using the Singular Value Decomposition (SVD)

The QR decomposition is our first experience with being able to calculate a matrix inverse. But there is a decomposition that could be considered the "ultimate" in terms of how easy its products are to invert as well as what it tells us about the channel. This is called the singular value decomposition (SVD).

SVD exists for any matrix of any size, whether or not it is invertible. The singular value decomposition breaks down a general matrix into three matrices. We are again going to consider mainly square matrices:

$$H = USV$$

Both U and V are unitary matrices. S is a diagonal matrix. This means inverse calculation is now trivial, because:

$$H^{-1} = (USV)^{-1} = V^*S^{-1}U^*$$

The hardware cost of the conjugate transpose is very small. Calculating the inverse of a diagonal matrix is also very simple, because it is a diagonal matrix with the reciprocal of diagonal elements:

$$S^{-1} = \left(\begin{bmatrix} S1 & 0 & 0 \\ 0 & S2 & 0 \\ 0 & 0 & S3 \end{bmatrix} \right)^{-1} = \begin{bmatrix} \dfrac{1}{S1} & 0 & 0 \\ 0 & \dfrac{1}{S2} & 0 \\ 0 & 0 & \dfrac{1}{S3} \end{bmatrix}$$

The natural question now is why bother with anything other than the SVD? Why even investigate the QR decomposition? Nothing could be better than SVD. The missing link here is that we are only considering the computational effort in calculating the inverse once the decomposition has happened. We have not yet considered how the decomposition itself is implemented in hardware. In Chap. 10 we will discover that the hardware cost of complex QR decomposition is much lower than that of complex SVD.

But the SVD is not only valuable because of how easy it makes inversion. Its real value lies in the insight it gives about the channel. SVD breaks down the channel into what we have always wished it was: multiple parallel SISO channels. Not only that but it also tells us which of these "SISO" channels is "strong" and which is "bad." The singular values in the S matrix are representative of the strengths of the different spatial streams.

The benefits of SVD are clearest when there is a feedback channel from the receiver to the transmitter. Most standards allow such a low bandwidth channel. This is not a reciprocal link where the transmitter and the receiver switch roles but rather a low throughput side channel where the receiver sends information, usually channel state information (CSIR), back to the transmitter.

Assume, for example, that the singular value matrix contains a null element at position ii. This means that antenna i is completely dead, perhaps because of the presence of a deep fade. If this information is known to both the transmitter and the receiver, they can agree not to transmit on said antenna. The MIMO order of the system would then be reduced by 1 for the duration of this fade; however, communication could still take place on the remaining antennas. Without SVD and the feedback channel, we would find the channel to be singular, fail in communication, and there would be no data for the whole period of outage.

It does not even need to be this extreme. Even with non-null singular values, the transmitter can make many interesting decisions based on SVD:

- With a total available power P_t, the transmitter may decide to distribute the power unequally among the transmit antennas. This usually involves putting more power into antennas with larger singular values to leverage their stronger channel. This is known as "water-filling."
- The transmitter may, if the standard allows it, use different modulation on different antennas. Stronger spatial streams can use larger QAM constellations to increase throughput, while streams with lower singular values use tighter constellations to reduce BER.
- The transmitter can decide to use different coding rates on different spatial streams, using higher rates on streams with higher singular values. Notice that most standards do not allow variations of MCS among transmit antennas, so the limit here is usually the standard.

One of the simplest ways that the transmitter can use CSI is to feedback the V matrix only to the transmitter. This allows the transmitter to transmit a distorted version of x:

$$\tilde{x} = V^*x$$

At the receiver, the observation vector is:

$$y = H\tilde{x} + n$$
$$y = HV^*x + n$$

And expanding the channel matrix:

$$y = USVV^*x + n = USx + n$$

To extract the estimates:

$$\hat{x} = S^{-1}U^*y$$

So, we have split the job of inverting the channel between the transmitter and the receiver by pre-distorting the transmitted signal. The SVD allowed us to do this by splitting the channel into easier components. Because the transmitter is usually less computationally loaded than the receiver, this allows the receiver to offload some of the computation to the transmitter. As with QR decomposition, we have not yet explained how to perform the decomposition nor how to implement it in hardware. The algorithm to perform SVD is explained in Sect. 9.14 and its implementation in hardware in Sect. 10.10.

9.11   Givens Rotations

We still have not described how to do either QR decomposition or SVD. In Sects. 9.9 and 9.10, we assumed that the decomposition was already performed and then used the desirable properties of the factors to invert the channel. In this section, we describe a very important *unitary transformation* called the Givens rotation. Givens rotations allow us to perform QR decomposition, but it also gives us an excellent visual representation of what happens during decomposition, which will form the basis of hardware implementation in Chap. 10.

A unitary transformation is the multiplication of a vector by a unitary matrix. Givens rotations are unitary transformations which rotate a point in the plane by a certain angle. The rotation is named after someone called Wallace Givens, so it is always spelled "Givens." There are two ways to understand Givens rotations, either analytically or graphically. We require an understanding of both.

A Givens rotation of a vector $\begin{bmatrix} C \\ D \end{bmatrix}$ by an angle θ is the following multiplication:

$$\begin{bmatrix} \cos\theta & \sin\theta \\ -\sin\theta & \cos\theta \end{bmatrix} \begin{bmatrix} C \\ D \end{bmatrix} = \begin{bmatrix} G \\ K \end{bmatrix}$$

If the angle θ is calculated so that it is the arctan of the vector, then the resulting vector will have a zero component:

$$\begin{bmatrix} \cos\theta & \sin\theta \\ -\sin\theta & \cos\theta \end{bmatrix} \begin{bmatrix} A \\ B \end{bmatrix} = \begin{bmatrix} E \\ 0 \end{bmatrix}, \quad \text{where } \theta = \tan^{-1}\left(\frac{B}{A}\right)$$

The rotation is best understood when seen graphically. In Fig. 9.7, we see two points in the complex plane, point (A, B) and point (C, D). The vector formed by point (A, B) is rotated by the same angle that it forms with the x-axis, causing it to rotate till it lies on the x-axis. This tells us that:

Fig. 9.7 Graphical
representation of Givens
rotations

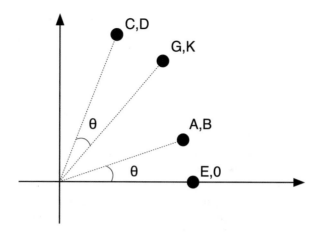

$$E = \sqrt{A^2 + B^2}$$

The vector formed by point (C, D) is also rotated by the same angle. Because this angle θ is not the angle that the vector (C, D) forms with the x-axis, the resulting point (G, K) does not lie on the x-axis. But the length of the vector remains the same during rotation:

$$\sqrt{C^2 + D^2} = \sqrt{G^2 + K^2}$$

We can see the unitary nature of the transformation in how it preserves vector length. Givens rotation is interesting in the context of QR decomposition because it nulls elements in vectors. If we combine the two vectors above, we end up with a unitary transformation of a matrix:

$$\begin{bmatrix} \cos\theta & \sin\theta \\ -\sin\theta & \cos\theta \end{bmatrix} \begin{bmatrix} A & C \\ B & D \end{bmatrix} = \begin{bmatrix} E & G \\ 0 & K \end{bmatrix}$$

The nulling of the off-diagonal element means that the matrix is now upper triangular. Thus:

$$\Psi \begin{bmatrix} A & C \\ B & D \end{bmatrix} = R$$

$$\begin{bmatrix} A & C \\ B & D \end{bmatrix} = \Psi^* R = QR$$

$$Q = \Psi^* = \begin{bmatrix} \cos\theta & -\sin\theta \\ \sin\theta & \cos\theta \end{bmatrix}$$

$$R = \begin{bmatrix} E & G \\ 0 & K \end{bmatrix}$$

There are two steps remaining to allow the Givens rotation to be used for QR decomposition of channel matrices:

- The channel matrix is complex; the above decomposition works only on real matrices. We will extend Givens rotations to complex 2×2 matrices in Sect. 9.12.
- The channel matrix is not generally 2×2. We will extend real Givens rotations to larger matrices below and use the exact reasoning to extend it for complex matrices in Sect. 9.12.

Let us extend Givens rotations to 3×3 matrices. The result will be immediately scalable for larger matrices:

$$\begin{bmatrix} A11 & A12 & A13 \\ A21 & A22 & A23 \\ A31 & A32 & A33 \end{bmatrix}$$

We begin by nulling element $A21$. To do this we start with a 3×3 identity matrix:

$$\begin{bmatrix} 1 & 0 & 0 \\ 0 & 1 & 0 \\ 0 & 0 & 1 \end{bmatrix}$$

Then calculate a Givens rotation angle using the diagonal element of the column where we want to null an element as a pivot. Because we want to null $A21$, we use $A11$ as a pivot, calculating the rotation angle as:

$$\theta_1 = \tan^{-1}\left(\frac{A21}{A11}\right)$$

A Givens rotation sub-matrix, identical to the 2×2 case, is inserted into the identity matrix at locations that reflect the element we want to null. If we want to null element ij, we insert the Givens factors into elements ii, jj, ij, and ji of the identity matrix. So, in this case, the unitary transformation matrix is:

$$\Psi_1 = \begin{bmatrix} \cos\theta_1 & \sin\theta_1 & 0 \\ -\sin\theta_1 & \cos\theta_1 & 0 \\ 0 & 0 & 1 \end{bmatrix}$$

And performing this unitary transformation, we rotate every element in row 1 with the corresponding element in row 2 and end up nulling element A21:

$$\begin{bmatrix} \cos\theta_1 & \sin\theta_1 & 0 \\ -\sin\theta_1 & \cos\theta_1 & 0 \\ 0 & 0 & 1 \end{bmatrix} \begin{bmatrix} A11 & A12 & A13 \\ A21 & A22 & A23 \\ A31 & A32 & A33 \end{bmatrix} = \begin{bmatrix} A11' & A12' & A13' \\ 0 & A22' & A23' \\ A31 & A32 & A33 \end{bmatrix}$$

Similarly, to null element 31, we calculate the angle:

$$\theta_2 = \tan^{-1}\left(\frac{A31}{A11'}\right)$$

And populate the unitary transformation matrix:

$$\Psi_2 = \begin{bmatrix} \cos\theta_2 & 0 & \sin\theta_2 \\ 0 & 1 & 0 \\ -\sin\theta_2 & 0 & \cos\theta_2 \end{bmatrix}$$

And nulling the element 31 will cause every element in the first row to be rotated with the corresponding element in the third row:

$$\begin{bmatrix} \cos\theta_2 & 0 & \sin\theta_2 \\ 0 & 1 & 0 \\ -\sin\theta_2 & 0 & \cos\theta_2 \end{bmatrix} \begin{bmatrix} A11' & A12' & A13' \\ 0 & A22' & A23' \\ A31 & A32 & A33 \end{bmatrix} = \begin{bmatrix} A11'' & A12'' & A13'' \\ 0 & A22' & A23' \\ 0 & A32' & A33' \end{bmatrix}$$

And finally, to null element 32, calculate the following angle. Note that we have moved to using element 22 as a pivot:

$$\theta_3 = \tan^{-1}\left(\frac{A32'}{A22'}\right)$$

$$\Psi_3 = \begin{bmatrix} 1 & 0 & 0 \\ 0 & \cos\theta_3 & \sin\theta_3 \\ 0 & -\sin\theta_3 & \cos\theta_3 \end{bmatrix}$$

Leading to the result:

$$\begin{bmatrix} 1 & 0 & 0 \\ 0 & \cos\theta_3 & -\sin\theta_3 \\ 0 & \sin\theta_3 & \cos\theta_3 \end{bmatrix} \begin{bmatrix} A11'' & A12'' & A13'' \\ 0 & A22' & A23' \\ 0 & A32' & A33' \end{bmatrix} = \begin{bmatrix} A11'' & A12'' & A13'' \\ 0 & A22'' & A23'' \\ 0 & 0 & A33'' \end{bmatrix}$$

And thus:

$$\Psi_3\Psi_2\Psi_1 \begin{bmatrix} A11 & A12 & A13 \\ A21 & A22 & A23 \\ A31 & A32 & A33 \end{bmatrix} = \begin{bmatrix} A11'' & A12'' & A13'' \\ 0 & A22'' & A23'' \\ 0 & 0 & A33'' \end{bmatrix} = R$$

$$A = (\Psi_3\Psi_2\Psi_1)^* R = QR$$

$$Q = \Psi_1^*\Psi_2^*\Psi_3^*$$

Notice that the order in which off-diagonal elements are nulled makes all the difference. There is no alternative to the order above. We null the elements in the first column top to bottom, then move to the next column top to bottom, and so on.

The reason for this is that any other order runs the risk of regenerating nulled elements. For example, assume we null element 21 first then move to element 32. Nulling element 32 requires rotating the second and third rows of the matrix. This involves rotating element 21 with element 31. Element 31 has not been nulled, and thus element 21 will become non-null. This would not have happened if we had nulled 31 first before moving on to element 32.

9.12   Complex Givens Rotations and Channel Inversion

The channel matrix is complex. Thus, we need to modify Sect. 9.11 to work with complex matrices. The complex Givens rotation for 2×2 matrices is fairly similar to real Givens rotation. But it has one limitation: it can only work with a matrix where the leading element, specifically element $(1, 1)$, is pure real.

In a 2×2 complex channel matrix H:

$$H = \begin{bmatrix} Ae^{j\phi_a} & Be^{j\phi_{b'}} \\ Ce^{j\phi_c} & De^{j\phi_d} \end{bmatrix}$$

The leading element is complex, and the complex Givens algorithm requires it to be pure real. Fortunately, we can use a very simple unitary transformation to take care of this:

$$Q1 = \begin{bmatrix} e^{-j\phi_a} & 0 \\ 0 & 1 \end{bmatrix}$$

$$Q1 * H = \begin{bmatrix} e^{-j\phi_a} & 0 \\ 0 & 1 \end{bmatrix} \begin{bmatrix} Ae^{j\phi_a} & Be^{j\phi_{b'}} \\ Ce^{j\phi_c} & De^{j\phi_d} \end{bmatrix} = \begin{bmatrix} A & Be^{j\phi_b} \\ Ce^{j\phi_c} & De^{j\phi_d} \end{bmatrix}$$

where:

$$\phi_b = \phi_{b'} - \phi_a$$

$Q1$ is a unitary transformation because:

$$\begin{bmatrix} e^{-j\phi_a} & 0 \\ 0 & 1 \end{bmatrix} \begin{bmatrix} e^{j\phi_a} & 0 \\ 0 & 1 \end{bmatrix} = \begin{bmatrix} e^{j\phi_a} & 0 \\ 0 & 1 \end{bmatrix} \begin{bmatrix} e^{-j\phi_a} & 0 \\ 0 & 1 \end{bmatrix} = \begin{bmatrix} 1 & 0 \\ 0 & 1 \end{bmatrix}$$

Once the matrix has a real leading element, we can use the complex Givens rotation to nullify element (2,1):

$$\begin{bmatrix} \cos\theta & \sin\theta e^{-j\phi} \\ -\sin\theta & \cos\theta e^{-j\phi} \end{bmatrix} \begin{bmatrix} A & Be^{j\phi_b} \\ Ce^{j\phi_c} & De^{j\phi_d} \end{bmatrix} = \begin{bmatrix} E & Ge^{j\phi_g} \\ 0 & Ke^{j\phi_k} \end{bmatrix}$$

The complex Givens rotation uses two phases θ and ϕ:

$$\phi = \phi_c$$

$$\theta = a\tan\left(\frac{C}{A}\right)$$

The complex Givens rotation is significantly more complicated than its real counterpart and requires division and arctans to find the various phases. However, in Chap. 10, we will find a very efficient way to implement it in hardware.

Only one question remains which is how to extend the complex Givens rotation to matrices larger than 2×2. We can and will take a cue from the real case in Sect. 9.11. However, there is a slight modification to take care of the necessity of having a real leading element.

As in the pure real case, we start with a general matrix A:

$$\begin{bmatrix} A11e^{j\phi_{11}} & A12e^{j\phi_{12}} & A13e^{j\phi_{13}} \\ A21e^{j\phi_{21}} & A22e^{j\phi_{22}} & A23e^{j\phi_{23}} \\ A31e^{j\phi_{31}} & A32e^{j\phi_{32}} & A33e^{j\phi_{33}} \end{bmatrix}$$

The first step is to transform element $A11$ to pure real. This will allow us to use complex Givens rotations to null the off-diagonal elements of the first column. Thus, the first unitary transformation is:

$$
U_1 \begin{bmatrix} A11e^{j\phi_{11}} & A12e^{j\phi_{12}} & A13e^{j\phi_{13}} \\ A21e^{j\phi_{21}} & A22e^{j\phi_{22}} & A23e^{j\phi_{23}} \\ A31e^{j\phi_{31}} & A32e^{j\phi_{32}} & A33e^{j\phi_{33}} \end{bmatrix} = \begin{bmatrix} e^{-j\phi_{11}} & 0 & 0 \\ 0 & 1 & 0 \\ 0 & 0 & 1 \end{bmatrix}
$$

$$
\times \begin{bmatrix} A11e^{j\phi_{11}} & A12e^{j\phi_{12}} & A13e^{j\phi_{13}} \\ A21e^{j\phi_{21}} & A22e^{j\phi_{22}} & A23e^{j\phi_{23}} \\ A31e^{j\phi_{31}} & A32e^{j\phi_{32}} & A33e^{j\phi_{33}} \end{bmatrix}
$$

$$
= \begin{bmatrix} A11 & A12e^{j\phi_{12}'} & A13e^{j\phi_{13}'} \\ A21e^{j\phi_{21}} & A22e^{j\phi_{22}} & A23e^{j\phi_{23}} \\ A31e^{j\phi_{31}} & A32e^{j\phi_{32}} & A33e^{j\phi_{33}} \end{bmatrix}
$$

The second unitary transformation nulls element 2,1 and is:

$$
U_2 = \begin{bmatrix} \cos\theta_1 & \sin\theta_1 e^{-j\phi_1} & 0 \\ -\sin\theta_1 & \cos\theta_1 e^{-j\phi_1} & 0 \\ 0 & 0 & 1 \end{bmatrix}
$$

$$
\begin{bmatrix} \cos\theta_1 & \sin\theta_1 e^{-j\phi_1} & 0 \\ -\sin\theta_1 & \cos\theta_1 e^{-j\phi_1} & 0 \\ 0 & 0 & 1 \end{bmatrix} \begin{bmatrix} A11 & A12e^{j\phi_{12}'} & A13e^{j\phi_{13}'} \\ A21e^{j\phi_{21}} & A22e^{j\phi_{22}} & A23e^{j\phi_{23}} \\ A31e^{j\phi_{31}} & A32e^{j\phi_{32}} & A33e^{j\phi_{33}} \end{bmatrix}
$$

$$
= \begin{bmatrix} A11' & A12'e^{j\phi_{12}''} & A13'e^{j\phi_{13}''} \\ 0 & A22'e^{j\phi_{22}'} & A23'e^{j\phi_{23}'} \\ A31e^{j\phi_{31}} & A32e^{j\phi_{32}} & A33e^{j\phi_{33}} \end{bmatrix}
$$

where:

$$
\phi_1 = \phi_{21}
$$

$$
\theta_1 = a\tan\left(\frac{A21}{A11}\right)
$$

Because the leading element will remain real, we can immediately move to nulling element 3,1:

$$U_3 = \begin{bmatrix} \cos\theta_2 & 0 & \sin\theta_2 e^{-j\phi_2} \\ 0 & 1 & 0 \\ -\sin\theta_2 & 0 & \cos\theta_2 e^{-j\phi_2} \end{bmatrix}$$

$$\begin{bmatrix} \cos\theta_2 & 0 & \sin\theta_2 e^{-j\phi_2} \\ 0 & 1 & 0 \\ -\sin\theta_2 & 0 & \cos\theta_2 e^{-j\phi_2} \end{bmatrix} \begin{bmatrix} A11' & A12'e^{j\phi_{12}''} & A13'e^{j\phi_{13}''} \\ 0 & A22'e^{j\phi_{22}'} & A23'e^{j\phi_{23}'} \\ A31e^{j\phi_{31}} & A32e^{j\phi_{32}} & A33e^{j\phi_{33}} \end{bmatrix}$$

$$= \begin{bmatrix} A11'' & A12''e^{j\phi_{12}'''} & A13''e^{j\phi_{13}'''} \\ 0 & A22'e^{j\phi_{22}'} & A23'e^{j\phi_{23}'} \\ 0 & A32'e^{j\phi_{32}'} & A33'e^{j\phi_{33}'} \end{bmatrix}$$

$$\phi_1 = \phi_{31}$$

$$\theta_1 = a\tan\left(\frac{A31}{A11'}\right)$$

As in the pure real case, we must move between elements in a very specific order to avoid the regeneration of null elements. We move to the second column to eliminate element $(3,2)$. To do this, we first make the diagonal element $(2,2)$ real:

$$U_4 \begin{bmatrix} A11'' & A12''e^{j\phi_{12}'''} & A13''e^{j\phi_{13}'''} \\ 0 & A22'e^{j\phi_{22}'} & A23'e^{j\phi_{23}'} \\ 0 & A32'e^{j\phi_{32}'} & A33'e^{j\phi_{33}'} \end{bmatrix} = \begin{bmatrix} 1 & 0 & 0 \\ 0 & e^{-j\phi_{22}'} & 0 \\ 0 & 0 & 1 \end{bmatrix}$$

$$\times \begin{bmatrix} A11'' & A12''e^{j\phi_{12}'''} & A13''e^{j\phi_{13}'''} \\ 0 & A22'e^{j\phi_{22}'} & A23'e^{j\phi_{23}'} \\ 0 & A32'e^{j\phi_{32}'} & A33'e^{j\phi_{33}'} \end{bmatrix}$$

$$= \begin{bmatrix} A11'' & A12''e^{j\phi_{12}'''} & A13''e^{j\phi_{13}'''} \\ 0 & A22' & A23'e^{j\phi_{23}''} \\ 0 & A32'e^{j\phi_{32}'} & A33'e^{j\phi_{33}'} \end{bmatrix}$$

We can proceed to perform complex Givens rotation to nullify element 3,2:

$$U_5 = \begin{bmatrix} 1 & 0 & 0 \\ 0 & \cos\theta_3 & \sin\theta_3 e^{-j\phi_3} \\ 0 & -\sin\theta_3 & \cos\theta_3 e^{-j\phi_3} \end{bmatrix}$$

$$
\begin{bmatrix} 1 & 0 & 0 \\ 0 & \cos\theta_3 & \sin\theta_3 e^{-j\phi_3} \\ 0 & -\sin\theta_3 & \cos\theta_3 e^{-j\phi_3} \end{bmatrix}
\begin{bmatrix} A11'' & A12'' e^{j\phi_{12}''''} & A13'' e^{j\phi_{13}''''} \\ 0 & A22' & A23' e^{j\phi_{23}''} \\ 0 & A32' e^{j\phi_{32}'} & A33' e^{j\phi_{33}'} \end{bmatrix}
$$

$$
= \begin{bmatrix} A11'' & A12'' e^{j\phi_{12}''''} & A13'' e^{j\phi_{13}''''} \\ 0 & A22'' & A23'' e^{j\phi_{23}''''} \\ 0 & 0 & A33'' e^{j\phi_{33}''} \end{bmatrix}
$$

$$
\phi_1 = \phi_{32}'
$$

$$
\theta_1 = a\tan\left(\frac{A32'}{A22'}\right)
$$

It is also often useful to produce an upper triangular matrix that has all-real diagonal elements. Thus, even though there are no elements in the third column that we need to nullify, we should make element $(3,3)$ pure real:

$$
U_6 \begin{bmatrix} A11'' & A12'' e^{j\phi_{12}''''} & A13'' e^{j\phi_{13}''''} \\ 0 & A22'' & A23'' e^{j\phi_{23}''''} \\ 0 & 0 & A33'' e^{j\phi_{33}''} \end{bmatrix} = \begin{bmatrix} 1 & 0 & 0 \\ 0 & 1 & 0 \\ 0 & 0 & e^{-j\phi_{33}''} \end{bmatrix}
$$

$$
\times \begin{bmatrix} A11'' & A12'' e^{j\phi_{12}''''} & A13'' e^{j\phi_{13}''''} \\ 0 & A22'' & A23'' e^{j\phi_{23}''''} \\ 0 & 0 & A33'' e^{j\phi_{33}''} \end{bmatrix}
$$

$$
= \begin{bmatrix} A11'' & A12'' e^{j\phi_{12}''''} & A13'' e^{j\phi_{13}''''} \\ 0 & A22'' & A23'' e^{j\phi_{23}''''} \\ 0 & 0 & A33'' \end{bmatrix}
$$

Thus:

$$
U_6 U_5 U_4 U_3 U_2 U_1 A = R
$$

$$
A = (U_6 U_5 U_4 U_3 U_2 U_1)^{-1} R
$$

$$
A = U_1^* U_2^* U_3^* U_4^* U_5^* U_6^* R
$$

And:

$$
Q = U_1^* U_2^* U_3^* U_4^* U_5^* U_6^*
$$

9.13 Sphere Decoding

In Sect. 9.8 we discussed the maximum likelihood solution as the optimal solution for MIMO decoding. The complexity of ML was prohibitive, but we had an inkling on how to approach this complexity from the SISO case. In SISO, if we drew a circle around the observation, we could measure the distance to only a fraction of the constellation points. In MIMO, the same concept applies, except by using a hypersphere.

Understanding this approach, called sphere decoding, requires a good understanding of QR decomposition, which we should have by now. The upper triangular nature of R allows us to restate the distance measurement condition in terms of successively more strict conditions relative to the hypersphere. This will allow us to state the problem in terms of a tree structure that can allow substantial simplifications.

To find the maximum likelihood solution, we measure the distance between the equalized observation $H^{-1}y$ and each of the possible constellation vectors x:

$$\left\| H^{-1}y - x \right\|$$

As we discussed in Chap. 5, we can measure the distance between the equalized observations and the sent symbols, or the distorted signals and the unequalized observations:

$$\left\| y - Hx \right\|$$

Because this latter distance does not involve the calculation of a matrix inverse, we will continue to use it.

Under the sphere decoding algorithm, we measure distances only for points that lie within a certain radius of a hypersphere around the observation:

$$\left\| y - Hx \right\| < r$$

But the x that minimizes the norm also minimizes the square of the norm:

$$\left\| y - Hx \right\|^2 < r^2$$

We can expand the channel matrix H using its QR decomposition:

$$\left\| y - QRx \right\|^2 < r^2$$

One important aspect of unitary matrices is that their norm is unity. Thus, multiplying by a unitary matrix does not change the norm of a vector. This allows us to multiply by the inverse of Q from the left without changing anything in the condition:

$$\|Q^*y - Rx\|^2 < r^2$$

And multiplying by -1 will also not change anything in the condition:

$$\|Rx - Q^*y\|^2 < r^2$$

Consider a particular example for a 3×3 matrix:

$$\left\| \begin{bmatrix} R11 & R12 & R13 \\ 0 & R22 & R23 \\ 0 & 0 & R33 \end{bmatrix} \begin{bmatrix} x_1 \\ x_2 \\ x_3 \end{bmatrix} - \begin{bmatrix} Q_{11}^* & Q_{21}^* & Q_{31}^* \\ Q_{12}^* & Q_{22}^* & Q_{32}^* \\ Q_{13}^* & Q_{23}^* & Q_{33}^* \end{bmatrix} \begin{bmatrix} y_1 \\ y_2 \\ y_3 \end{bmatrix} \right\|^2 < r^2$$

Each row by column multiplication produces a scalar value. Let us number them p_1 through p_3, allowing us to write the square norm as:

$$|p_1|^2 + |p_2|^2 + |p_3|^2 < r^2$$

where:

$$p_1 = R11.x_1 + R12.x_2 + R13.x_3 - Q_{11}^*.y_1 - Q_{21}^*.y_2 - Q_{31}^*.y_3$$
$$p_2 = R22.x_2 + R23.x_3 - Q_{12}^*.y_1 - Q_{22}^*.y_2 - Q_{32}^*.y_3$$
$$p_3 = R33.x_3 - Q_{13}^*.y_1 - Q_{23}^*.y_2 - Q_{33}^*.y_3$$

When we considered ML in Sect. 9.8, we found that we had to calculate an inordinate number of distances to find the solution. In the inequality above, not all variables change as we carry out the different tests. The matrix H is constant and its components Q and R are also constant, at least for the duration of the packet. We also have a single observation that we are trying to decode; thus, the vector y is constant for all the tests. The vector x is only one of the possible sent vectors. This will definitely change across the different tests and is solely responsible for the large number of "nodes" we need to visit. Thus, it is the elements of x that will be our focus for the rest of this section.

Notice that p_3 contains only x_3, while p_2 contains only x_2 and x_3, and p_1 contains all elements of x. This comes from the upper triangular nature of the matrix R, which allows us to perform successive pruning of the number of nodes that need to be visited.

The tree structure is the best way to visualize the reduction in complexity introduced by sphere decoding (SD). Figure 9.8 shows the tree for 3×3 QPSK. Each level of the tree represents an antenna, and thus a new entry into the vector x. Thus, at the first level of the tree, we are dealing with antenna number 3. There are four possible values for x_3, specifically $1 + j$, $1 - j$, $-1 + j$, and $-1 - j$. We are ignoring the normalization factor $1/\sqrt{2}$ without loss of generality because it does not

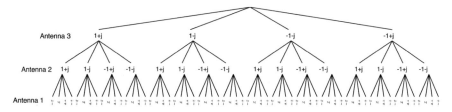

Fig. 9.8 Decision tree for a 3 × 3 QPSK system

change the solution. Remember that the solution we are seeking is the sent vector x that minimizes the distance to the observation.

At the second level of the tree, each value of x_3 could lead to four values of x_2. At the third level of the tree, we are finally visiting x_1. There are 16 nodes at the second level and 64 nodes at the third level. In general, at level P of a k-QAM system, there will be k^P nodes, leading to $k^{N_{rx}}$ nodes at the lowest level. The lowest level of the tree represents a complete vector to which we must measure the distance. Thus, there are 64 distances to calculate in a 3 × 3 QPSK system. For each final level node, the vector x is the collection of values of x_1 through x_3 met on the path to the final node.

Now, at the highest level of the tree, assume that we perform the following check:

$$|p_3|^2 < r^2$$

This check involves only p_3 and thus involves only x_3. If this condition is not satisfied, there is no way that the final condition for inclusion of the node is satisfied. Recall that the final condition is::

$$|p_1|^2 + |p_2|^2 + |p_3|^2 < r^2$$

And since any $|p|^2$ is a positive number, the condition $|p_3|^2 < r^2$ is a necessary but not sufficient prerequisite for $|p_1|^2 + |p_2|^2 + |p_3|^2 < r^2$. Assume we perform $|p_3|^2 < r^2$ at the level of x_3 and find that it is not satisfied for $x_3 = 1 + j$, $1 - j$; this allows us to prune all the branches of the tree under these values. This is shown in Fig. 9.9.

As shown in Fig. 9.9, we now know that 32 of the 64 final nodes have no chance of satisfying the final condition and thus do not need to be visited. By "visited," we mean calculate the distance for. At the second level, we need to visit 8 nodes instead of 16. We perform the following check:

$$|p_2|^2 + |p_3|^2 < r^2$$

And again, this check must pass for the final condition to be true. Thus, we can prune the tree again. Figure 9.10 shows the second level where only three of the eight nodes "pass."

This means that at the final level there are only 12 nodes that need to be visited. This is three possibilities for x_2 each leading to four possibilities for x_3. So,

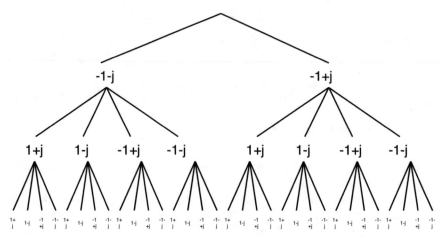

Fig. 9.9 Tree with nodes for $x_3 = 1 + j$, $1 - j$ pruned

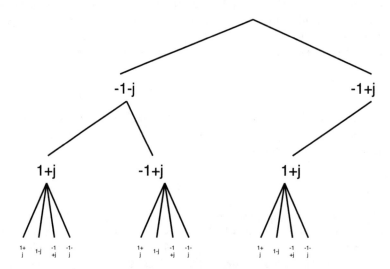

Fig. 9.10 Tree with five out of eight candidates from the second antenna pruned

originally, we needed to calculate the distance to 64 vectors. Testing the p-metric for the third antenna, we remove 32 of the 64 vectors from consideration. At the level of the second antenna, we know we may need to go out of 8*4 branches, but we prune 5*4, and need to only visit 12 vectors.

For the 12 vectors that satisfy the condition, we calculate the distance for each and pick the candidate with the smallest distance. Notice, however, that calculating the distance, or rather the norm square, is as simple as calculating the sum of p norm squares for the candidates. At the second level, we already have the sum of p_3 and p_2. So, we only need to calculate the square of p_1 for each candidate to compare their "distances" to the observation.

We can intuitively see when it helps to use SD and when it does not. SD is best when pruning happens earlier in the tree. It is also best when the radius r is chosen wisely. In fact, the two are not unrelated. For the condition at the top level to be violated often, and thus allow a lot of pruning to happen early, r needs to be chosen as small as possible. This allows p_3 to quickly outpace r.

But what stops us from using an arbitrarily small r to cut down on nodes as early as possible? The problem is that we do not know if a path is a possibility unless we reach the lowest level. Assume we choose a very small r; in that case, we will prune a lot early, but there is also a distinct possibility that when we reach toward the lower levels of the tree, *all* candidates fail. In that case, we have no recourse but to go back from the beginning and try a larger r. But if we increase r, do we really need to retrace from the very beginning? Why not take only one step back and keep working with the last survivors at the last level? The problem is this will not yield the ML solution. It is very possible that a path that incurred a large metric early on will incur smaller metrics at lower levels and end up winning, except that having chosen a radius that was too small, we killed this candidate branch too early.

To better understand this point, we should examine the performance of the SD algorithm as described above. By "performance," we mean BER-SNR. The BER curve of SD as described above will be identical to the ML BER curve. This is because the algorithm will *always* find the optimal solution that the ML algorithm dictates. However, the problem with the above algorithm is that it has non-constant throughput. It might need to visit a few nodes for the current sample, but it might visit a lot of nodes for the next one. It all depends on the radius used and the amount of noise the sample was exposed to. While there are ways to find an average complexity for the algorithm, it is practically impossible to expect any reliable time-invariant throughput.

There are alternatives that allow the SD algorithm to produce constant through-put, or at least to impose a floor on throughput. However, all such algorithms will produce sub-ML performance. The best known sub-optimal SD algorithm is the K-best algorithm. Under the K-best regime, we are not forced to accommodate all nodes that pass at any certain level; we only dig deeper for the K-best nodes.

Assume, for example, we have a 4×4 64 QAM system. At the first level, we might have a situation where all 64 children nodes pass, forcing us to test for 64×64 nodes at the second level. In the next cycle, we might have only one node passing at the first level, delivering results much faster, thus the non-constant throughput.

But in a K-best system, we choose a number K, and we do not have to accept any more successes at any level above this level K. So let us assume that we pick $K = 7$. At the first level, 30 nodes pass the test, but we will only accept the 7 best, that is to say the 7 with the least distance metric $|p|$. At the second level, we could have 7*64 successful candidates, of which we will again accept only the 7 best, and so on. The K-best algorithm guarantees a constant throughput, because it guarantees a maximum number of nodes we need to visit. Specifically, we need to visit:

$$k + \sum_{i=0}^{N_{\text{rx}}-2} K.k$$

where k is the QAM modulation order and K is the maximum number of visitations per level. Note that the K-best algorithm might occasionally deliver faster than promised, which happens whenever one or more levels have less than K successes. But it will never deliver slower than promised.

The K-best algorithm has a severe cost: it does not deliver ML BER because it has a distinct chance of missing the ML solution. Recall that the ML solution is the solution that minimizes the total distance, which is the summation of all p square metrics. With the K-best algorithm, we could disqualify the path that ends up being the optimal path early on.

This could happen because we are not preserving all paths that satisfy the condition; we are preserving only the K-best. Thus, we can and will often throw away a path at some level because it has a bad metric early on, when this metric would have preserved a good metric deeper into the tree, causing it to be optimal. This would never happen with vanilla SD, because SD will always either preserve the winner or will inform you at some level that you should have used a larger radius.

There are major design choices involved in the K-best algorithm. Specifically, we need to be judicious about the choice of K and its interplay with the radius. A higher K will bring us closer to ML performance but will also reduce throughput. One great idea is to allow variable values of K to be used at different levels. At higher levels of the tree, high values of K can be used to prevent potential winners from being prematurely killed. At lower levels of the tree, as the chance of killing a winner decreases, we can use a smaller K.

9.14   Two-Sided Jacobi for Singular Value Decomposition

The SVD algorithm most suitable for hardware adaptation is two-sided Jacobi. It uses a couple of unitary transformations to diagonalize a general matrix and is very suitable for parallel implementation. However, it has two major shortcomings: there is no complex version, and strictly speaking it only works for 2×2 matrices.

We can get over the first limitation of Jacobi by using a bunch of unitary transformations to transform any complex matrix into a real matrix. Because the transformations are unitary, they can all be rolled in with the U and V matrices. We begin with a general 2×2 complex matrix H:

$$H = \begin{bmatrix} Ae^{j\phi_a} & Be^{j\phi_{b\prime}} \\ Ce^{j\phi_c} & De^{j\phi_d} \end{bmatrix}$$

First, we use a unitary transformation to make element $(1, 1)$ pure real:

$$U_1 \begin{bmatrix} Ae^{j\phi_a} & Be^{j\phi_{b'}} \\ Ce^{j\phi_c} & De^{j\phi_d} \end{bmatrix} = \begin{bmatrix} e^{-j\phi_a} & 0 \\ 0 & 1 \end{bmatrix} \begin{bmatrix} Ae^{j\phi_a} & Be^{j\phi_{b'}} \\ Ce^{j\phi_c} & De^{j\phi_d} \end{bmatrix} = \begin{bmatrix} A & Be^{j\phi_b} \\ Ce^{j\phi_c} & De^{j\phi_d} \end{bmatrix}$$

We then use complex a Givens rotation to nullify element $(2, 1)$:

$$U_2 \begin{bmatrix} A & Be^{j\phi_b} \\ Ce^{j\phi_c} & De^{j\phi_d} \end{bmatrix} = \begin{bmatrix} \cos\theta_1 & \sin\theta_1 e^{-j\phi_1} \\ -\sin\theta_1 & \cos\theta_1 e^{-j\phi_1} \end{bmatrix} \begin{bmatrix} A & Be^{j\phi_b} \\ Ce^{j\phi_c} & De^{j\phi_d} \end{bmatrix} = \begin{bmatrix} E & Ge^{j\phi_g} \\ 0 & Ke^{j\phi_k} \end{bmatrix}$$

The complex Givens rotation uses two phases theta and phi:

$$\phi_1 = \phi_c$$

$$\theta_1 = a\tan\left(\frac{C}{A}\right)$$

We use another unitary transformation to make element $(2, 2)$ pure real. Because element $(2, 1)$ is null, it will also remain real even as it rotates by the phase of element $(2, 2)$:

$$U_3 \begin{bmatrix} E & Ge^{j\phi_g} \\ 0 & Ke^{j\phi_k} \end{bmatrix} = \begin{bmatrix} 1 & 0 \\ 0 & e^{-j\phi_k} \end{bmatrix} \begin{bmatrix} E & Ge^{j\phi_g} \\ 0 & Ke^{j\phi_k} \end{bmatrix} = \begin{bmatrix} E & Ge^{j\phi_g} \\ 0 & K \end{bmatrix}$$

We are in the homestretch. If we can get rid of the phase of element $(1, 2)$ using unitary transformations, the matrix will become pure real. Once the matrix is pure real, we can use the Jacobi algorithm to diagonalize it. The following transformation will cause the matrix to become real:

$$U_4 \begin{bmatrix} E & Ge^{j\phi_g} \\ 0 & K \end{bmatrix} U_5 = \begin{bmatrix} e^{j\phi_g/2} & 0 \\ 0 & e^{-j\phi_g/2} \end{bmatrix} \begin{bmatrix} E & Ge^{j\phi_g} \\ 0 & K \end{bmatrix} \begin{bmatrix} e^{-j\phi_g/2} & 0 \\ 0 & e^{j\phi_g/2} \end{bmatrix}$$

$$= \begin{bmatrix} E & G \\ 0 & K \end{bmatrix}$$

There are two things to notice here. First, we are multiplying by unitary matrices on both sides. This was not necessary when doing QR decomposition. While performing SVD, it will be necessary at some point because the decomposition must break the matrix into two unitary matrices U and V multiplied on both sides; thus, we will need unitary transforms on both sides.

The second major note is that the above transformation does not actually remove the phase of element $(1, 2)$. In reality, it exchanges the phases of elements $(2, 1)$ and $(1, 2)$. Element $(2, 1)$ is pure real; thus, when it gives element $(1, 2)$ its phase, the phase will end up being zero. But element $(2, 1)$ is also null; thus, when it is given the non-trivial phase of element $(1, 2)$, it will still be zero. In other words, the above transformation will not work to make the matrix pure real if element $(2, 1)$ were not zero.

We finally reach a point where we can apply the real two-sided Jacobi algorithm. Notice that this works with any 2×2 real matrix. The fact that element $(2, 1)$ is null is irrelevant for Jacobi but was necessary to allow us to make the original complex matrix real. The Jacobi algorithm is called two-sided because it multiplies by unitary matrices on both sides.

We start by calculating two phases:

$$\theta_{\text{sum}} = a \tan \left(\frac{G}{K - E} \right)$$

$$\theta_{\text{diff}} = a \tan \left(\frac{G}{K + E} \right)$$

We calculate two further phases from these two:

$$\theta_{\text{sum}} = \theta_l + \theta_r$$

$$\theta_{\text{diff}} = \theta_r - \theta_l$$

$$\theta_r = \frac{\theta_{\text{sum}} + \theta_{\text{diff}}}{2}$$

$$\theta_l = \frac{\theta_{\text{sum}} - \theta_{\text{diff}}}{2}$$

And doing the final unitary transformations:

$$U_6 \begin{bmatrix} E & G \\ 0 & K \end{bmatrix} U_7 = \begin{bmatrix} \cos \theta_r & -\sin \theta_r \\ \sin \theta_r & \cos \theta_r \end{bmatrix} \begin{bmatrix} E & G \\ 0 & K \end{bmatrix} \begin{bmatrix} \cos \theta_l & \sin \theta_l \\ -\sin \theta_l & \cos \theta_l \end{bmatrix} = \begin{bmatrix} S_1 & 0 \\ 0 & S_2 \end{bmatrix}$$

We can now gather all the matrices involved in the diagonalization:

$$U_6 U_4 U_3 U_2 U_1 H U_5 U_7 = S$$

$$H = (U_6 U_4 U_3 U_2 U_1)^* S (U_5 U_7)^* = USV$$

Thus, we have performed SVD on a complex 2×2 matrix. Now to extend it to larger matrices, we can take a lead from QRD in Sects. 9.11 and 9.12. To extend to larger matrices, we perform the same steps we performed above repeatedly. For example, to perform SVD on a 3×3 matrix, we first pick two off diagonal elements and null them. So, start by nulling elements $(2, 1)$ and $(1, 2)$. To perform this, we multiply by unitary matrices that impact the first two rows marked by x below:

$$\begin{bmatrix} x & 0 & x \\ 0 & x & x \\ - & - & - \end{bmatrix}$$

We then move to nulling elements $(3, 1)$ and $(1, 3)$, which impact elements marked by x below:

$$\begin{bmatrix} x & x & 0 \\ - & - & - \\ 0 & x & x \end{bmatrix}$$

This is where we hit a wall. To null $(3, 1)$ and $(1, 3)$, we end up affecting element $(1, 2)$ which we had nulled in the preceding step. We could have hit the same problem with QRD, but as discussed in Sect. 9.11, there is an order of nulling that avoids regenerating null values because we end up rotating two null values. This is impossible with SVD. For example, above, element $(1, 2)$ will rotate with element $(3, 2)$ which is non-null. Thus element $(1, 2)$ will end up regenerating.

This is inevitable, but all is not lost. In the example above, element $(1, 2)$ will regenerate but to a potentially smaller value than its original value because the vector formed by $(1, 2)$ and $(3, 2)$ in the second step has a smaller amplitude than the original vector of $(1, 2)$ and $(2, 2)$.

Thus, Jacobi is iterative. In the 3×3 example above, we null elements in the following order: $(1, 2)$ and $(2, 1)$; $(1, 3)$ and $(3, 1)$; $(3, 2)$ and $(2, 3)$. When we finish, we will find that all the elements we nulled earlier have regenerated, albeit to lower amplitudes. So, we perform another iteration, nulling off-diagonal pairs in the same order again. This time when we finish, the off-diagonal elements would have regenerated to even lower values while the diagonals start to converge to singular values. The number of iterations we perform will depend on the size of the matrix and the wordlength of the fixed-point system. We should stop once the off-diagonal elements are small enough that they appear to be nulls in the wordlength of registers used in the system.

9.15   Comparison of MIMO Decoding Algorithms and Algorithm as Part of Cognition

We discussed multiple MIMO decoding algorithms in this chapter. They show a tradeoff between complexity and performance. Optimal performance is afforded by ML, which is matched by some implementations of SD (not K-best). Maximum likelihood has the highest complexity because the number of nodes that must be visited grows exponentially with the number of antennas. This is particularly disastrous for large QAM constellations.

SVD is relatively complicated. The only hardware-friendly algorithm for SVD works only on real 2×2 matrices. The two-sided Jacobi algorithm is itself non-trivial in hardware, but it is in its extension to large complex matrices that the complexity comes from. To extend SVD to complex matrices, multiple unitary transformations are required before Jacobi can be applied. Large matrices also require multiple Jacobi iterations before the factors converge to the true decomposition. However, SVD provides incredible insight into CSIR that allows us to play games, especially when there is a feedback channel available to the transmitter.

MMSE and ZF are the least complex of the approaches we discussed; they also deliver the worst performance. MMSE is slightly more complex than ZF but offers better performance at low SNR by avoiding noise maximization. Both ZF and MMSE can be used with successive interference cancellation techniques to improve performance.

In Sect. 7.14, we discussed how the multitude of modes (MCS) available to the system allow it to adapt to different channel conditions. This is contingent on the hardware being able to support all the different modes. This incentivizes reconfigurable hardware and flexible solutions. It also requires cognition that allows the system to make decisions about which MCS to use on the fly.

This chapter suggests another layer of adaptation: the adaptation of algorithm. All the algorithms discussed here perform the same thing: MIMO decoding. You give them an observation vector and channel estimates; they give you their best guess about the sent vector. However, they do so at differing performance and consuming different power. If you find a hardware solution that could switch between those algorithms on the fly, this could allow you to trade off performance and power based on channel conditions.

For example, if the channel is good, perhaps we can get the QoS we need from something as simple as MMSE. So, we switch to a low power mode where we perform MMSE decoding. If the channel is in dire straits and we still need good QoS, then we need to squeeze as many bits as possible out no matter how much power we burn in the decoder, so we switch to SD. If there is a directional interferer, maybe we need to beamform, and we need SVD to give us more insight into the channel. But if the interferer is not directional, maybe ZF-SIC could help without consuming too much power.

The degree of cognition we need here is significantly beyond that discussed in Sect. 7.14. The same applies to the level of hardware flexibility needed. And this discussion can extend to any complex subsystem in any large system. Hardware implementations that afford this level of flexibility are often called accelerators.

Accelerator architecture is like that of processors (Fig. 9.11). Accelerators contain processing units arranged in a generic manner (Sect. 8.8), memory banks to keep intermediate results (Sect. 8.9), a micro-coded controller (Sect. 8.12), and switching structures that allow the PUs to be reconfigured.

The design and implementation of said architectures is often very challenging and requires a thorough understanding of the range of algorithms supported and the hardware used. Some of the challenges include:

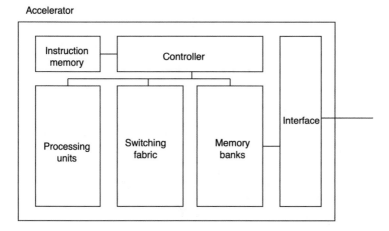

Fig. 9.11 Accelerator architecture. The interface communicates between the accelerator and a main processor

- What processing units are used. This is particularly difficult in problems interesting enough to require accelerators because the PUs are rarely simple multiply-accumulate units. Also, because the PU needs to support a lot of different (though related) algorithms. So, there must be a high degree of reconfigurability.
- There will inevitably be a degree of time-sharing (hardware reuse) involved in an accelerator. This is because it is impossible to support disparate algorithms with a direct implementation of PUs. This will require intelligent use of memories to store inputs, outputs, and particularly intermediate results.
- The way PUs talk to each other will differ substantially from algorithm to algorithm; this requires programmable and efficient routing structures.
- Even though the algorithms are doing the same thing, they do it in fundamentally different ways. This means a fundamentally different sequence of events, thus requiring very flexible control.

Chapter 10
Advanced Issues in Migrating to Hardware: MIMO Decoders as Case Studies

10.1 Naïve Implementation of Unitary Transformations

The equation below implements a real Givens rotation:

$$\begin{bmatrix} \cos\theta & \sin\theta \\ -\sin\theta & \cos\theta \end{bmatrix} \begin{bmatrix} A & C \\ B & D \end{bmatrix} = \begin{bmatrix} E & G \\ 0 & K \end{bmatrix}$$

Figure 10.1 shows a direct implementation of this equation using multipliers and lookup tables. We use a divider to obtain the tan of the phase of rotation. The arctan is obtained from a lookup table. This gives us the phase θ, which is then used to look up values for cosine and sine from lookup tables. These sinusoid values are then used to multiply the original vectors to obtain the result vectors. Figure 10.1 shows only the transformation of (C, D) to (G, K); but the transformation of (A, B) to $(E, 0)$ can be done similarly.

The implementation in Fig. 10.1 is extremely inefficient. One reason for this is the use of multipliers. Multipliers are area and power-hungry and they dominate the critical path of most circuits. But multipliers are only the tip of the iceberg. We also use a divider, which is much worse than a multiplier in almost all respects. Trigonometric and inverse trigonometric functions require the use of three lookup tables.

We can easily see that the whole arrangement in Fig. 10.1 is going to perform miserably in power and delay. But its dynamic range is even worse. The multipliers in Fig. 10.1 will double the wordlength of their inputs, and the divider will be even worse. But Givens rotation has a very good dynamic performance. As shown in Fig. 9.7, any operation that involves the rotation of a vector will have good dynamic performance because the vector is only rotated and its x and y components can only have a maximum value equal to the length of the vector. This reduces the requirements on the size of the output of multipliers significantly.

For example, assume that inputs to the multipliers in Fig. 10.1 are 7-bits long. This means the noiseless output of the multipliers will be 14-bits long. But if we are

© The Author(s), under exclusive license to Springer Nature Switzerland AG 2023
K. Abbas, *From Algorithms to Hardware Architectures*,
https://doi.org/10.1007/978-3-031-08693-9_10

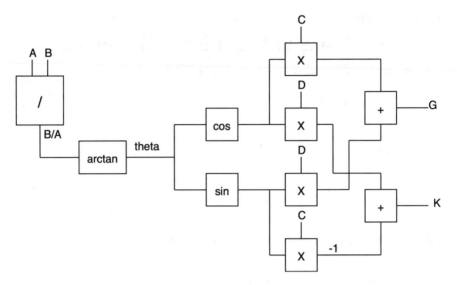

Fig. 10.1 A naïve implementation of Givens rotations. The multiply/add section can be duplicated to calculate $(E, 0)$

aware of the graphical interpretation of Givens rotation, we can reduce this output significantly. C and D both have a maximum value of 127 (we are assuming unsigned registers). This means that the vector (C, D) has a maximum length of $\sqrt{127^2 + 127^2} = 179.6$. Thus, the outputs of the multipliers will not exceed 8 bits.

Even the divider in Fig. 10.1 has more lenient dynamic conditions than a run of the mill divider. The output of the divider in Fig. 10.1 is immediately and exclusively fed to an inverse trigonometric function. This allows us to limit the output of the divider to the range that causes the arctan output to be bound in the range $-\pi$, π within the wordlength of the system.

10.2   CORDIC

Figure 10.2 shows two distinct operations which we will call vectoring and rotation. In the vectoring operation, a point (x, y) is rotated till it reaches the x-axis. The outputs of the vectoring operation are the magnitude of the original vector as well as the phase that the original point formed with the x-axis. Thus, the inputs to the vectoring operation are:

$$x, y$$

And the outputs are:

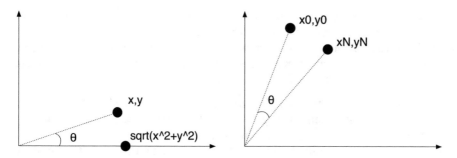

Fig. 10.2   Vectoring operation, left. Rotation operation, right

Fig. 10.3 CORDIC
through microrotations. The
initial and final points are
shown as solid circles

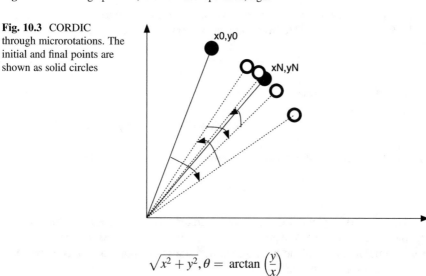

$$\sqrt{x^2 + y^2}, \theta = \arctan\left(\frac{y}{x}\right)$$

The rotation operation takes a point and rotates it by a given phase to produce a new point. Thus, the inputs are x, y, and θ, while the output is a new point which will generally not be at the x-axis.

The vectoring operation implements the part of Givens rotation which rotates the vector (A, B) to point $(E, 0)$. The rotation operation implements the part of Givens rotation which takes point (C, D) to point (G, K).

CORDIC is a processing unit that allows us to perform many transcendental functions efficiently. When applied to Givens rotations, it allows us to perform both vectoring and rotation operations using shifters and adders only. When we get deeper into the CORDIC algorithm, we will find that it needs some more auxiliary hardware, but it will still be much more efficient than the naïve implementation in Sect. 10.1.

CORDIC is an acronym, and as far as acronyms go, it is not excellent. It stands for COordinate Rotation DIgital Computer. The name suggests how closely connected it is to Givens rotations. Figure 10.3 shows the principle behind CORDIC. In Fig. 10.3 we use the CORDIC algorithm to rotate point (x_0, y_0) to point (x_N, y_N).

But instead of doing this in a single rotation, we reach it through a consecutive series of smaller rotations, each called a microrotation.

Figure 10.3 might seem counterintuitive at first. Why would performing many rotations be more efficient than a single rotation. We will build up to the answer, but there are two main reasons CORDIC works:

- The microrotation phases are the same regardless of the total angle we want to rotate by. Distinguishing between one total rotation angle and another is only through the directions of the microrotations.
- There must be some useful property for the microrotation angles that makes them desirable.

Both points will become much clearer when we look at the problem mathematically. Consider the scalar G from the matrix rotation in Sect. 9.11:

$$G = C.\cos\theta + D.\sin\theta$$

Taking the cosine common:

$$G = \cos\theta(C + D.\tan\theta)$$

Similarly, the scalar value K can be obtained as:

$$K = \cos\theta(D - C.\tan\theta)$$

To more clearly parallel the geometric interpretation of the Givens rotation, we rename (C, D) as (x_0, y_0). We also rename (G, K) as (x_N, y_N). This indicates that both pairs are points in the plane with the unitary transformation being a rotation. And thus, the two rotation equations are:

$$(x_0, y_0) \rightarrow (x_N, y_N)$$
$$x_N = x_0 \cos\theta + y_0 \sin\theta = \cos\theta(x_0 + y_0.\tan\theta)$$
$$y_N = y_0 \cos\theta - x_0 \sin\theta = \cos\theta(y_0 - x_0.\tan\theta)$$

These equations can also be used to represent the microrotations because there is nothing about them that limits them to a certain rotation angle. When we use these equations to represent a single microrotation, in other words a single iteration, we use the indices i and $i + 1$ to indicate rotation by angle θ_i:

$$x_{i+1} = \cos\theta_i(x_i + y_i.\tan\theta_i)$$
$$y_{i+1} = \cos\theta_i(y_i - x_i.\tan\theta_i)$$

But again, why would we do multiple microrotations instead of a single rotation? Let us count the operations involved in this microrotation. We need a lookup table for the tan function, a lookup table for the cosine function, a multiplication by the tan, an addition, and a final multiplication by the cosine.

Ignore the cosine for now; we will come back to it later. We are free to choose the set of θ_i used for the microrotations. We will choose phases for which $\tan\theta_i = 2^{-i}$. This makes $y_i\tan\theta_i = y_i2^{-i}$. So, for each microrotation, the whole equation is now reduced to a shift and an addition (ignoring the cosine).

10.3   Vectoring CORDIC

The vectoring operation starts with a point in the plane (x_0, y_0) and ends with a point on the x-axis $(x_N, 0)$. While rotating, the phase that the original point makes with the x-axis is also calculated as $\theta_N = \text{atan}\left(\frac{y_0}{x_0}\right)$. We start with three registers carrying the initial x and y values, as well as a null phase register:

$$x = x_0, y = y_0$$
$$\theta_0 = 0$$

On every iteration i following this, the values of x, y, and the phase are updated according to:

$$x_{i+1} = \left(x_i + a_i.y_i.2^{-i}\right)$$
$$y_{i+1} = \left(y_i - a_i.x_i.2^{-i}\right)$$
$$\theta_{i+1} = \theta_i + a_i\text{atan}2^{-i}$$

These equations are run for N iterations, beginning with $i = 0$ and up to $i = N - 1$. Ostensibly, at the end of the iterations, we should end up with $x_N = \sqrt{x_0^2 + y_0^2}$ and $y_N = 0$. Also, we should end up with $\theta_N = \text{atan}\left(\frac{y_0}{x_0}\right)$. Now, we figure out how and why this is true. Particularly, we should figure out what the factor a_i means, why it makes sense, and how it is calculated.

The factor a_i simply indicates the direction in which the current microrotation takes place. Notice that all phases are broken down into the same set of microrotation phases. The phases start very large at $\theta_0 = \text{atan}1 = \frac{\pi}{4}$. The phases then get progressively smaller as their tans also get progressively smaller. As the number of iterations gets larger, we allow the algorithm to fine-tune the rotation even more.

We can guess that at the end of the N iterations, there will be a residual left in the registers. The more the iterations, the less the residue left. The number of iterations needed is a matter of trial and error but recall that all numbers in registers are fixed-point. Thus, the number of iterations needed is only the number necessary to reduce the residue left in the registers below their resolutions. In general, having N be four more than the size of the registers is enough to ensure good performance.

Table 10.1 Evolution of the x, y, and θ registers with each step of the CORDIC algorithm

Step	X	Y	a	Theta	Magnitude	Ratio
Initialization	7	3	0	0	7.61577	1
0	10	−4	1	0.78540	10.77033	1.41421
1	12	1	−1	0.32175	12.04159	1.58114
2	12.25	−2	1	0.56673	12.41219	1.62980
3	12.5	−0.46875	−1	0.44237	12.50879	1.64248
4	12.52930	0.3125	−1	0.37996	12.53319	1.64569
5	12.53906	−0.07904	1	0.41120	12.53931	1.64649
6	12.54030	0.11688	−1	0.39557	12.54084	1.64669
7	12.54121	0.01891	1	0.40338	12.54122	1.64674
8	12.54128	−0.03008	1	0.40729	12.54132	1.64676
9	12.54134	−0.00558	−1	0.40534	12.54134	1.64676
10	12.54135	0.00666	−1	0.40436	12.54135	1.64676
11	12.54135	0.00054	1	0.40485	12.54135	1.64676
12	12.54135	−0.00252	1	0.40509	12.54135	1.64676
13	12.54135	−0.00099	−1	0.40497	12.54135	1.64676

What distinguishes one total phase from another is the directions of microrotations. In every iteration, we must decide if we are rotating clockwise or counterclockwise. So a_i is either $+1$ or -1 causing the microrotation to be clockwise or counterclockwise. The factor a_i should simply be the sign of the last value of y we calculated:

$$a_i = \text{sign}(y_i)$$

This is made much clearer through an example. Table 10.1 shows an initial point $(7, 3)$ being vectored by the CORDIC algorithm. The first column of the table is the index of the step (iteration), with the first row representing the initial loadouts of the registers. The second and third columns are the X and Y elements of the point at the iteration, in other words x_i and y_i. Theta is the angle at the current step θ_i. The "a" column indicates the direction of rotation in the current iteration a_i. The magnitude column is $\sqrt{x_i^2 + y_i^2}$. The ratio column is the ratio between the length of the vector at iteration i and the length of the initial vector, in other words $\frac{\sqrt{x_i^2 + y_i^2}}{\sqrt{x_0^2 + y_0^2}}$

We start from a point in the top half-plane with a positive value for Y. In iteration 0, the microrotation pushes the point into the bottom half-plane, which is obvious from the fact that Y is -4. In vectoring, we want to end up at the x-axis, which means that Y being negative shows that we have overshot, and in iteration 1, we must rotate back counterclockwise. Thus, "a" for step 1 is the sign of Y in step 0. Y in step 1 becomes $+1$; the fact that it is positive means we must rotate clockwise in step 2. At each step, the microrotation angle gets progressively smaller, and the only decision we make is the direction of said microrotation.

What is also very clear is that the longer we go, meaning the more steps we take, the closer Y gets to 0, and the closer theta converges toward a number. We can see that this number is $\operatorname{atan}\left(\frac{3}{7}\right)$. The number of steps (iterations) needed depends on how many significant digits we want to converge, which directly relates to the wordlength of the registers.

There is only one problem, x_N is supposed to converge to $\sqrt{x_0^2 + y_0^2}$ as y_N converges to 0. This clearly does not happen. In fact, the last two columns of the table show this explicitly. The original amplitude of the vector is 7.6158. The CORDIC algorithm converges to a vector amplitude of 12.54. So why does this happen and what are we going to do about it?

At the beginning of this section, we used this equation to update values of X:

$$x_{i+1} = \left(x_i + a_i.y_i.2^{-i}\right)$$

But the original equation for the Givens rotation was:

$$x_{i+1} = \cos\theta_i(x_i + y_i.\tan\theta_i)$$

The tan term is replaced with $a_i.\ y_i.\ 2^{-i}$. But what about the cosine term? We completely ignored it without justification. The cosine term, or in fact the accumulation of multiple cosine terms from each of the microrotations, will lead to a deviation of the amplitude of the vector from its original value.

This might seem like a very strong limitation on the CORDIC algorithm. But things are not as bad as they initially seem. If we run the CORDIC algorithm for the same number of iterations for point $(5, 5)$, we will find that the initial amplitude of the point is 7.0711. The final value of the amplitude after 13 iterations is 12.54. The ratio between the two is 1.647. If we run again with the initial point $(2, 10)$, the original amplitude is 10.198, the final amplitude is 16.973, and their ratio is 1.647.

The bad news is none of the vectors converge toward their amplitude. The good news is the vectors' amplitude is scaled by the exact same number 1.647 regardless of the initial point. This means that we have not actually failed to obtain the amplitude of the vectors; we obtained them scaled by a constant factor that is only a function of the number of iterations, not the inputs, which means we can use a single constant multiplier to obtain the actual length of the vector.

In fact, the factor is:

$$\text{Factor} = \prod_{i=0}^{N-1} \cos\theta_i = \prod_{i=0}^{N-1} \cos\left(\operatorname{atan}2^{-i}\right)$$

Because the phases of the microrotations are constants, the factor also converges toward a constant value. More iterations mean the factor gets closer to this value. Therefore, the scaling factor is the same for all starting points. But why did we end up with a constant for the cosine factor, but had to perform shifts and adds for the tan factor? Because cosine is an even function and tan is an odd function. The microrotation phase magnitudes are fixed; however, the direction that each takes will depend on the starting point. But because the cosine function is even, it will give the same result regardless of the direction of each microrotation. The same cannot be said for tan.

Let us take account of the hardware cost of the vectoring CORDIC now that we have figured out each of its constituents. We will need one lookup table containing the arctan values for each of the microrotations. We will need a shifter and an adder for each of x and y, plus another adder for the θ register. We will ostensibly also need two constant multipliers to account for the scaling factor. But we only need one for the X register. Recall that in vectoring, Y ends up null, so there is no need to scale it.

CORDIC is obviously doing a much better job than the naïve implementation we discussed in Sect. 10.1. It uses no full multipliers and no dividers. Even the LUT used for the arctans is significantly smaller than the LUTs in Sect. 10.1. In fact, this LUT will only have as many entries as the number of iterations.

10.4   Rotation CORDIC

The rotation CORDIC is incredibly similar to the vectoring CORDIC even though it performs a fundamentally different operation. As explained in Sects. 9.11 and 10.2, a rotation operation takes a point (x_0, y_0) and rotates it by a given angle θ_0 to a new coordinate point (x_N, y_N). Thus, it differs from the vectoring CORDIC in both its inputs and outputs. A rotation CORDIC accepts an angle to rotate by as an input rather than produce it as an output. Additionally, the final point resulting from rotation will not be on the x-axis (at least not generally), and thus its y_N output will not be null.

The ultimate results of the rotation are:

$$x_N = x_0 \cos \theta_0 + y_0 \sin \theta_0 = \cos \theta_0 (x_0 + y_0 . \tan \theta_0)$$
$$y_N = y_0 \cos \theta_0 - x_0 \sin \theta_0 = \cos \theta_0 (y_0 - x_0 . \tan \theta_0)$$

The update equations for the rotation CORDIC are:

$$x_{i+1} = \left(x_i + a_i . y_i . 2^{-i}\right)$$
$$y_{i+1} = \left(y_i - a_i . x_i . 2^{-i}\right)$$
$$\theta_{i+1} = \theta_i + a_i \text{atan2}^{-i}$$

At each iteration, we add or remove the microrotation angle from the θ register. This register is initially loaded with θ_0. Eventually, we want this register to be nullified. If the phase register becomes negative, then we have overshot, and we should rotate in the opposite direction. Thus:

$$a_i = -\text{sign}(\theta_i)$$

This is a very similar reasoning to what we did with the y register in vectoring CORDIC. We start with $i = 0$. This allows us to perform a first rotation at an angle $\text{atan}(1) = \frac{\pi}{4}$. We use progressively smaller phases to get the phase register as close to zero as possible.

The rotation CORDIC suffers from the same constant amplitude scaling issue that affects the vectoring CORDIC. In other words, the common cosine factors lead to a distortion of the magnitude of registers that is only a function of the number of iterations. As in the vectoring CORDIC, we can address this by multiplying by the constant scaling factor at the end. In this case, however, we scale both the x and y registers because neither of them will end up being null.

The same considerations affect the number of iterations in the rotation CORDIC as in the vectoring CORDIC. In fact, both types of CORDIC are so incredibly similar that they can be implemented using the same hardware. The only difference between the two is how the direction of rotation (the "a" register) is determined. In hardware, it will be determined through the sign bit of either the y or θ registers. So, to use the same hardware to implement rotation and vectoring CORDIC, we use a multiplexer to pick "a" as either the sign bit of y or of θ. We will look at the hardware implementation of both types of CORDIC in more detail in Sect. 10.5.

One limitation of the CORDIC algorithm as stated above for both vectoring and rotation modes is that it is limited to the right half-plane. In other words, it is limited to phases of plus or minus 90 degrees. This is a fundamental result of the limited set of available microrotations, the biggest of which is $\frac{\pi}{4}$. We can extend the range of the rotation CORDIC by doing the following simple initial modification:

$$\text{If } x_0 < 0$$

$$d = -\text{sign}(y_0)$$

$$x_0 = -d.y_0$$

$$y_0 = d.x_0$$

$$\theta_0 = \theta_0 + d.\frac{\pi}{2}$$

And for vectoring CORDIC:

$$\text{If } x_0 < 0$$

$$d = \text{sign}(y_0)$$

$$x_0 = -x_0$$

$$y_0 = y_0$$

$$\theta_N = d.(\pi - d.\theta_N)$$

10.5   Iterations, Wordlength, and Hardware Implementation of CORDIC

The discussion of CORDIC in this chapter has assumed that all inputs, outputs, and arithmetic operations use floating-point registers. To implement rotation or vectoring CORDIC, we need the following fixed-point hardware blocks:

- Adders
- Shifters
- Lookup tables

Figure 10.4 shows a hardware implementation of vectoring CORDIC. The implementation is regular, with each stage being identical to the next. There are N stages, where N is the number of iterations. Every stage, x_i and y_i, with one of them being shifted, are added and subtracted in two adders/subtractors. Whether addition or subtraction happens depends on the sign of y_i. The sgn hardware block extracts the MSB of y_i.

Before addition in the calculation of x_{i+1}, y_i is shifted by as many bits as the stage we are in; this implements the multiplication by 2^{-i}. y_{i+1} is calculated similarly. Simultaneously and if needed, the angle θ is calculated. θ starts out at null. Addition or subtraction takes place at every stage based on the same control bit a_i used for the calculation of x and y. There is an atan lookup table where the values of atan(2^{-i}) are stored. Each stage, the corresponding looked up value, is either added or subtracted from the θ register.

The implementation in Fig. 10.4 is unpipelined. The critical path is the y-path or x-path going through N shifters and N adders. Thus, the critical path delay is $N(t_{shift} + t_{add})$. Shifters in Fig. 10.4 are hardwired, so their delay is negligible next to adder delay and adders dominate the performance of CORDIC.

If the implementation in Fig. 10.4 is internally pipelined between all stages, clock speed shoots up by N and critical path delay drops to single stage delay ($t_{shift} + t_{add}$),

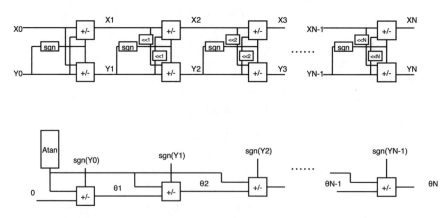

Fig. 10.4 Parallel vectoring CORDIC

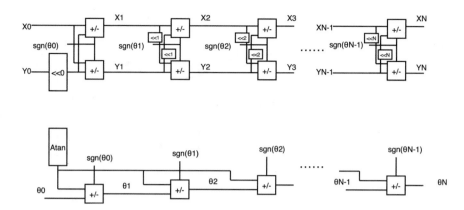

Fig. 10.5 Parallel rotation CORDIC

again predominantly t_{add}. Because adder delay is directly proportional to wordlength, we should thoroughly consider the wordlength we use.

Figure 10.5 shows the parallel implementation of rotation CORDIC. It is incredibly similar to Fig. 10.4; in fact it seems nearly identical. The main difference is in the θ path. The register θ starts with an initial value instead of null and the direction of rotation is determined by the sign of θ rather than y. Rotation and vectoring CORDIC are so similar that in many cases it would make sense to design a single CORDIC and decide whether to use it as either on the fly.

Figure 10.6 shows a time-shared implementation of CORDIC. It trades off performance for lower area. A single stage is used, and an intermediate memory is included. Each iteration, x_i and y_i, are read from the right-hand side memory ports and used as input to the stage. The stage produces x_{i+1} and y_{i+1} which are stored in memory through the write ports. Because the PU in CORDIC is relatively simple, time-sharing can be controlled easily, and memory need not be large. The CORDIC in Fig. 10.6 has a critical path of $(t_{shift} + t_{add})$, but will take N cycles to produce a single output. Time-sharing can be useful when the number of iterations is large, especially when wordlength is also large. These results show that there is no reason to use an unpipelined parallel CORDIC. We should use either a pipelined implementation or a time-shared CORDIC.

And this raises the question of the difference between iterations and wordlength and whether the two are related. Iterations define the number of times we perform add/shifts. As seen in the algorithm in Sects. 10.3 and 10.4, this assumes that inputs, outputs, and operations are all floating point. In hardware in Figs. 10.4 and 10.5, iterations define the number of stages in each CORDIC.

Wordlength considers what happens when adders and shifters, and thus all registers, in CORDIC are fixed-point. In other words, how many bits the adders in Figs. 10.4 and 10.5 are. This is, at least to a first order, independent from the number of iterations. So, to restate this in clear terms, the number of iterations defines the number of stages, while wordlength defines the size of adders in the stages.

Fig. 10.6 Time-shared
rotation CORDIC

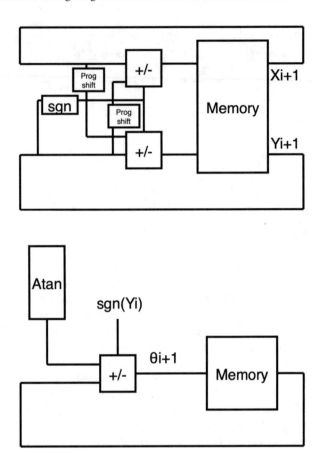

But are the two truly independent? Consider Table 10.1 in Sect. 10.3. We see that the more iterations we perform, the closer the output is to the ideal output. So, at which point is the result "close enough?" The answer is when the result is close enough to the ideal output that we cannot tell the difference within the system's bus size. For example, in Table 10.1, notice that the last four rows show identical values for amplitude. This is because the precision used in the table does not allow us to see decimal places where the distinction is clear. If the adders and registers in the system only accommodate five decimal points, it would make no sense to go beyond iteration 10.

The discussion above is complicated by the fact that additions in Table 10.1 were calculated as floating-point operations. Making an informed decision about the number of iterations in CORDIC thus requires experience, trial and error, and simulations. But a general rule of thumb is that it never makes sense to have fewer iterations than the wordlength of the inputs and that four extra iterations above the input wordlength are usually enough to preserve the outputs relatively quantization noise-free.

Fig. 10.7 Implementation
of real Givens rotations
using CORDIC

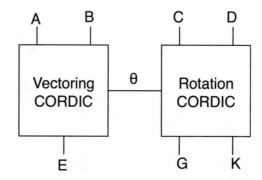

10.6   CORDIC for Givens Rotations

Givens rotations were discussed in detail in Sects. 9.11 and 9.12. Again, there are
two types of operations related to the following unitary transformation:

$$\begin{bmatrix} \cos\theta & \sin\theta \\ -\sin\theta & \cos\theta \end{bmatrix} \begin{bmatrix} A & C \\ B & D \end{bmatrix} = \begin{bmatrix} E & G \\ 0 & K \end{bmatrix}$$

We have shown in Sect. 10.3 that the transformation of vector $\begin{bmatrix} A \\ B \end{bmatrix}$ to $\begin{bmatrix} E \\ 0 \end{bmatrix}$ is
called a vectoring operation and can be implemented using a vectoring CORDIC.
The rotation of vector $\begin{bmatrix} C \\ D \end{bmatrix}$ into vector $\begin{bmatrix} G \\ K \end{bmatrix}$ is a rotation operation that can be
implemented using a rotation CORDIC.

Figure 10.7 shows how CORDIC processors can be used to implement real
Givens rotation. Elements A and B are fed to X and Y in the vectoring CORDIC,
respectively. This produces an angle θ and a magnitude, which will be element E.
The angle θ is fed to a rotation CORDIC with elements C and D being fed to X and Y.
This produces outputs G and K on the X and Y outputs, respectively.

10.7   Super CORDIC: Handling Complex Numbers

The complex vectoring and rotation CORDIC operations were discussed in detail in
Sect. 9.12. They are repeated below for easy reference:

$$\begin{bmatrix} \cos\theta & \sin\theta\, e^{-j\phi} \\ -\sin\theta & \cos\theta\, e^{-j\phi} \end{bmatrix} \begin{bmatrix} A & B e^{j\phi_b} \\ C e^{j\phi_c} & D e^{j\phi_d} \end{bmatrix} = \begin{bmatrix} E & G e^{j\phi_g} \\ 0 & K e^{j\phi_k} \end{bmatrix}$$

$$\phi = \phi_c$$

$$\theta = \text{atan}\left(\frac{C}{A}\right)$$

As with the real case, there are two operations shown in the matrix multiplication above. The first is a "vectoring" operation of the vector formed by the real element A and the complex element C. This is analogous to vectoring in the real case, with two major differences: one of the original points is complex, and there are two phase outputs from the operation.

Figure 10.8 shows the implementation of the "super vectoring CORDIC," the hardware unit used to implement this first operation. It consists of two real vectoring CORDICs and is very easy to reverse engineer. The complex number $Ce^{j\phi_c}$ is fed to the first vectoring CORDIC. The real part of $Ce^{j\phi_c}$ is fed to the x input, and the imaginary part is fed to the y input. It will thus produce ϕ_c and C at the output. The phase ϕ_c is one of the two phase outputs needed from the super CORDIC. C is fed as the y input of the second vectoring CORDIC, with A being the x input. This allows the second vectoring CORDIC to calculate $E = \sqrt{A^2 + C^2}$ and $\theta = \text{atan}\left(\frac{C}{A}\right)$, completing the function of super vectoring.

The second part of the matrix multiplication is complex Givens rotation, which in the real case is handled by a single rotation CORDIC. As with the vectoring operation, we use a super rotation CORDIC to handle the complex case. The super rotation CORDIC performs the following operation:

Fig. 10.8 Using a "super vectoring CORDIC" to obtain the rotation phases and magnitude

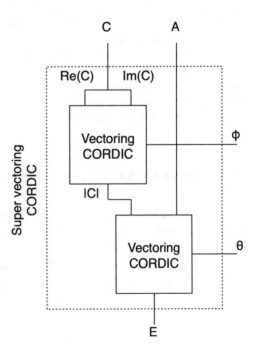

$$\begin{bmatrix} \cos\theta & \sin\theta e^{-j\phi} \\ -\sin\theta & \cos\theta e^{-j\phi} \end{bmatrix} \begin{bmatrix} Be^{j\phi_b} \\ De^{j\phi_d} \end{bmatrix} = \begin{bmatrix} Ge^{j\phi_g} \\ Ke^{j\phi_k} \end{bmatrix}$$

The complex rotation CORDIC accepts two complex inputs $Be^{j\phi_b}$ and $De^{j\phi_d}$ and produces two complex outputs $Ge^{j\phi_g}$ and $Ke^{j\phi_k}$. It also needs two phases as inputs, which it obtains from a super vectoring CORDIC.

Figure 10.9 shows how a super rotation CORDIC can be implemented using three real rotation CORDICs. It takes a little more algebra than the vectoring case to prove that this circuit corresponds to the equation, but it is straightforward. Real CORDIC numbered I performs the following operations:

$$\begin{bmatrix} \cos\phi & \sin\phi \\ -\sin\phi & \cos\phi \end{bmatrix} \begin{bmatrix} D_r \\ D_i \end{bmatrix} = \begin{bmatrix} P_1 \\ P_2 \end{bmatrix}$$

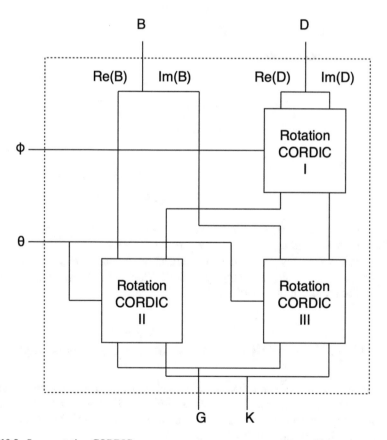

Fig. 10.9 Super rotation CORDIC

$$P_1 = D_r \cos \phi + D_i \sin \phi$$
$$P_2 = -D_r \sin \phi + D_i \cos \phi$$

The CORDIC numbered *II* performs:

$$Q_1 = B_r \cos \theta + P_1 \sin \theta$$
$$Q_2 = -B_r \sin \theta + P_1 \cos \theta$$

The CORDIC numbered *III* performs:

$$R_1 = B_i \cos \theta + P_2 \sin \theta$$
$$R_2 = -B_i \sin \theta + P_2 \cos \theta$$

And:

$$G = Q_1 + jR_1$$
$$K = Q_2 + jR_2$$
$$G = B_r \cos \theta + P_1 \sin \theta + j(B_i \cos \theta + P_2 \sin \theta)$$
$$G = B \cos \theta + \sin \theta (D_r \cos \phi + D_i \sin \phi + j(-D_r \sin \phi + D_i \cos \phi))$$
$$G = B \cos \theta + \sin \theta \left(D_r e^{-j\phi} + jD_i e^{-j\phi} \right)$$
$$G = B \cos \theta + D \sin \theta e^{-j\phi}$$

And similarly:

$$K = -B\sin\theta + D \cos \theta e^{-j\phi}$$

There is one last issue that we must address. Complex Givens rotation is always contingent on the leading element being pure real. This is not generally true for a random matrix. However, making an element in an array real can be easily implemented using a single vectoring CORDIC and a single rotation CORDIC. We will discuss this in detail in Sect. 10.10.

10.8   CORDIC for QR Decomposition and Systolic Arrays

Forming super CORDIC units that can be reused is one thing, using them to build a large circuit that performs anything useful is different. We saw how algorithms can be moved to hardware in Chaps. 3 and 8. But here, we will discover a new architecture when we directly implement QRD, namely, systolic arrays.

The problem statement is as follows: we have a complex 4×4 matrix and we want to apply unitary transforms to transform it into an upper triangular matrix. Note that the angles obtained from these transformations also implicitly provide the Q matrix. We will also assume that the diagonal elements of the matrix are pure real when needed. This is necessary to allow super CORDICs to function (Sect. 10.7). Any conclusions we make about the architecture can be readily and without complication extended to larger matrices.

The order at which off-diagonal elements are nullified is (2,1), (3,1), (4,1), (3,2), (4,2), (4,3). The implementation is shown in Fig. 10.10. Rotation CORDICs accept phases horizontally and pass them along to adjacent rotation CORDICs. The phases are generated from vectoring CORDICs at the head of the row. The first row nullifies the first column. Four CORDICs are necessary. The second row nullifies the second column. Because the first element is already null by the time we move to the second row, we need only three CORDICs. The same logic applies to the third row which contains only a single rotation CORDIC.

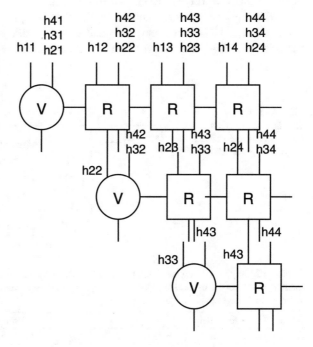

Fig. 10.10 Systolic array performing upper-triangulization of a 4×4 matrix. Circles are vectoring super CORDIC. Squares are rotation super CORDIC

10.9   Collapsed Arrays and Time-Sharing: Accelerators as an Approach to Implementation

The fully expanded architecture in Sect. 10.8 does not make sense in many respects. Firstly, it usually produces too much throughput for the system. Secondly, every row operates for one fewer cycle than the row before it. Additionally, lower rows must wait for upper rows to finish processing first. Notice that for the second row to start nulling element 3,2, it must operate on a matrix where the second row is fully updated from nulling the entirety of the first column.

For these reasons, CORDIC arrays as used for QRD readily lend themselves to time-sharing. In Fig. 10.11 we use a single row of CORDIC reused over three cycles to finish the decomposition. The processing unit is thus the upper row from the systolic array of Sect. 10.8. A memory bank is used to read all the inputs of the CORDIC and write their intermediate results. An address generator, usually part of a larger controller, is used to determine the addresses for read and write every cycle.

Table 10.2 shows the data read and written every cycle. Data is marked by matrix indices. The order of data in Fig. 10.11 and Table 10.2 is identical. So, in cycle 1, the inputs to the vectoring CORDIC are h11 and h21, the inputs to the first rotation CORDIC are h12 and h22, and so on. The same applies to outputs. There are potentially eight PU inputs per cycle and seven significant outputs. Note that the vectoring CORDIC will always produce only one non-null output.

Fig. 10.11 Collapsed CORDIC array. Squares are rotation CORDIC; circles are vectoring CORDIC

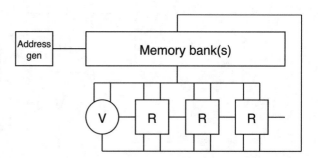

Table 10.2 Inputs and outputs to the PUs of the collapsed array by cycle. These will also dictate memory contents per cycle. Note don't cares increasing every cycle to indicate the rows getting shorter and PUs getting less busy

Cycle	Read data	Write data
1	h11, h21, h12, h22, h13, h23, h14, h24	h11, h12, h22, h13, h23, h14, h24
2	h11, h31, h12, h32, h13, h33, h14, h34	h11, h12, h32, h13, h33, h14, h34
3	h11, h41, h12, h42, h13, h43, h14, h44	h11, h12, h42, h13, h43, h14, h44
4	h22, h32, h23, h33, h24, h34, X, X	h22, h23, h33, h24, h34, X, X
5	h22, h42, h23, h43, h24, h44, X, X	h22, h23, h43, h24, h44, X, X
6	h33, h43, h34, h44, X, X, X, X	h33, h34, h44, X, X, X, X

X's indicate don't cares. In the read data column, don't cares indicate the corresponding CORDIC will not need these inputs; thus, they could be anything. In the write data column, don't cares indicate the output of the corresponding CORDIC is garbage and need not be written to memory.

In cycles where the array is nulling the first column of the array, none of the PUs are free and all memory ports are engaged (cycles 1–3). When nullifying the second column, one of the rotation CORDICs is redundant (Sect. 10.8), and thus some memory inputs and outputs from memories are don't cares. While nulling element 4,3, two rotation CORDICs are not used, leading to four inputs and outputs being don't cares in cycle 6.

Figure 10.11 suggests that the memory is multiport. In fact, it suggests that there are eight read ports and seven write ports. Memories with such a huge number of ports are unrealistic. We should not count on more than two ports in a realistic RAM. The contention of memory and the huge requirements on memory bandwidth with hardware reuse is not a problem unique to MIMO decoding. In fact, we discussed it in detail in Sect. 8.9.

If we cannot have this huge number of ports, we need to overclock the memory. Fig. 8.11 suggests we should clock the memory at 15 times the clock of the PUs if the memory is single port, or eight times if the memory is two-port. This is again impractical since the PU is a CORDIC, which has a relatively short critical path.

We discussed ways to address memory contention with hardware reuse in Sect. 8.9 including memory bank partitioning. The order of memory partitioning reduces the clock frequency by the same order. For example, use of two memory partitions requires a two-port memory to be clocked at four times the rate of CORDIC, which is realistic.

Notice that effective partitioning is not easy because it requires us to choose assignment of data while writing so that no more than four inputs need to be read from the same bank in the following cycle. Partitioning also requires the use of switching structures (Sect. 8.8) and a relatively complicated controller to manage memory addressing.

10.10 Generic Use of CORDIC and Two-Sided Jacobi in Hardware

CORDIC can be used to implement transcendental functions other than Givens rotations. This is used in applications beyond MIMO decoding and to perform flexible unitary transformation such as those used in Jacobi.

Table 10.3 lists some functions that can be supported by CORDIC. The first column is the function produced. The R/V column determines whether a rotation or vectoring CORDIC is used. The output column indicates the significant output to look at. For vectoring CORDIC, X indicates you should examine the magnitude, and θ indicates you should only look at the phase.

Table 10.3 Miscellaneous uses of CORDIC

Function	x	y	θ	R/V	Output
$|A|$	Re(A)	Im(A)	–	V	X
Angle(A)	Re(A)	Im(A)	–	V	θ
Dist(A, B)	Re(A) $-$ Re (B)	Im(A) $-$ Im (B)	–	V	X
atan(b)	1	b	–	V	θ
cos(θ)	1	0	θ	R	X
sin(θ)	0	1	θ	R	X

For example, if you examine the rotation CORDIC equation in Sect. 10.2 and use $x = 1$ and $y = 0$, then the x output would be the cosine of the θ input. This allows us to use the rotation CORDIC as an implementation for cosine, which is often more efficient than LUTs when the number of phases to look up is high. The functions listed in the table are, in order, absolute value of a complex number, phase of a complex number, distance between two points in the complex plane, inverse tan, cosine, and sine.

We can now use this understanding to implement a complicated decomposition such as two-sided Jacobi from Sect. 9.14. Two-sided Jacobi involves multiple steps, some of which require a lot of preprocessing. Thus, if we can support two-sided Jacobi, we can probably support any decomposition.

The first part of two-sided Jacobi is a complex Givens rotation to nullify element $(2, 1)$. Since we discussed how this can be done in detail in Sects. 10.7 and 10.8, we can start at the next step:

$$
U_4 \begin{bmatrix} E & G^{j\phi_g} \\ 0 & K \end{bmatrix} U_5 = \begin{bmatrix} e^{j\phi_g/2} & 0 \\ 0 & e^{-j\phi_g/2} \end{bmatrix} \begin{bmatrix} E & Ge^{j\phi_g} \\ 0 & K \end{bmatrix} \begin{bmatrix} e^{-j\phi_g/2} & 0 \\ 0 & e^{j\phi_g/2} \end{bmatrix}
$$

$$
= \begin{bmatrix} E & G \\ 0 & K \end{bmatrix}
$$

This transformation is applied on both sides of the matrix and at first glance seems very different from anything we have done so far. However, it can still be easily implemented using CORDIC. This is shown in Fig. 10.12. First, the phase of $G^{j\phi_g}$ is obtained through a (real) vectoring CORDIC by feeding the real and imaginary parts of $G^{j\phi_g}$ to the inputs of the vectoring CORDIC. This generates the phase of the complex number according to Table 10.3. This phase is fed to a shifter where it is divided by 2.

There are eight rotation operations in the matrix multiplications by U_4 and U_5. These are implemented using eight rotation CORDICs in Fig. 10.12. The upper row of CORDIC perform the left rotation where E and $Ge^{j\phi_g}$ are rotated by a positive phase and F and K by a negative phase. F is the null element in the matrix.

In the second row of rotation CORDIC, we apply the right rotation. Because this is a right rotation it acts on columns instead of rows, so we send F and G each to the

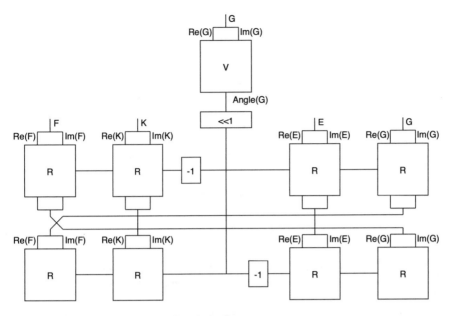

Fig. 10.12 Diagonal exchange transform in hardware

opposite side in Fig. 10.12. Notice that the order of the right and left rotations does not matter, and they can be interchanged. Thus, the upper and lower row of rotation CORDICs in Fig. 10.12 can be exchanged.

One might object that there is no F in the equation. In fact, F is a null, and its rotations can be removed to save hardware. To take this even further, if we go back to the phase exchange equation, we find that it can be implemented using a single real vectoring CORDIC instead of the entirety of Fig. 10.12. Notice that the diagonal exchange transform does the following:

$$\begin{bmatrix} E & G^{j\phi_g} \\ 0 & K \end{bmatrix} \rightarrow \begin{bmatrix} E & G \\ 0 & K \end{bmatrix}$$

So, nothing changes except that G becomes its absolute value. Thus, we only need a vectoring CORDIC to calculate the absolute value of G instead of the entirety of Fig. 10.12. However, notice that this simplification works only for 2×2 matrices. According to Sect. 9.14, we need to apply these transformations to entire rows in larger matrices, and thus we must be ready to apply them to nontrivial cases where none of the elements are null or even real.

To perform the real Jacobi algorithm, we begin by calculating the sum and difference phases:

Fig. 10.13 Sum and difference phases

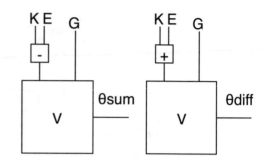

$$\theta_{\text{sum}} = \text{atan}\left(\frac{G}{K - E}\right)$$

$$\theta_{\text{diff}} = \text{atan}\left(\frac{G}{K + E}\right)$$

The hardware implementation is straightforward using adders and vectoring CORDIC according to Fig. 10.13. The right and left phases can be obtained as:

$$\theta_r = \frac{\theta_{\text{sum}} + \theta_{\text{diff}}}{2}$$

$$\theta_l = \frac{\theta_{\text{sum}} - \theta_{\text{diff}}}{2}$$

And their hardware implementation is trivial using adders and hardwired shifters. Finally, we can perform the two-sided Jacobi transformation:

$$\begin{bmatrix} \cos\theta_r & -\sin\theta_r \\ \sin\theta_r & \cos\theta_r \end{bmatrix} \begin{bmatrix} E & G \\ 0 & K \end{bmatrix} \begin{bmatrix} \cos\theta_l & \sin\theta_l \\ -\sin\theta_l & \cos\theta_l \end{bmatrix} = \begin{bmatrix} S_1 & 0 \\ 0 & S_2 \end{bmatrix}$$

The implementation in Fig. 10.14 needs some explanation. The left transformation can be implemented as a Givens rotation using two rotation CORDICs according to Sect. 10.6. However, the matrix is slightly different from a Givens rotation because the negative sign on the sin is in element $(1, 2)$ instead of $(2, 1)$. This can be amended by using a negative sign with the input phase to the rotation CORDIC. Since cosine is even and sine is odd, this takes care of the left rotation.

The right rotation also needs this sign flip. But additionally, we need to act on columns rather than rows. We can already see a pattern where left transforms act on rows and right transforms act on columns. As with the diagonal exchange transform, it does not matter if we perform the right rotation or the left rotation first.

At several points in QRD and SVD, we performed transforms of the form:

Fig. 10.14 Two-sided
Jacobi in hardware

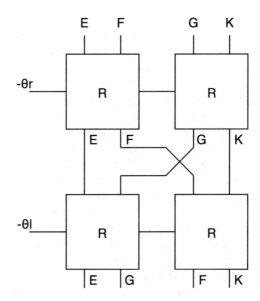

$$\begin{bmatrix} e^{j\phi} & 0 & 0 \\ 0 & 1 & 0 \\ 0 & 0 & 1 \end{bmatrix} \begin{bmatrix} Ae^{j\phi_a} & Be^{j\phi_b} & Ce^{j\phi_c} \\ De^{j\phi_d} & Ee^{j\phi_e} & Fe^{j\phi_f} \\ Ge^{j\phi_g} & He^{j\phi_h} & Ke^{j\phi_k} \end{bmatrix} = \begin{bmatrix} Ae^{j(\phi_a-\phi)} & Be^{j(\phi_b-\phi)} & Ce^{j(\phi_c-\phi)} \\ De^{j\phi_d} & Ee^{j\phi_e} & Fe^{j\phi_f} \\ Ge^{j\phi_g} & He^{j\phi_h} & Ke^{j\phi_k} \end{bmatrix}$$

This happened as part of both decompositions when we turned leading elements of matrices real. This transformation can be performed using three rotation CORDICs. These CORDICs will rotate the first row of elements in the matrix. The second and third rows remain unchanged. To rotate the first row, feed ϕ to the phase inputs of the rotation CORDICs. The x and y inputs of each CORDIC are the real and imaginary parts of A, B, and C. In cases where $\phi = \phi_a$, the first rotation CORDIC is replaced with a vectoring CORDIC which calculates A and ϕ_a.

10.11 Is a Naïve Implementation Always Bad?

In Sect. 10.1 we discussed what we termed a naïve implementation of unitary transforms. For the rest of the chapter, we discussed how CORDIC can be used to implement a host of transcendental functions including unitary transformations more efficiently. The implementation in Sect. 10.1 was "naïve" because it implemented the arithmetic of the equations directly without any adaptation. This led to hardware that uses many multipliers and lookup tables. Thus, Fig. 10.1 has a power-hungry PU, a complicated critical path, and many challenges in fixed-point design.

Implementations using CORDIC use only adders, shifters, and a simple LUT for atan. This LUT is not even necessary. Go back to Sect. 10.4; the rotation CORDIC looked at the sign of θ_i to determine the direction of the current microrotation. Once this was decided, the value of the phase needed to be updated by adding or subtracting atan(2^{-i}) to allow the direction of the following microrotation to be determined, thus the need for the LUT.

But examine how rotation CORDICs are used for Givens rotations in Sects. 10.8 and 10.9. The phase input of every rotation CORDIC is obtained from a vectoring CORDIC. All the rotation CORDICs in a certain row will rotate by this same phase. The discussion here and below applies only when the rotation CORDIC phase input is the unprocessed phase output of a vectoring CORDIC, and thus does not apply to Jacobi, for example (Sect. 10.10).

In Fig. 10.11, the vectoring CORDIC calculates a direction a_i during every microrotation. This is used to compound a value for θ. This value of θ is then handed to the row of rotation CORDIC, where they break it down to obtain directions a_i. We could (and should) have handed the directions determined by the vectoring CORDIC a_i directly to the row of rotation CORDIC. This negates the need for any CORDIC to even have a θ path, which would also remove all the lookup tables from all CORDICs. In other words, why force the vectoring CORDIC to add up theta so that the rotation CORDICs can proceed to break it down in exactly the reverse way. Instead, allow the vectoring CORDIC to inform rotation CORDICs of which direction to microrotate in every iteration.

Another reason CORDIC implementations are efficient is they do not show the wordlength growth typical of multipliers. This is because even though the raw equations for unitary transforms involve multiplications, these multiplications are by sinusoids with limited magnitude. In the visual sense of Fig. 10.1, the outputs of a CORDIC may be larger than the inputs, but they cannot be larger than the amplitude of the vector formed by the original point. This is because what we are doing is ultimately a rotation, an operation with a unit norm.

CORDIC is thus very popular in applications that involve multiplication by sinusoids, including phasors. Some people use CORDIC to implement twiddle factor multiplication in FFT. And CORDIC is very often used to implement frequency offset estimation and correction (Sect. 10.6).

Having said all this, multiplier-based implementations of unitary transformations are not unknown. This is because such implementations are direct and easy and do not require special hardware. After all, CORDIC is special, and requires someone familiar with its inner workings. But another reason multiplier-based applications are coming into vogue is that multipliers are starting to become cheap. This is particularly true on FPGA, where dedicated arithmetic slices are becoming plentiful, fast, and power efficient. If you do not use these multipliers for something, they still exist.

10.12 Clocks, Architectures, Speed, Throughput, and Latency: Sorting Through the Lingo

Of all the performance metrics we have discussed in this book, "speed" is the most confusing. We took a stab at the definitions in Sect. 3.9 and tried to connect throughput to clock implicitly in Sects. 7.14 and 7.15. By this point, we have a comprehensive view of the transceiver chain and a good understanding of hardware architectures. This allows us to finally address "speed" in a systematic manner. We will first define different terms used to refer to speed, namely, bitrate, clock speed, symbol rate, throughput, and latency. We will then focus on how throughput and clock rate are related, and how this can be used to plan hardware in a receiver.

Bitrate This is the rate of flow of bits through a subsystem, block, or system as measured in Mbps. For the transmitter, for example, bitrate will refer to the uncoded data rate as calculated from the MCS in Sect. 7.14. It could also refer to the raw data rate as radiated out of the antenna. Or maybe it is the goodput as defined in Sect. 2.7. When applied to subsystems, it could refer to the equivalent bitrate in a symbol-domain block. In bit-domain blocks, it usually refers to the coded data rate. As you can tell from the above wishy-washy definition, bitrate is highly underdefined. If you are given specifications or requirements in terms of bitrate, make sure you ask as many clarifying questions as you need.

Symbol Rate The rate at which QAM symbols are radiated out of the radio, or the rate symbols are flowing through a symbol-domain subsystem. Symbol rate is measured in MSps. On a system level in 802.11n, the symbol rate is identical to bandwidth. At the subsystem level, symbol rate only applies to symbol-domain blocks; thus, it is the rate at which blocks like the FFT and the MIMO decoder produce outputs. For example, in a 20 MHz SISO receiver, the FFT receives and should produce outputs at the rate of 20 MSps if it is not accumulating latency.

Throughput This is the closest term to the "speed" we refer to when discussing performance. Throughput is the rate at which useful outputs are produced out of a subsystem. It is measured in samples per second (more likely Mega samples per second). What constitutes a "sample" or useful output is related to the application. But when defining throughput, you need to be very specific and clear.

For example, assume we are performing a 32-point FFT. If we say that throughput is 100 MSps, does that means that frequency-domain samples are produced at this rate, or that frames of 32 samples are produced at this rate? If it is the former, then $X(k)$ is produced at a rate of 100 MHz, but if it is the latter, then $X(k)$ is produced at a rate of 3.2 GHz. This is a 32-fold difference and should not be left to individual interpretation. You can state throughput in MSps, but you need to be explicit about what a "sample" is.

Latency This is the time (in seconds or full clock cycles) that it takes for the first sample to be produced out of the block or system. When pipelining is used, latency is a direct result of the depth of the pipeline. But in some cases, latency is fundamentally related to the algorithm. For example, both the frequency-domain interleaver and FFT have innate latency. Both require an entire block of data to be available before the first output can be produced. This is clearest in FFT where the DFT equation shows that 64 (in 20 MHz mode) time-domain samples must be received before we begin to calculate any of the frequency-domain samples.

Assuming time-domain samples arrive at a rate of 20 MHz, it takes 3.2 μsec to receive 64 samples. During this time, which is the duration of an OFDM symbol, the FFT is sitting idle. Notice that this needs only happen in the first OFDM symbol, especially if you use internal pipelining, because on subsequent symbols, we will be processing the old signal while accumulating inputs for the new one, which is why this is latency rather than just generic delay.

This inevitable latency might seem small, but we must see it on a system-wide basis. For example, upper layers usually require the receiver to acknowledge receipt within a time-period or the transmitter will score a timeout. While 3.2 μsec might not sound like much, if we allow every subsystem to accumulate delay, we might run into a wall.

But this latency is easy to absorb. In some ways, it is hard not to absorb it. Processing units are usually fast enough that we can operate them so fast that they compensate for latency. For example, if FFT has accumulated a latency of 3.2 μsec, and the MIMO decoder needs to finish an OFDM symbol in 3.2 μsec to introduce zero latency, what happens if MIMO decoding finishes in 1 μsec? The total latency through the system is now reduced to $(3.2 - (3.2 - 1))$ μsec. This is the concept of latency budgeting, where we can allow some subsystems to accumulate latency if we force others to make it up.

At this point, it is worth pointing out that the rate at which data arrives is often a limiting factor for symbol-domain blocks. This is because their PUs can operate at high rates, but data input is only available at 20 or 40 MHz. This is not as big of a problem in bit-domain blocks because bitrates are higher than symbol rates by the modulation order.

Clock Rate It is the clock speed of the subsystem. This is often related to but not identical to throughput, bitrate, and symbol rate. This is because systems may use parallelism to accept and process multiple samples at a time; it could also have hardware reuse where it needs to run for multiple cycles before producing a single output.

For example, assume we are designing a symbol-domain block that produces outputs at a rate of 20 MSps. This rate is the required throughput. When we implement the block, we find that the PU can be clocked at a rate of 200 MHz; we thus time-share the PU over ten cycles to produce a single output. So, in this case, the clock rate is 200 MHz, while the throughput is 20 MSps because we produce a useful output every ten clock cycles. Thus, the clock speed and the throughput are different, although they are related through the order of hardware reuse.

As another extreme example, assume that when we try to implement the block, we find that it can only be clocked at 2 MHz, producing output at a rate of 2 MSps. In such case, we'll use parallelism to obtain the required 20 MSps. Specifically, we apply parallelism of order 10. In such case, the clock is 2 MHz and throughput is 20 MSps. So, because we can have hardware reuse or parallelism, we cannot equate clock speed and throughput, although the two are usually related.

Planning Clocks and Rates

We need to bring all of this together. How can we plan ahead and make decisions about the architecture of all our blocks? In this section, we will describe a systematic way to determine rates, speeds, and throughputs throughout a receiver. This can be extended to any similarly complicated system. We start with the MCS and bandwidth of the system and calculate its throughput in Mbps. We trace this back through the system, determining the throughput requirements of every block. This allows us to decide their hardware implementation, and finally decide their clock rates.

Consider a 2×2 transmitter in 20 MHz mode. The IFFT produces output at 20 MSps; the constellation mapper also produces a throughput of 20 MSps (notice we are including pilots and nulls). The input to the mapper is in the bit-domain and is thus at a rate of $20k$ Mbps, where k is the number of bits in a QAM symbol. This is also the throughput of each of the interleavers. The spatial parser gathers these two streams for an input throughput of $20k \times 2$ Mbps. This rate is the output throughput of the puncturer, which means that the input rate to the encoder is $40k/r$ Mbps. Thus, the input to the puncturer is at a rate of $40 \times 2k$ Mbps.

Knowing these throughput requirements will then inform our hardware design decisions. For each block in the chain, we know the output rate and the input rate. We know the algorithm the block implements and can break it down into basic operations. Through experience or by running the synthesizer, we figure out the clock speed of the PU used to implement the algorithm. Based on the required throughput from the block and the speed of the PU, we can make a decision about the architecture. According to Chap. 8, if throughput is higher than PU bandwidth, we resort to parallelism or pipelining. If PU throughput is higher than throughput, we implement hardware reuse. Once the hardware is set, we calculate the latency of the block and find its clock rate from the critical path.

But this is where a major complication arises. If we follow the approach described above to design the system, we will end up with a different clock rate for each block. Do we allow each block to operate at its own clock? Clock rates are the result of critical path analysis on a per block basis. Thus, they can vary wildly between blocks. For example, the critical path in the MIMO decoder is likely much longer than in the interleaver.

So, do we let every block operate at an independent clock? If we do, we will face multiple problems:

- How can you make sure that data is produced in a way and at a rate that the next block accepts? If the interleaver finishes in a burst of very high-rate data, how can you make sure that the mapper can accept data of this nature?
- But more importantly, if every block operates at a different clock, how do we hand data between them?

Moving data across clock domains is not simple. It is a major problem that is beyond the scope of this book. If data is moved ad hoc between registers clocked at different rates, a phenomenon called metastability occurs. The receiver register will fail catastrophically on a regular basis and the system will not work. To move data safely, special interfaces must be used. These could be synchronization circuits that use handshaking or asynchronous FIFOs. But in all cases, this is a major issue that cannot be taken lightly, and it is not something that can be done at will.

But even ignoring metastability, clocks are not free. Clocks are not treated by placement and routing tools as regular signals, because they are not. Clock distribution networks use dedicated metal layers to minimize skew and jitter (variable delay and phase noise). Thus, it is very hard to find a routing solution for a design with two clocks, let alone one with half a dozen.

On the other hand, if we enforce the use of a single clock through the whole system, we would be forcing all the blocks to have a single critical path. This is the slowest path in the slowest block. Everyone else will then have to slow down to the rate of this weakest link.

The answer is usually something in the middle. For example, you should not use multiple clocks between the encoder and the puncturer. They are not that far-off in their rate and the puncturer is very simple to begin with. But perhaps you should allow the FFT to use its own clock because it is a high intensity block that accumulates a lot of latency by virtue of algorithm.

Chapter 11
Synchronization

11.1 802.11n Packet Structure

Figure 11.1 shows the packet structure for 802.11n. The packet consists of a payload and a header. The payload is shown as HT-DATA on the figure and the header is everything else. The figure is not drawn to scale and the header is typically shorter than the payload.

The payload is the data that the packet delivers. This data has a modulation and coding scheme dictated by the transmitter and chosen from the list of allowed MCS in the standard. The payload is usually the MAC MPDU as handed down to the physical layer but in some cases, it can be a service or sounding packet.

The header consists of preamble fields (L-STF, L-LTF, HT-STF, and HT-LTF) and signal fields (L-SIG and HT-SIG). The preamble contains known data and is used to estimate channel state information and receiver non-idealities. The signal part of the header contains information that the transmitter needs to tell the receiver about the payload.

The header contains an L copy and an HT copy of every field. L stands for legacy and HT stands for high throughput. SIG stands for signal. LTF stands for long training field. STF stands for short training fields. All training fields are preambles. The first three fields of the header are the legacy header, which is identical to the header of 802.11a. Figure 11.1 is in time domain, and the duration of each field clearly shows that most of them are not standard OFDM fields. Recall that the OFDM field in 802.11n has a length of either 3.6 or 4 μsec depending on the length of GI.

The header performs the following functions:

- Informs the receiver about decisions the transmitter made about the payload, like which modulation coding scheme they will be using, coding technique, type of payload, etc.
- Allows packet edge detection (Sect. 11.8). This is usually performed using the L-STF since it is the first field in the packet.

© The Author(s), under exclusive license to Springer Nature Switzerland AG 2023   373
K. Abbas, *From Algorithms to Hardware Architectures*,
https://doi.org/10.1007/978-3-031-08693-9_11

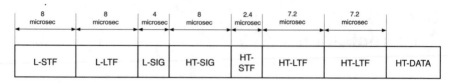

Fig. 11.1 802.11n packet structure

- Allows automatic gain control (Sect. 7.15). This is usually done using the HT-STF.
- Allows frequency offset estimation (Sect. 11.6). Can be performed by any training field, but usually either short training fields due to their periodicity.
- Allows timing offset estimation (Sect. 11.7). Usually performed using long training fields, especially the L-LTF.
- Channel estimation (Sect. 11.5). Done using long training fields. In systems with spatial multiplexing, this must be the HT-LTF. In setups with SISO communication, both the L-LTF and the HT-LTF can be used.

11.2   The Legacy Header

The legacy header is understood by both legacy and high throughput terminals. It consists of a preamble formed of a short training field and a long training field. It also contains a rather misleading signal field. It is used by high throughput terminals to aid in offset estimation and perform packet edge detection. The legacy header also serves to keep legacy terminals off the channel for the duration of the packet, thus reducing collisions in CSMA-CA.

11.2.1   L-STF

The legacy short training field is the first part of the packet. Its structure is identical to the short training field in 802.11a. It serves a lot of purposes. As part of the legacy header, it engages legacy devices and informs them that there is a packet over the air. The legacy terminal will then try to decode the packet to see if it is directed to them.

The L-STF is also used for packet edge detection by high throughput terminals (Sect. 11.8). It is used to generate coarse frequency offset estimates (Sect. 11.6). It is also used to begin automatic gain control (Sect. 7.15).

Most of the functions that employ the L-STF happen before the channel estimates are available. In most high throughput terminals, channel estimates are only available after the HT-LTF. Thus, even though the receiver knows the values in the L-STF at the transmitter, it does not know what values to expect at the receiver. So, how can we use the L-STF to do packet edge detection or frequency offset estimation, for example? Notice that all these functions are contingent on having known data *at the receiver*.

The *L*-STF has a structure that we can leverage to make estimations before inverting the channel. The following is the frequency-domain symbol of the *L*-STF in 20 MHz:

$$
\begin{aligned}
S_{-26,26} = \sqrt{\frac{13}{6}} \{ & 0,0,1+j,0,0,0,\,-1-j,0,0,0,1+j,0,0,0,\,-1-j,0,0,0, \\
& -1-j,0,0,0,1+j,0,0,0,0,0,0,0,\,-1-j,0,0,0,\,-1-j,0,0,0, \\
& 1+j,0,0,0,1+j,0,0,0,1+j,0,0,0,1+j,0,0 \}
\end{aligned}
$$

Only subcarriers -26 through 26 are listed because subcarriers on the periphery are null. The *L*-STF uses QPSK, but there is a strange preponderance of null carriers. The nulls are plentiful and are distributed regularly throughout the whole symbol. This is not how a QPSK OFDM symbol in payload looks like (Sect. 7.13).

Also, the normalization factor of the symbol does not fit with the traditional normalization factor for QPSK in the payload (Sect. 5.4). Because of the plentiful nulls, the symbol has a lower average energy than the typical payload symbol. To raise its power, we increase the normalization factor. But even accounting for the nulls, *L*-STF has much higher energy than a payload QPSK symbol. As discussed above, the *L*-STF is used for a lot of critical synchronization functions before the channel is known. The high energy helps make the symbol more resistant to noise, and thus improves the outcomes of these functions.

The *L*-STF symbol in 40 MHz mode is:

$$
\begin{aligned}
S_{-58,58} = \sqrt{\frac{13}{6}} \{ & 0,0,1+j,0,0,0,\,-1-j,0,0,0,1+j,0,0,0,\,-1-j,0,0,0, \\
& -1-j,0,0,0,1+j,0,0,0,0,0,0,0,\,-1-j,0,0,0,\,-1-j,0,0,0, \\
& 1+j,0,0,0,1+j,0,0,0,1+j,0,0,0,1+j,0,0,0,0,0,0,0,0,0, \\
& 0,0,0,0,0,j-1,0,0,0,\,-j+1,0,0,0,j-1,0,0,0-j+1,0,0, \\
& 0,\,-j+1,0,0,0,j-1,0,0,0,0,0,0,0,\,-j+1,0,0,0,\,-j+1,0,0, \\
& 0,j-1,0,0,0,j-1,0,0,0,j-1,0,0,0,j-1,0,0 \}
\end{aligned}
$$

The lower half of the symbol is a copy of the 20 MHz symbol. The upper half, representing the upper half band, is a 90-degree rotated version of the lower half band. Legacy devices do not have a 40 MHz mode; thus, when receiving a 40 MHz packet, they will only observe the lower 64 subchannels. Notice that the subchannel bandwidth in 20 and 40 MHz modes is equal because 40 MHz mode has exactly double the number of non-null subcarriers.

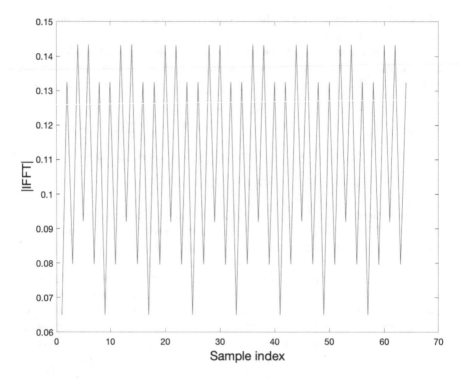

Fig. 11.2 Time-domain amplitude of the *L*-STF showing 16-sample periodicity

The preponderance of frequency-domain nulls in the *L*-STF symbol means that when we convert it to time domain, we will not obtain 64 unique samples. Instead, we will get 16 unique samples repeated four times. This periodicity is exactly the kind of feature we need for packet edge detection (Sect. 11.8), and particularly for frequency offset estimation (Sect. 11.6), where the low period allows discerning larger offsets. The *L*-STF in time domain is shown in Fig. 11.2. The periodicity is clearly visible.

The 16 samples have a duration of:

$$\frac{16}{20}\mu\text{sec} = 0.8\,\mu\text{sec}$$

This is repeated ten times without a guard interval to give an *L*-STF duration of:

$$0.8 * 10\,\mu\text{sec} = 8\,\mu\text{sec}$$

Why can the *L*-STF get away without a guard interval? Mainly because it comes at the beginning of the packet. Guard intervals come in the form of prefixes rather than suffixes in 802.11n. Thus, the *L*-STF will not introduce anything meaningful by adding a prefix. If there is any data before the L-STF that can introduce ISI, then there is a collision and the whole packet is lost anyway.

11.2.2 L-LTF

The legacy long training field is the second part of the legacy header. It continues the tradition of the L-STF in fooling legacy terminals into believing there is a legacy packet over the air. The L-LTF may be used to improve frequency offset estimates. In legacy devices, it is used to obtain channel estimates. In 802.11n and especially in multi-antenna modes, it cannot be used to estimate the channel, and is mainly used for timing offset estimation (Sect. 11.7). It is perhaps a little interesting that the L-LTF and the L-STF are both the same duration (Fig. 11.1). But notice that the "short" in L-STF refers to the 16-sample periodicity observed in the time-domain signal rather than the duration of the whole field.

The L-LTF is a BPSK field. This helps fortify it against the channel. In 20 MHz mode, it is:

$$L_{-26,26} = \{1, 1, -1, -1, 1, 1, -1, 1, -1, 1, 1, 1, 1, 1, 1, 1, -1, -1, 1, 1, 1, -1, 1, 1, -1, 1, 1,$$
$$1, 1, 1, 0, 1, -1, -1, 1, 1, -1, 1, 1, -1, 1, 1, -1, -1, -1, -1, -1, -1, 1, 1, 1, -1, -1, 1,$$
$$1, -1, 1, -1, 1, 1, 1, 1, 1\}$$

Clearly, the field is not sparse like the L-STF. It will thus not have any form of periodicity when converted to time domain. However, the L-LTF field is formed of two time-domain symbols attached to each other; each is 3.2 μsec long. This is 6.4 μsec. It also has a special very long combined GI CP right after the L-STF which is 0.8*2 μsec. The total time of the L-LTF is thus 8 μsec.

For 40 MHz devices, the L-LTF in frequency domain is:

$$L_{-58,58} = \{1, 1, -1, -1, 1, 1, -1, 1, -1, 1, 1, 1, 1, 1, 1, 1, -1, -1, 1, 1, 1, -1, 1, 1, -1, 1, 1, 1, 1, 1, 0,$$
$$1, -1, -1, 1, 1, -1, 1, 1, -1, 1, 1, -1, -1, -1, -1, -1, 1, 1, 1, -1, -1, 1, 1,$$
$$-1, 1, -1, 1, 1, 1, 1, 1,$$

$$0, 0, 0, 0, 0, 0, 0, 0, 0, 0, 0, 0, 0$$

$$j, j, -j, -j, j, j, -j, j, -j, j, j, j, j, j, j, j, -j, -j, j, j, j, -j, j, j, -j, j, j, j, j, j, 0,$$
$$j, -j, -j, j, j, -j, j, j, -j, j, j, -j, -j, -j, -j, -j, j, j, j, -j, -j, j, j,$$
$$-j, j, j, j, j\}$$

The lower band is a copy of the 20 MHz mode, allowing legacy devices to keep up. The upper half band is a 90-degree rotated version of the lower half band. This strange Q-only BPSK signal is unknown to legacy devices, but legacy devices will not see this half band anyway.

11.2.3 L-SIG

The legacy signal field. This is the last part of the legacy header. The symbol is BPSK and has a format that would be very familiar in the payload. The symbol is 3.2 μsec long and has a GI of 0.8 μsec for a total duration of 4 μsec. In 802.11a, this field gives the receiver critical information from the transmitter. In 802.11n, the *L*-SIG field is only used to tell legacy terminals how long the channel will be occupied. And we do it in a very roundabout way.

The format of the *L*-SIG field is shown in Fig. 11.3. The field is BPSK, and coding rate is ½. This is a feature of all signal fields. They use the smallest possible constellation and the lowest possible coding rate. This is because the information in signal fields is critical, and we want it to arrive safely.

Most of the information in the *L*-SIG field is irrelevant except for length and rate. Rate is comparable to MCS in 802.11n; thus, it informs the legacy receiver about the modulation and coding rate. This does *not* have to be the true modulation and coding used in the high throughput payload. The *p* field is a parity bit that allows the receiver to check the integrity of the field. Length is very interesting. It is an indication of the length of the packet, but it is calculated in a very special way:

- Length is reported in units of bits.
- The time duration of these bits is interpreted in terms of the modulation in the "rate" field. Again, this does not have to be the true modulation used in the payload.
- Length is calculated for the summation of the duration of the high throughput header and the high throughput payload.

So, length is length of the packet if the packet were legacy. Because the packet is *not* legacy, this length is then a very special number. In the next section, we will see that legacy devices will abandon high throughput packets as soon as the legacy header ends. The "length" field helps these legacy terminals determine how long they must wait before attempting to transmit over the channel. This may allow the high throughput transmission to conclude without collisions from legacy devices.

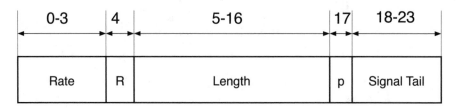

0-3	4	5-16	17	18-23
Rate	R	Length	p	Signal Tail

Fig. 11.3 Format of the *L*-SIG field. Numbers are bit indices not field lengths

11.3   The High Throughput Header

The high throughput header is expected to be understood only by high throughput terminals. Legacy terminals will (and must) fail to understand it. It contains a preamble composed of a short training field and a long training field. It also contains two signal fields.

11.3.1   HT-SIG1 and HT-SIG2

The high throughput signal field includes two OFDM symbols with a long guard interval each for a total duration of 8 μsec. Each symbol carries a number of information bits for high throughput terminals. The field uses BPSK modulation and coding rate ½.

The BPSK used in the HT-SIG field is 90-degree rotated so that the symbols used are j and -j instead of 1 and − 1. High throughput terminals know this, expect it, and can handle it. Legacy devices are not aware of the high throughput header or of the possibility of rotated BPSK. Instead, they think the payload has started. They will try to decode the message based on the rate in the *L*-SIG field. They will immediately fail CRC checks and will give up trying to receive the packet. However, because they know the length of the rest of the packet (provided by the *L*-SIG field), they know that the channel is busy till the high throughput packet ends. They will not try to transmit for the duration of the packet. This reduces the probability of collision under CSMA regime.

The first HT-SIG1 field is shown in Fig. 11.4. There are two pieces of information that the transmitter hands to the receiver in this field. The first is the length of the high throughput payload. This is measured in bits after the end of the high throughput header as interpreted through the MCS. The MCS is the modulation coding scheme used in the payload, which must be drawn from the list of allowed MCS in the standard (Sect. 7.14). Both pieces of information are critical. But the MCS is particularly important because it allows the baseband decoder, the depuncturer, and the deparser to function properly.

The HT-SIG2 field is shown in Fig. 11.5. It carries more diverse information than the first field. The ADV CODING bit indicates whether we are using LDPC (low-density parity-check) or convolutional coding at the transmitter. LDPC is a more powerful coding technique than convolutional coding that can optionally be used in 802.11n. The next bit is reserved for future use. The SOUNDING bit is true

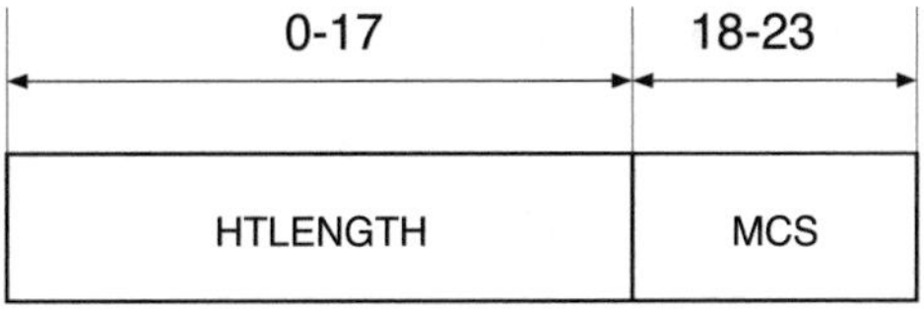

Fig. 11.4 First HT-SIG field

0-17	18-23
HTLENGTH	MCS

0	1	2	3-4	5	6	7-8	9	10-17	18-23
ADV CODING	Reserved	Sounding	Num HT-LTF	Short GI	Aggregat	Scrambler init	20/40	CRC	Signal Tail

Fig. 11.5 Second HT-SIG field

when the packet does not carry data payload but is used by the transmitter to carry out sounding of the channel. This allows the transmitter and the receiver to paint a more detailed picture of channel state information than that provided by plain channel estimation.

Before the end of this section, we will discover that the HT-LTF field can be repeated several times to provide noise immunity. NUMBER HT-LTF informs the receiver about how many such repetitions there are. SHORT GI is a single bit which is true when a 400 ns GI is used. Note that the default ("0") state is a long GI of 800 ns. The AGGREGATE bit is true when the packet is part of several packets that need to be aggregated to form a single MPDU before handing over to the MAC. SCRAMBLER INIT is the initialization seed of the scrambler.

The 20/40BW field indicates the bandwidth and thus the number of subcarriers used. Notice that the legacy header does not carry any unique data on the upper half band, so a terminal that thinks the bandwidth is 20 MHz till the HT-SIG2 field will not end up missing anything. The CRC bits provide cyclic redundancy check for the header, allowing the receiver to decide early if it is useful to continue decoding the packet. The SIGNAL TAIL bits provide the number of zeros attached to the end of the packet at the transmitter to make it into a whole number of OFDM symbols.

11.3.2   HT-STF

The high throughput short training field is a single OFDM symbol with a long GI. Its format in frequency domain in 20 MHz mode is:

$$HTS_{-26,26} = \sqrt{13/12}\{0, 0, 1+j, 0, -1-j, 0, 1+j, 0, -1-j, 0, -1-j, 0, -1-j, 0, 1+j$$
$$0, 1+j, 0, 1+j, 0, 1+j, 0, -1-j, 0, -1-j, 0, 0, 0,$$
$$1+j, 0, -1-j, 0, -1-j, 0 \ 1+j, 0, -1-j, 0, -1-j, 0, -1-j, 0,$$
$$-1-j, 0, 1+j, 0, -1-j, 0, -1-j, 0, 1+j, 0, 0\}$$

For the second antenna:

$$\sqrt{13/12}\{0, 0, \ -1-j, 0, 1+j, 0, 1+j, 0, 1+j, 0, \ -1-j, 0, \ -1-j, 0, \ -1-j$$
$$0, \ -1-j, 0, 1+j, 0, \ -1-j, 0, \ -1-j, 0, 1+j, 0, 0, 0,$$
$$1+j, 0, \ -1-j, 0, \ -1-j, 0 \ 1+j, 0, \ -1-j, 0, \ -1-j, 0, \ -1-j, 0,$$
$$-1-j, 0, 1+j, 0, 1+j, 0, 1+j, 0, \ -1-j, 0, 0\}$$

For the third antenna:

$$\sqrt{13/12}\{0, 0, \ -1-j, 0, \ -1-j, 0, 1+j, 0, 1+j, 0, 1+j, 0, 1+j, 0, 1+j$$
$$0, 1+j, 0, 1+j, 0, \ -1-j, 0, \ -1-j, 0, \ -1-j, 0, 0, 0,$$
$$-1-j, 0, \ -1-j, 0, \ -1-j, 0, 1+j, 0, 1+j, 0, 1+j, 0, 1+j, 0,$$
$$1+j, 0, 1+j, 0, 1+j, 0, 1+j, 0, \ -1-j, 0, 0\}$$

And for the fourth antenna:

$$\sqrt{13/12}\{0, 0, \ -1-j, 0, \ -1-j, 0, 1+j, 0, 1+j, 0, \ -1-j, 0, \ -1-j, 0, 1+j$$
$$0, 1+j, 0, \ -1-j, 0, \ -1-j, 0, \ -1-j, 0, \ -1-j, 0, 0, 0,$$
$$-1-j, 0, \ -1-j, 0, \ -1-j, 0, \ -1-j, 0, 1+j, 0, 1+j, 0, \ -1-j, 0,$$
$$-1-j, 0, 1+j, 0, 1+j, 0, \ -1-j, 0, \ -1-j, 0, 0\}$$

The 40 MHz mode is formed by 90-degree rotation of the 20 MHz symbol in the upper half of the tones. As with the legacy STF, when transformed to time domain, the HT-STF exhibits periodicity. In practice, the HT-STF is typically used to perform or improve AGC. The periodicity is over 16 samples. In 20 MHz mode, the HT-STF symbol is formed of two periods with a long GI. Each period is $\frac{3.2}{4} =$ 0.8 μsec long, and thus the duration of the HT-STF is $0.8 * 2 + 0.8 = 2.4$ μsec.

11.3.3   HT-LTF

The high throughput long training field is very special. It is a BPSK field, but it is interleaved between antennas and formed in a very specific way. Its main function is to allow channel estimation, especially in MIMO channels. This will be discussed in detail in Sect. 11.5. The HT-LTF is based on a primary sequence that can be drawn from several options provided in the standard. The following are samples for 20 MHz and 40 MHz:

$HTL_{-26,26} = \{-1,1,-1,1,1,1,-1,-1,-1,-1,1,1,1,1,1,-1,1,-1,1,-1,-1,1,1,1,1,-1,1,1,$

$\qquad 0,-1,1,1,-1,-1,1,1,-1,-1,1,1,-1,-1,1,1,-1,-1,1,1,1,1,1,$
$\qquad \quad -1,1,1,1,1,1,1,1,1\}$

$HTL_{-58,58} = \{-1,1,1,1,1,-1,1,1,-1,-1,-1,-1,1,1,1,-1,1,1,1,-1,1,1,1,-1,1,1,-1,1,1,-1$

$-1,1,1,-1,-1,1,1,1,-1,-1,-1,1,1,-1,1,1,-1,1,1,1,1,-1,-1,1,1,-1,$
$\quad -1,-1,-1,-1,-1,1,1,-1,1,1,-1,1,1$

$0,0,0,-1,-1,-1,-1,-1,-1,-1,-1,-1,-1,-1,-1,1,1,1,1,-1,1,1,$
$\quad -1,1,1,$

$-1,-1,1,1,-1,1,1,1,-1,1,1,-1,1,1,-1,-1,-1,-1,1,1,1,-1,1,1,-1,-1,1,$
$\quad -1,-1,$

$\qquad -1,1,1,-1,1,1,1,1,1,-1,1,1,1,-1,1,1,-1,1,1,1,1\}$

The aim now, especially in MIMO, is to obtain nonoverlapping sets on different antennas from this basic sequence. The number of HT-LTF symbols and the way they are distributed can be very confusing, but we will take it a step at a time.

Figure 11.6 shows the setup in 2×2 MIMO. There are two HT-LTFs sent on each antenna. Each consists of the repetition of a set. Each set is nonoverlapping with the other so that null carriers in Set1 correspond to occupied carriers in Set2. The same also applies to 3×3 MIMO in Fig. 11.7 and 4×4 MIMO in Figs. 11.7 and 11.8.

These sets are obtained from the master HTL sequence provided above. In 20 MHz 2×2 MIMO, for example:

$$\text{Set1} = [-26:2:-2],[+2:2:+26]$$
$$\text{Set2} = [-25:2:-1],[+1:2:+25]$$

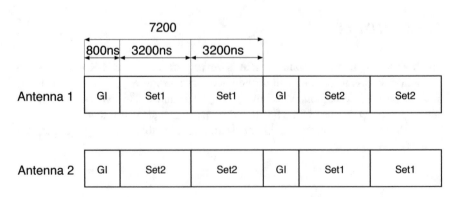

Fig. 11.6. 2×2 HT-LTF

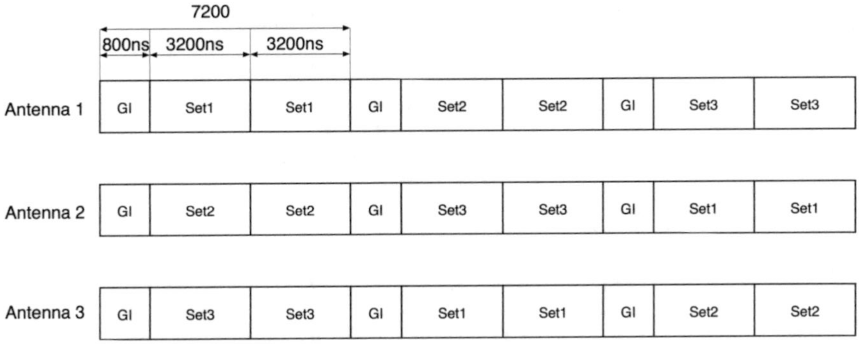

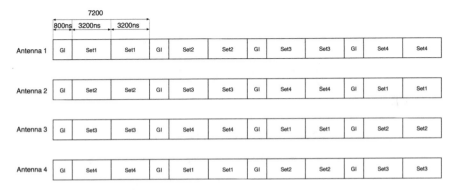

Fig. 11.7 3 × 3 HT-LTF

Fig. 11.8. 4 × 4 HT-LTF

The sets indicate the indices from HTL that are occupied in each set. Other indices are null. So, for example, Set1 has non-null subcarriers at indices −26, −24, −22, ..., −2 and 2, 4, 6, ..., 26. Set2 has non-null subcarriers at −25, −23, −21, ..., −1 and 1, 3, 5, ..., 25. The non-null subcarriers in the two sets are obviously nonoverlapping, which is the reasoning behind their design.

For 3 × 3 MIMO, three sets are derived as follows:

$$\text{Set1} = [-26:3:-2], [+2:3:+26]$$
$$\text{Set2} = [-25:3:-1], [+3:3:+24]$$
$$\text{Set3} = [-24:3:-3], [+1:3:+25]$$

Again, the three sets are designed so that they do not have overlapping non-null subcarriers. For 4 × 4 MIMO, four sets are derived as:

$$Set1 = [-26 : 4 : -2], [+3 : 4 : +23]$$
$$Set2 = [-25 : 4 : -1], [+4 : 4 : +24]$$
$$Set3 = [-24 : 4 : -4], [+1 : 4 : +25]$$
$$Set4 = [-23 : 4 : -3], [+2 : 4 : +26]$$

And for 40 MHz 2×2:

$$Set1 = [-58 : 2 : -2], [+2 : 2 : +58]$$
$$Set2 = [-57 : 2 : -3], [+3 : 2 : +57]$$

40 MHz 3×3:

$$Set1 = [-58 : 3 : -2], [+3 : 3 : +56]$$
$$Set2 = [-57 : 3 : -3], [+4 : 3 : +57]$$
$$Set3 = [-56 : 3 : -2], [+1 : 3 : +58]$$

40 MHz 4×4:

$$Set1 = [-58 : 4 : -2], [+5 : 4 : +57]$$
$$Set2 = [-57 : 4 : -5], [+2 : 4 : +58]$$
$$Set3 = [-56 : 4 : -4], [+3 : 4 : +55]$$
$$Set4 = [-55 : 4 : -3], [+4 : 4 : +56]$$

It should be clear what we are doing in Figs. 11.6, 11.7, and 11.8. For example, in Fig. 11.6, Set1 is first sent on antenna 1, while Set2 is simultaneously sent on antenna 2. The two are nonoverlapping and will thus not interfere with each other in frequency domain, allowing us to perform channel estimation in Sect. 11.5. Each set is sent twice on each antenna to allow the receiver to average out noise while performing channel estimation. Each set is 3.2 μsec and a single long GI of 800 ns is used for a total period of 7.2 μsec. The roles are then reversed, and everything is repeated except with Set1 on antenna 2 and Set2 on antenna 1. This allows all subcarriers to be estimated on the two antennas without interpolation.

The same logic applies to the 3×3 case in Fig. 11.7 and 4×4 in Fig. 11.8. Every set must be experienced by every antenna, but no two antennas can have the same set simultaneously. This allows all antennas to estimate all subcarriers while ensuring that different antennas do not interfere with each other during channel estimation. The number of times that each set is repeated on each antenna is determined by the transmitter and indicated in the HT-SIG2 field.

11.4   Scrambling, Padding, and CRC

A few fields from the HT-SIG field have not yet been explained. They mostly pertain to the scrambler and CRC. The scrambler and descrambler are identical hardware blocks shown in Fig. 11.9. It operates on bit streams at both the transmitter and the receiver. The scrambler/descrambler is a simple LFSR.

The purpose of the scrambler is to make the output signal look unintelligible to terminals that are not equipped with a corresponding descrambler. The aim here is not encryption, privacy, or data protection. The scrambler is very simple and any attempt to break its code will immediately succeed. Scrambled data is supposed to look unintelligible to benign terminals that are not trying to break it but that are not equipped with an appropriate descrambler. The objective is to allow terminals operating on a different standard in the same band to quit trying to decode while informing them the channel is busy. The HT-SIG 2 field contains bits for initializing the scrambler at the start of operation.

The scrambler/descrambler in 802.11n, Fig. 11.9, is known as an additive scrambler. It is very popular in wireless communication systems, and one of its advantages is that the scrambler and the descrambler are identical. The LFSR is a ring with positive feedback. One bit from the LFSR is used as an output; this bit is XOR'ed. with the input data stream to obtain an output stream. It is imperative that the receiver and the transmitter use the same sequence to scramble and descramble, which is why they need to exchange information about scrambler initialization.

Notice that in the additive scrambler, the LFSR and the scrambling operation are independent. This means that the bit sequence out of the LFSR is data-independent and is predictable based on the initialization. This means that one can replace the LFSR with a small memory containing the sequence of the LFSR. Since we only need the MSB of the LFSR, the memory will be small, especially since the polynomial is short.

CRC is short for cyclic redundancy check. CRC is an involved method of data validation, like a super charged parity check. The idea is to pass data through an LFSR and then look at the remainder left in the register. If the transmitter provides the receiver with an "expected" remainder, then the receiver can take a primary stab at error detection.

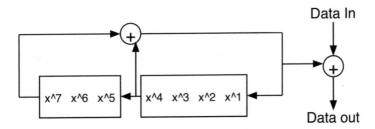

Fig. 11.9 The scrambler/descrambler

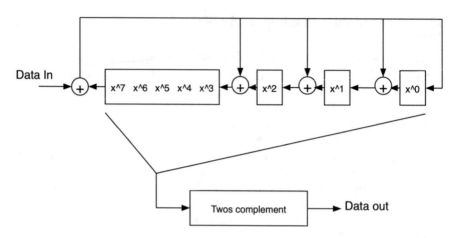

Fig. 11.10 CRC calculation in 802.11n

CRC is often used to check the integrity of message bodies. But in 802.11n, it is used independently to check the integrity of the header. This allows the receiver to abandon the packet early if critical parts of the header have not been received correctly. For example, if the signal field is corrupted, we should know this, and we should abandon the packet as early as possible, simply because we cannot even tell the MCS being used, the length of the packet, the number of HT-LTF, and a host of other critical information.

CRC on the header is performed as follows:

- Form a bit stream to be checked by concatenating L-SIG(0–16), HT-SIG1(0–23), and HT-SIG2(0–9).
- Initialize the LFSR in Fig. 11.10 to all 1's. Feed the bit stream above to the data input of the setup in Fig. 11.10.
- Take the remainder in the LFSR.
- Take the 2's complement of the remainder.

This output is then compared to bits 10–17 of HT-SIG2, which contains the 8-bit CRC as calculated at the transmitter. If the CRC calculated by the receiver matches the CRC calculated at the transmitter, we have good confidence that the three signal fields were received correctly and we can go deeper into the packet.

11.5 MIMO Channel Estimation: Easier Than You Think

In Sect. 11.3, we saw how the HT-LTF can be derived from nonoverlapping sets. The frequency-domain structure of the HT-LTF is very conducive to MIMO channel estimation. We will explore a simple case below. This is a 2×2 system where channel estimates are performed in frequency domain. The same concept can be extended to any size channel matrix and to time domain.

In the first slot of Fig. 11.6, Set1 is sent on antenna 1 and Set2 on antenna 2. This means that for the even-numbered tones, only transmitter 1 is active, and thus:

$$\begin{bmatrix} y_1 \\ y_2 \end{bmatrix} = \begin{bmatrix} h_{11} & h_{12} \\ h_{21} & h_{22} \end{bmatrix} \begin{bmatrix} x_1 \\ 0 \end{bmatrix} + \begin{bmatrix} n_1 \\ n_2 \end{bmatrix}$$

This allows the receiver to estimate even numbered subchannel coefficients:

$$\widehat{h_{11}}\Big|_{even} = \frac{y_1}{x_1}$$

$$\widehat{h_{21}}\Big|_{even} = \frac{y_2}{x_1}$$

These must be estimates because there is noise added to the observations. Recall that the HT-LTF field is repeated. The estimate is obtained by averaging these repetitions. The more repetitions, the better the estimate. Meanwhile, antenna 2 is transmitting on odd-numbered subcarriers. This allows the receiver to estimate the following channel coefficients:

$$\begin{bmatrix} y_1 \\ y_2 \end{bmatrix} = \begin{bmatrix} h_{11} & h_{12} \\ h_{21} & h_{22} \end{bmatrix} \begin{bmatrix} 0 \\ x_2 \end{bmatrix} + \begin{bmatrix} n_1 \\ n_2 \end{bmatrix}$$

$$\widehat{h_{12}}\Big|_{odd} = \frac{y_1}{x_2}$$

$$\widehat{h_{22}}\Big|_{odd} = \frac{y_2}{x_2}$$

According to Fig. 11.6, this is then followed by an exchange of transmitted fields, with antenna 1 transmitting Set2 and antenna 2 transmitting Set1. This allows us to calculate $\widehat{h_{12}}\big|_{even}$, $\widehat{h_{22}}\big|_{even}$, $\widehat{h_{11}}\big|_{odd}$, and $\widehat{h_{21}}\big|_{odd}$. Thus, all coefficients can be estimated for all subcarriers. This will always be true for 802.11n regardless of the number of antennas. In cellular systems (Sect. 12.4), pilots are used instead of a training field. This will allow estimates to be obtained for only a subset of subcarriers. Remaining subcarrier channel estimates must be obtained by interpolation.

Why are we assuming that the channel model is a scalar multiplication of channel coefficients with sent symbols? In other words, are we doing this estimation in frequency domain or in time domain? The answer is it does not matter; it will work for both. If we are doing this in frequency domain, then the fact that we are multiplying constant channel coefficients means that we are assuming the channel is flat within the subchannel, which is the assumption we generally make for OFDM (Sect. 7.1) and is also the reason we had to make an independent estimate for every subcarrier above. If we are doing this in time domain, then the corresponding assumption is that the channel is flat per subcarrier, which makes it single path, which allows us to transform convolution in time domain to multiplication.

11.6   Frequency Offset Estimation and Correction: What OFDM Would Die Without

In Sect. 7.3, we discussed why frequency offsets are particularly devastating for OFDM systems. In Sect. 5.5, we discussed the impact of frequency offsets on the constellation of a received QAM signal. In symbols with a large time duration, which OFDM signals are, the impact of frequency offsets cannot be ignored. In this section, we will see how different parts of the header and the packet can be used to find an initial estimate for the frequency offset, and then keep improving it.

To reiterate where frequency offset comes from, the passband signal coming out of the transmitter is:

$$x_p(t) = x(t)e^{j\omega_{ct}t}$$

And down converting at the receiver, we obtain a receiver baseband signal of:

$$x_r(t) = x(t)e^{j\omega_{ct}t}e^{-j\omega_{cr}t}$$

Ideally ω_{ct} and ω_{cr} should be equal. But due to inevitable variability in oscillators, their difference produces a frequency offset term:

$$x_r(t) = x(t)e^{j\Delta\omega t}$$

This phasor is a time-evolving phase difference that causes the constellation of the baseband signal $x(t)$ to revolve in time. This is particularly dangerous in OFDM, where the duration of the symbol is very long, allowing the phase to grow substantially by the end of the symbol.

How can we estimate the amount of frequency offset, and once we estimate it, how can we correct this offset in incoming signals? Estimating offsets requires using known data. Because we know what phase to expect from this data, we can examine the incoming phase and observe its time evolution to estimate the offset.

There are two sets of known data that can be used by the receiver: the preamble and the pilots. The preamble occurs in the header and is a one-time event. The pilots occur within the packet. Both are used in frequency offset estimation and in very similar ways. The preamble (particularly the L-STF) is used to produce an initial estimate of the offset to allow us to start receiving the packet, while the pilots allow us to perform continuous improvement.

There is a particularly problematic element to using the preamble to do anything: we usually do not have channel estimates yet. As discussed in Sect. 11.5, the channel estimates are only available by the time the HT-LTF is received. Frequency offsets need to already have been calculated for channel estimation to even begin. So how can we use data that has been corrupted by the channel to estimate offsets when we also need to estimate offsets to understand channel state information?

The answer for frequency offsets will also be the answer for packet edge detection: use a feature of the header to perform the needed task, rather than the specific header samples. So, we will use properties of the header that do not get affected by the channel. We will use periodicity of the short training fields to get coarse estimates for frequency offset quickly and efficiently.

In Sect. 11.2, we understood that there is a time-domain periodicity in the L-STF. The L-LTF is also repeated as a whole, allowing us to think of it as periodic in its own way, albeit with a period of 64 samples in 20 MHz mode. Thus, we can use either to do the task. To understand which is better, we must examine how the estimates are extracted and study the impact of the period.

Assuming that we have a sample at time nT, its baseband frequency offset value is:

$$x_r(nT) = x(nT)e^{j\Delta\omega nT}$$

The signal D samples later is:

$$x_r(nT + DT) = x(nT + DT)e^{j\Delta\omega(n+D)T}$$

And the product of the two samples is:

$$x_r(nT + DT)x_r^*(nT) = x(nT + DT)e^{j\Delta\omega(n+D)T}\, x^*(nT)e^{-j\Delta\omega nT}$$

But if the signal is cyclic with a cycle of D samples, then regardless of the channel (but assuming low enough Doppler):

$$x(nT + DT) = x(nT)$$
$$x_r(nT + DT)x_r^*(nT) = |x(nT)|^2 e^{j\Delta\omega DT}$$

And taking this over a window to average out the effect of noise:

$$z = \sum_{i=0}^{L-1} x_r(iT + DT)x_r^*(iT) = \sum_{i=0}^{L-1} |x(iT)|^2 e^{j\Delta\omega DT} = e^{j\Delta\omega DT} \sum_{i=0}^{L-1} |x(iT)|^2$$

Notice that we were able to reduce the product of the two received signals into a pure real power because the signal is periodic. This allows us to completely get rid of the phase of the received baseband signal because the product is pure real. In turn, the only phase component of the correlation above is due to frequency offset. This allows us to obtain the frequency offset as:

$$Z = e^{j\Delta\omega DT} \sum_{i=0}^{L-1} |x(iT)|^2$$

$$\text{Angle}(Z) = \Delta\omega DT$$

$$\Delta\omega = \frac{\text{angle}(Z)}{DT}$$

The hardware implementation of the above equation may initially seem intimidating. After all, we need to calculate an *arctan* from the equation to be able to extract the phase of the complex number. However, in Sect. 10.3, we described how the use of a vectoring CORDIC can readily produce the phase of a complex number.

Figure 11.12 shows the implementation of frequency offset estimation. The multiplier-adder combo calculates the summation of the correlation calculation over L samples. The clear signal is used to clear the contents of the accumulator. Notice that we are not calculating energy in a sliding window; we know where the correlation should start and where it ends. The result of accumulation is Z; this is then fed to a vectoring CORDIC, with $Re(Z)$ going to the x input of the CORDIC and $Im(Z)$ going to the y input. The phase output of the CORDIC is a scaled version of the frequency offset; the magnitude output is discarded.

There is one limitation of the above method for frequency offset estimation that is inescapable: the limits of the phase. Any phase can only lie between π and $-\pi$. This will in turn limit the amount of frequency offset we can calculate. Specifically:

$$-\pi < \text{angle}(Z) < \pi$$

And expanding the expression of the angle:

$$-\pi < \Delta\omega DT < \pi$$

Which leads to:

$$-\frac{\pi}{DT} < \Delta\omega < \frac{\pi}{DT}$$

And thus, the frequency offset that can be calculated is at an absolute maximum of $\frac{\pi}{DT}$ either as a positive offset or a negative offset.

In 802.11n, the time of a sample in 20 MHz bandwidth is:

$$\frac{1}{20 * 10^6} = 50 \text{ ns}$$

Obtaining the limits of linear frequency offset:

$$-\frac{1}{2DT} < \Delta f < \frac{1}{2DT}$$

In other words:

$$| \Delta f | < \frac{1}{2DT}$$

And using the above symbol duration:

$$|\Delta f| < \frac{10}{D} \text{MHz}$$

We would like this limit to be relatively high. It sets an upper bound on the amount of frequency offset we can resolve. If the limit is low and the offset is high, we cannot use this method to estimate the offset.

Obviously, the only parameter affecting this resolution is D, the periodicity of the signal we are using to extract the offset from. In Sect. 11.2, we learned that L-LTF symbols are repeated twice across two different OFDM symbols. Thus, there is a periodicity of an entire OFDM symbol, which for 20 MHz is 64. Thus, the discernible frequency offset is:

$$|\Delta f| < \frac{10}{64} \text{ MHz}$$

$$|\Delta f| < 156 \text{ kHz}$$

On the other hand, L-STF has a periodicity of 16, and thus the upper limit on offset that can be discerned is:

$$|\Delta f| < \frac{10}{16} \text{MHz}$$

$$|\Delta f| < 625 \text{ kHz}$$

So, are these limits too restrictive or too permissive? Can we use L-LTF or not? To answer this, we must relate everything back to 802.11n. Assume we are transmitting in the 5.2 GHz band. This is the band that will impose a worse condition on frequency offsets. Offsets are measured in parts per million of the carrier. An offset of just 50 ppm from this carrier will lead to an offset of 265 kHz at each side of the system. This is an offset of over 0.5 MHz between the transmitter and the receiver in the worst case. The bandwidth of the system is 20 MHz, so a mere offset of 0.5 MHz may not seem so staggering. However, in 20 MHz, the channel is divided into 64 subchannels, each of which is thus 312.5 kHz wide. The offset is in fact much larger than the subchannel. This will lead to the situation in Fig. 11.11, where sampling happens off center, leading to enormous, accumulated ICI.

The L-LTF can calculate offsets up to 156 kHZ. For the 5.2 GHz carrier, this is an offset of 30 ppm. The L-STF can distinguish offsets up to 120 ppm. This is one reason not to use the L-LTF in frequency offset estimation, and it is very good on its own. The other reason is that the L-STF comes first, allowing us to use it to estimate offsets very early in the packet.

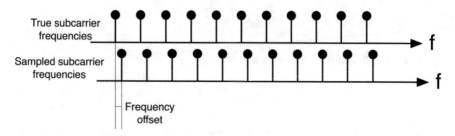

Fig. 11.11 Frequency offset in the frequency domain

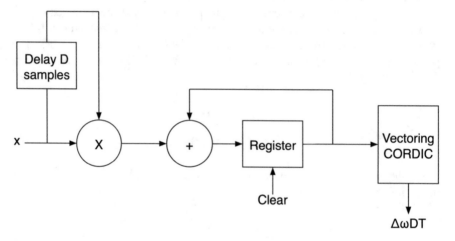

Fig. 11.12 Obtaining the phase and calculating the frequency offset using CORDIC

Once a frequency offset has been estimated, how do we correct incoming signals? This is simple if we consider the baseband signal with offset:

$$x_r(nT) = x(nT)e^{j\Delta\omega nT}$$

To correct, we multiply by a phasor with the opposite phase:

$$x_r(nT)e^{-j\Delta\omega nT} = x(nT)e^{j\Delta\omega nT}e^{-j\Delta\omega nT} = x(nT)$$

This can be performed by feeding the offset signal to a rotation CORDIC where the rotation phase is:

$$\theta = \Delta\omega nT$$

This is shown in Fig. 11.13. Note that the phase should not be allowed to increase without a limit. The phase should rotate back after 2π, and thus there will be a finite

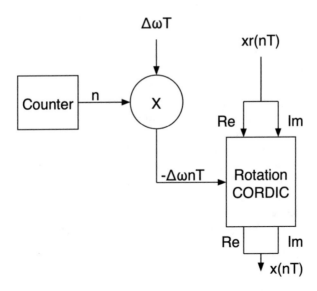

Fig. 11.13 Using phase offset to correct frequency offset using rotation CORDIC

number of phases that we rotate through. The number will depend on the amount of frequency offset relative to the sampling time. By expanding angular frequency in terms of linear frequency:

$$\theta = 2\pi \Delta f n T$$

And the phase reaches its maximum value of 2π:

$$2\pi = 2\pi \Delta f n T$$

And thus, the maximum distinct number of samples of phase we have is:

$$N = \text{ceil}\left(\frac{1}{\Delta f T}\right)$$

And the counter in Fig. 11.13 should only count up to this N.

Frequency offset is estimated once at the beginning of the packet using the *L*-STF. The short period of the *L*-STF allows a large offset to be estimated. This takes care of the bulk of offset early on, allowing channel estimation to proceed. Some residual errors remain. These still must be handled within the packet to maintain good PER.

Within the payload, we only have access to pilots as carriers of known data. Pilots will repeat every OFDM symbol. This means the period of the pilot is either 64 or 128. The period is very high relative to the *L*-STF meaning that only very small offsets can be corrected. However, this is fine because large offsets would have already been corrected by using the *L*-STF in the header, and pilots are only needed to perform fine correction.

Notice that whether we use 64 subcarriers or 128 subcarriers, the amount of frequency offset we can correct using pilots will be the same. This is because 64 subcarriers are always associated with 20 MHz, while 128 subcarriers are always associated with 128 subcarriers. This means that with 64 subcarriers $D = 64$ but $T = 50$ ns and thus $DT = 3.2$ µsec. In 40 MHz mode, $D = 128$ but $T = 25$ ns and thus $DT = 3.2$ µsec.

There is one problem here. In Sect. 7.13, we see that pilots exist in the frequency domain. In time domain, every sample contains information from all subcarriers. Pilot-driven frequency offset estimation must thus be calculated in frequency domain after the FFT. To understand how offset estimation can take place in frequency domain, assume we have a time-domain sample r_n. The time-domain sample exactly one OFDM symbol (4 µsec) later will be:

$$r_{n+N} = r_n e^{j2\pi\Delta fT}$$

where T is the OFDM symbol duration (4 µsec) and N is the number of time-domain samples in an OFDM symbol. This would be 64 in 20 MHz bandwidth and 128 in 40 MHz bandwidth. We are assuming a periodicity of an entire OFDM symbol, which is true for all training fields as well as pilots.

Using inverse DFT, we can obtain an expression of r_n in terms of its frequency-domain constituents:

$$r_n = \frac{1}{N} \sum_{k=0}^{N-1} X(k)H(k)e^{j2\pi n(k+\Delta f)/N}$$

And similarly, we can estimate a time-domain sample N samples later:

$$r_{n+N} = \frac{1}{N} \sum_{k=0}^{N-1} X(k)H(k)e^{\frac{j2\pi(n+N)(k+\Delta f)}{N}}$$

$$= \frac{1}{N} \left\{ \sum_{k=0}^{N-1} X(k)H(k)e^{\frac{j2\pi n(k+\Delta f)}{N}} \right\} e^{j2\pi\Delta f}$$

$$r_{n+N} = r_n e^{j2\pi\Delta f}$$

Now we use DFT to calculate the frequency-domain of the first N samples:

$$R_{1,k} = \sum_{n=0}^{N-1} r_n e^{\frac{-j2\pi kn}{N}} \quad k = 0, 1, \dots, N-1$$

The following OFDM symbol N samples later in frequency domain can be calculated as:

$$R_{2,k} = \sum_{n=0}^{N-1} r_{n+N} e^{\frac{-j2\pi kn}{N}} \quad k = 0, 1, \ldots, N-1$$

But we already obtained a relation between r_{n+N} and r_n, so:

$$R_{2,k} = e^{j2\pi\Delta f} \sum_{n=0}^{N-1} r_n e^{\frac{-j2\pi kn}{N}}$$

$$R_{2,k} = e^{j2\pi\Delta f} R_{1,k}$$

This means that the two frequency-domain OFDM symbols have a constant frequency offset on a sample-by-sample basis. This means we can use a very similar approach to that we used in time domain to calculate an average frequency offset. Specifically, we define a correlation between the two frequency-domain symbols:

$$z = \sum_{k=0}^{L-1} R_{1,k} R_{2,k}^*$$

$$z = \sum_{k=0}^{L-1} R_{1,k} \left(e^{j2\pi\Delta f} R_{1,k}\right)^*$$

$$z = e^{-j2\pi\Delta f} \sum_{k=0}^{L-1} R_{1,k} R_{1,k}^*$$

Because the amount inside the summation is an autocorrelation, it has zero phase:

$$z = e^{-j2\pi\Delta f} \sum_{k=0}^{N-1} |R_{1,k}|^2$$

And we can finally obtain the frequency offset as:

$$\Delta f = -\frac{1}{2\pi} \text{angle}(z)$$

Notice that this gives values of normalized frequency, meaning that offset will lie between 0 and 1. This is because the DFT equation uses normalized frequency indices k. To convert it to absolute frequency, divide by the total sampling time. In other words:

$$\Delta f \rightarrow \frac{\Delta f}{NT}$$

$$\Delta f = -\frac{1}{2\pi NT} \text{angle}(z)$$

The maximum frequency offset that can be corrected is:

$$|\Delta f| < \frac{\pi}{2\pi NT}$$

Notice that $\frac{1}{NT}$ is the subcarrier spacing if $T = T_s$ and $N = 64$ *or* 128. This means that the maximum absolute frequency offset that can be corrected is half the subcarrier spacing. This is comparable to the time-domain approach where L-LTF was used but is less capable than the achievable correction using L-STF.

Note the following about frequency offset estimation and correction:

- The L-STF in time domain is used first to correct the bulk of frequency offset. This is aided by the large cap on correctable offset stemming from the smaller period of the L-STF.
- Still in time domain, the L-LTF can be used to improve frequency offset estimates from L-STF. The large period of the L-LTF has a smaller cap on correctable offset, but this is fine when it follows L-STF.
- During the packet, pilots are used in the frequency domain to keep up with frequency offset estimation and correction.
- The need to keep performing frequency offset estimation does not stem from a time-variant nature of frequency offset. For any transmitter-receiver pair, the offset is deterministic. However, we often communicate with multiple terminals, and any estimate will always be nonideal due to noise.
- In the frequency domain, we calculated signals over a window. When used with pilots, the summation will be on all the pilots in the OFDM symbol, which will not be in a continuous window, but will be in the positions indicated in Sect. 7.13.

11.7   Timing Offset Estimation and Correction: What OFDM Does Not Really Care About

In Sect. 11.8, we will discuss packet edge detection. This is the process of deciding whether there is a packet over the air. It is the very first thing we do while synchronizing the packet. We will find that this is a dirty and rough process. The packet edge detector produces one bit indicating whether it thinks there is a packet or not. There are fundamental trade-offs involved in making sure the receiver does not fire up in response to noise and that no packets are missed.

What the packet edge detector will *definitely not do* is tell us when exactly the packet starts. It will fire up somewhere in the L-STF, and will probably catch on very early, but when exactly is impossible to tell. Somewhere down the receiver chain, we perform FFT. FFT must be done on a complete OFDM symbol of 64 samples in 20 MHz mode. We need someone to help us recognize which time-domain sample is 0 in one OFDM symbol, which synchronizes the OFDM symbols for the rest of the packet. This is timing offset estimation.

Timing offset estimation and correction is necessary but not particularly critical in OFDM systems. It is certainly nowhere near as critical as frequency offset estimation and correction. From Sect. 7.3, we understand why OFDM systems are a lot more resistant to timing offsets than they are to frequency offsets. It has to do with the long symbol time and the presence of a cyclic prefix.

In fact, we can miss timing synchronization by as many samples as there are in the GI and still be fine. The reason is the GI is filled with a cyclic prefix, which allows us to perform FFT with an offset and still get correct results. Notice, however, that any samples we miss in timing offset synchronization are samples that are taken away from the GI, and thus are samples that can no longer be used to protect against ISI.

We use the L-LTF for timing offset estimation. We know the structure of the field at the transmitter, and thus we kind of know what we should see at the receiver. If we perform correlation between the sent signal and the signal we receive, we can figure out how many samples of offset we have. So, first we calculate the correlation:

$$\sum_{i=0}^{L-1} x(n-i)T(n-i-t)$$

The sequence T is the L-LTF sequence as sent from the transmitter. L is the length of the L-LTF. The signal x is the time-domain part of the received header signal we think is the L-LTF. Notice that timing offset estimation occurs after a packet edge has been detected. This means we already know there is a packet over the air. We sort of know where the L-LTF starts, but we do not know exactly.

The variable t is a relative shift between the known sequence T and the header preamble x. Our objective is to find the value of t that maximizes the above correlation. So, we will sweep t and pick the value that gives us the maximum absolute correlation. This t will then be the number of samples by which the L-LTF in x is shifted relative to the correct training field T. This is also the shift that the entire packet observes relative to the sample at which packet edge detection fired. In other words:

$$\max_{t}\left\{\left|\sum_{i=0}^{L-1} x(n-i)T(n-i-t)\right|\right\}$$

We are trying to correlate to the known *values* of the training field (T) at the transmitter. x has passed through the channel, and when we perform timing offset estimation, we do not yet have channel estimates. This might not work because correlation might be destroyed by the effect of the channel. In fact, we talked a lot in

Sect. 11.6 about how it is not acceptable to correlate the transmitted values of the
L-STF in frequency offset estimation; instead, we had to rely on periodicity.

Unfortunately, the *L*-LTF does not have periodicity. But even if it did, periodicity
would not help. Periodicity was useful with frequency offset estimation because it
introduced a phase difference between periods that could be extracted independent
of the received sequence. Timing offset estimation cannot be distilled from phase.
Fortunately, the *L*-LTF is BPSK. So even if it suffers from a linear channel, the
sequence will still have the same pattern. This will preserve a high absolute value of
correlation at the receiver. In other words, for correlation to be high, we are not
looking for the sequence at the receiver to be exactly 1,1,-1,-1,1,1,-1,... Instead, we
are looking for any sequence where the received sequence is:

$$\vartheta, \vartheta, \vartheta e^{j\pi}, \vartheta e^{j\pi}, \vartheta, \vartheta, \vartheta e^{j\pi}, \ldots$$

where ϑ itself could be a complex number containing a phase.

A possible hardware implementation for timing offset estimation and correction is
shown in Fig. 11.14. The multiplier, adder, and memory perform the correlation.
Every cycle a new product sample is added to the contents of memory. The product
is calculated between the incoming stream x and a shifted version of *T*. The shift
register is variable. Once the sum of *L* samples is calculated, the value is stored in
memory. The shift value of the variable shift register is incremented, and we start
calculating a new sum from sample 0.

Assume we sweep *K* values of *t*. Once we have *K* sums stored in memory, we
compare their absolute values. This completes timing offset estimation. The
t corresponding to the highest amplitude then sets the shift value of another variable
shift register. We pass the incoming packet samples through this shifter, which
completes timing offset correction.

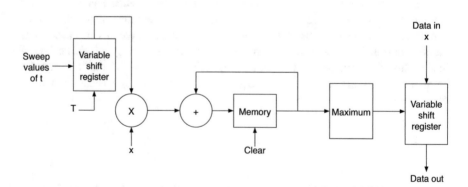

Fig. 11.14 Implementation of timing offset estimation

11.8   Packet Edge Detection: Waking the Receiver

Packet edge detection is a very rough operation that answers a very simple question: is there a packet over the air? This may sound like CCA in Sect. 7.15, but the two are distinct. In CCA, we sniff any form of energy on the air before we feel free to transmit. In packet edge detection, we are detecting if there are 802.11n packets over the air before we try to detect them. In other words, the difference is that CCA will detect *any* energy, while packet edge detection is looking for standard-specific packets.

The packet edge detection problem will produce a single bit. The bit indicates that we detected a packet on the channel, or we did not. There are four possible outcomes of this hypothesis test:

Detect. This means there is a packet, and we manage to detect its presence. This is a desirable outcome. The probability that this occurs is P_d.

False alarm. This means there is no packet over the air, yet we say there is one. This is a bad outcome we want to avoid. Its probability is P_f.

Miss. This means there was a packet, but we did not detect its presence. This is the worst outcome and should aim to minimize it. This has the probability P_m.

No packet. Means there is no packet, and we maintain there are none. This is a desirable outcome whose probability is P_n.

As with CCA, packet edge detection is an operation that needs us to constantly compare some metric to some threshold. If we exceed the threshold, we judge a packet to be present. If we do not, we judge there is no packet. Thus, the choice of threshold will make a huge difference to the probabilities we described above.

These probabilities have intrinsically contradictory requirements. For example, to reduce the probability of missing a packet, we should reduce the threshold. But reducing the threshold will lead to an increase in false alarms. However, if we must choose between the two, we should choose false alarms instead of missing. Missing a packet is disastrous because it means there is data that is going to waste and will require a retransmission. A false alarm will cause the receiver to start decoding in the absence of meaningful data, which it will soon realize by failing CRC and will stop trying. It is wasteful and annoying, but it is not disastrous.

We already have a hint about how to do packet edge detection from CCA. As with CCA, the first thing we should try is to detect energy over the air. This is called the single window approach. Energy can be measured by calculating the autocorrelation of the incoming signal over a window. Thus, our metric will be:

$$m = \sum_{r=0}^{L-1} x(n-r)x^*(n-r)$$

The length of the window is L. We need a window to average out the effect of noise. The longer the window, the less the effect of noise. As every new observation $x(n-r)$ becomes available, we do not have to recalculate the entire summation.

Instead, the metric is calculated in a sliding window. After a full window is calculated, every new sample, we perform only one multiplication, one addition, and one subtraction. Assume the newest sample is x_{new} and the oldest sample is x_{old}. Every step, we update the metric by performing:

$$m + x_{new}x_{new}^* - x_{old}x_{old}^*$$

The multiplication in $x_{old}x_{old}^*$ can be avoided because it was already performed when the old sample was the newest.

Figure 11.15 shows the arrival (in time domain) of an 802.11n packet. Before the packet, there is only noise. Packet arrival is obvious from the sudden jump in amplitude. Within the first portion of the packet, we see substantial oscillations. These are not due to noise but due to the periodic nature of the L-STF.

Figure 11.16 shows the metric m against time. This is measured along the same timescale as Fig. 11.15. The metric jumps once the packet starts because of the rise in energy. It is very difficult to predict how many samples into the packet this jump occurs, which is why we still need to perform timing offset estimation. The curve in Fig. 11.16 is much smoother than in Fig. 11.15 because it is averaged over a window, reducing the impact of both noise and L-STF variations.

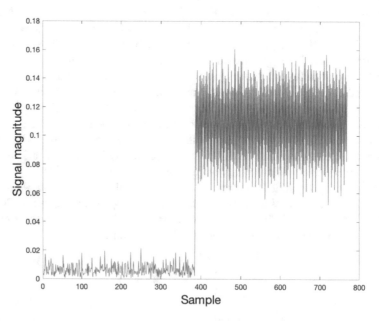

Fig. 11.15 Noisy start of packet. Most of the variation inside the packet is due to L-STF structure rather than noise

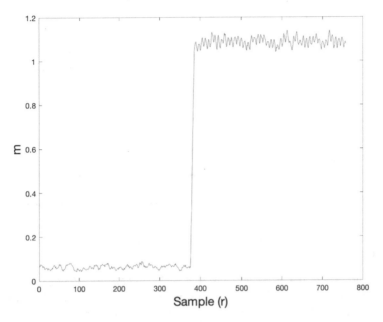

Fig. 11.16 Results of using energy in a single window for packet edge detection. Window size is 10

To decide on whether there is a packet, we compare this metric to a threshold:

$$m \gtrless \text{Threshold}$$

How should we choose the threshold? If we choose the threshold too low, we increase the rate of false alarms. If we choose the threshold too high, we increase the rate of misses. But assume that through trial and error we find a threshold that is just perfect, we will still have a major problem if any terminal moves.

If we optimize the threshold based on the levels in Fig. 11.16 and the transmitter and the receiver move away from each other, the rate of misses will rise. This is because the energy in the window drops, while the threshold remains the same. So, the threshold level must be adaptive based on the amount of path loss experienced.

We might protest the above by saying that AGC is performed on the incoming signal. This guarantees that signal level is about the same regardless of path loss (Sect. 1.3), and thus no adjustment is necessary for the threshold. There are multiple problems with this argument:

- Control over VGA gain is not continuous. There are discrete settings for amplifier gain, and AGC decides when to jump from one to another. This means that signal amplitude can exhibit significant variation before amplifier gain has to be adjusted.
- AGC happens once we know there is a packet. In fact, in 802.11n AGC normally does not conclude before HT-STF (Sect. 11.3). Packet edge detection is the very

Fig. 11.17 Single window energy

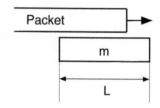

Fig. 11.18 Double window energy measurement for packet edge detection

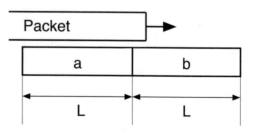

first function of the receiver chain. It is performed using L-STF before anything else in the receiver happens. When we perform packet edge detection, the AGC does not even know that there is a packet to adjust the gain for.

- Even if we assume that the AGC is ideal and can adjust gain before a packet edge is detected, the single window approach still fails. If the transmitter and receiver move further apart, gain is increased which will magnify noise. If the threshold is not adjusted, noise will cause the rate of false alarms to jump.

Our approach so far has involved a single sliding window (Fig. 11.17). Much of our trouble with this approach is because the metric is not normalized. We do not know what to refer the energy in the window to. If distance changes, energy will change. We do not know what level of energy is "normal" and thus, we do not know where to set the threshold.

Figure 11.18 shows an extension of the single window approach. This is called double window packet edge detection. There are two windows, each measuring autocorrelation the same way the single window did above. The two windows can be of different length, but there is no advantage to doing this, so we will assume they are of equal length. The two windows must be touching; in other words, there are no gaps between them. The two energies can be calculated as:

$$a = \sum_{r=0}^{L-1} x(n-r)x^*(n-r)$$

$$b = \sum_{r=0}^{L-1} x(n-r-L)x^*(n-r-L)$$

Window a spans samples from n to $n - L + 1$. Window b spans samples from $n - 2L + 1$ to $n - L$. The metric is neither energy a nor energy b, but rather their quotient:

$$m = \frac{a}{b}$$

The main contribution of this method is that it introduces normalization to the metric. The metric has a "normal" value of 1. It will jump up significantly if we detect a packet. But it will always go back down to 1. The knowledge that the metric being 1 means something is very comforting and allows us to pick the threshold intelligently.

Figure 11.19 shows the metric for the double window approach as we enter the edge of an 802.11n packet. Outside the packet, both windows are calculating noise energy. Because noise is a time-invariant process for the duration of the two windows, the two windows will carry roughly equal energy and the metric will be unity. The wider the window, the closer to unity the metric will be. Within the packet the two windows will be measuring packet energy. Thus again, the metric will be close to unity. At the packet edge, however, window "a" will start encroaching into the packet and will calculate packet energy. Window "b," on the other hand, will still be calculating noise energy. Thus, the metric will jump. The metric will drop back down to unity once both windows are comfortably within the packet.

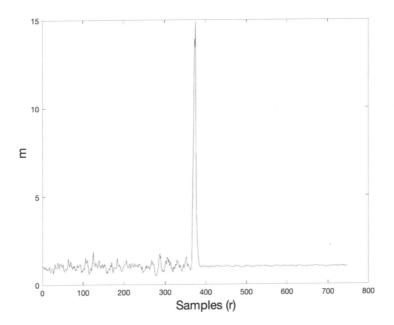

Fig. 11.19 Results for double window packet edge detection

In Fig. 11.19, the metric hovers around unity during in both noise and the packet. However, it shows much less variation inside the packet. This is because when both "a" and "b" are within the packet, they are both calculating L-STF energy. Even though these samples are noisy, the high amplitude of the training field samples drowns out the noise. Outside the packet, noise variations are more pronounced because they do not have a large signal to drown them out.

Both window approaches, and in fact any approach that uses energy, have a major issue. These approaches will detect any energy as a packet. This is particularly wasteful and counterproductive in the 2.4 GHz band. This band is busy and full of interference from non-Wi-Fi devices. Every energy spike from these devices will cause the receiver to think there is a packet. This is not as bad as it sounds because the receiver will soon recognize that it cannot synchronize the packet or perform CRC on the header and will give up. However, it is still bad because it causes receivers to waste energy. Notice that this is very different from CCA, because in CCA we want to know if there is *any* energy on the channel regardless of the standard of the transmitter.

What we are missing here is that we are looking at the packet as if we do not know anything about it. We do know the structure of the preamble at the transmitter. And while we do not know how the channel impacts the header before the HT-LTF (Sect. 11.3), we do know that the L-STF is periodic with a period of 16 (Sect. 11.2).

This allows us to formulate a new metric based on sliding correlation between two samples $D = 16$ samples apart in a window L samples long:

$$s = \sum_{r=0}^{L-1} x(n-r)x^*(n-r-D)$$

This correlation will be a complex number. For noise it will be a small number because noise is decorrelated. For the L-STF of a Wi-Fi header, the correlation will be high because of the periodicity. But more critically, the correlation will be low for high-energy transmissions that do not exhibit this particular periodicity, significantly reducing the rate of false alarms due to detection of nonstandard transmissions. Because the correlation is complex, we need to calculate its absolute value:

$$|s| = \left| \sum_{r=0}^{L-1} x(n-r)x^*(n-r-D) \right|$$

And to introduce normalization, we also calculate the energy of the current window:

$$p = \sum_{r=0}^{L-1} x(n-r)x^*(n-r)$$

Note that p is a real number because it is an autocorrelation; thus, there is no need to calculate its absolute value.

The metric is the quotient of $|s|^2$ and p, and the test is:

$$\frac{|s|^2}{p} \gtrless \text{Threshold}$$

The implementation is shown in Fig. 11.20. In the bottom branch, samples are delayed D cycles in a shift register. The absolute value of the delayed sample is calculated using a vectoring CORDIC. This value is squared using a multiplier. The

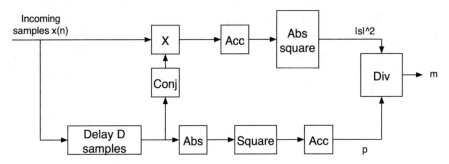

Fig. 11.20 Correlation of L-STF for packet edge detection

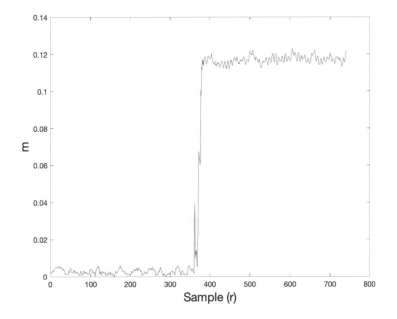

Fig. 11.21 Results for packet edge detection using correlation of periodic samples

accumulator adds a new square value every cycle and removes the oldest square value, thus calculating p.

In the top branch, a multiplier calculates the product of the sample and the delayed sample. The accumulator adds this complex number to the stored total and subtracts the oldest sample from the window. The absolute squared value of this complex accumulation is calculated to find $|s|^2$.

We can also avoid the use of the divider in Fig. 11.20 if we change the test from the above to:

$$|s|^2 \gtrless \text{Threshold} * \mathsf{p}$$

This allows us to use a multiplier instead of a divider, which is certainly welcome. The metric $\frac{|s|^2}{p}$ is shown in Fig. 11.21. The approach is normalized due to division by energy; this gives it immunity from path loss variations. It will also not detect high-power interference. When the packet starts, the metric spikes briefly before it starts to climb into a consistently high value. This initial spike is due to partial inclusion of L-STF periods as the packet starts.

Chapter 12
Cellular Radios

12.1 What Does a Cellular System Look Like?

In this chapter, we will explore 4G LTE as an example of the cellular wireless network physical layer. Our focus will be to contrast LTE to the 802.11n PHY layer. There is nothing special about either standard, which is why they are suitable as prototypes to explore differences and similarities between Wi-Fi and cellular.

Wi-Fi and cellular networks both perform wireless communication, but they seek different goals in different environments. Therefore, their hardware will sometimes look similar and at other times radically different. We have dedicated most of this book to Wi-Fi, and it is a little unfair that we are dedicating a single chapter to cellular. But this will be enough to stress how the concepts explored in this book can be extended to other systems.

Figure 12.1 shows the main reason LTE and 802.11n are different: they serve radically different network architectures. Wi-Fi creates connections between user terminals and access points or routers. These connections are symmetric. This means that the uplink and downlink are basically indistinguishable. In other words, if the AP is transmitting and the user terminal is receiving, communication is identical to the case where the user terminal is transmitting, and the AP is receiving. They have the same spectral mask, use the same MCSs, and perform the same modulation. In fact, the standard will specify very limited distinguishing features between an uplink and a downlink connection.

This is good for Wi-Fi and meshes well with the ad hoc nature of the Internet. Access points are diverse and are set up individually by users. There is no central authority to control who installs what where. Thus, because the network can be expanded or compressed ad hoc, it does not make sense to specify asymmetry in the connections. This is because a useful asymmetric downlink requires prior knowledge of where access points will be placed and how many of them there are.

Figure 12.2 shows how cellular networks typically look and why they are called cellular. This is a conceptual view that does not apply to a specific standard. The

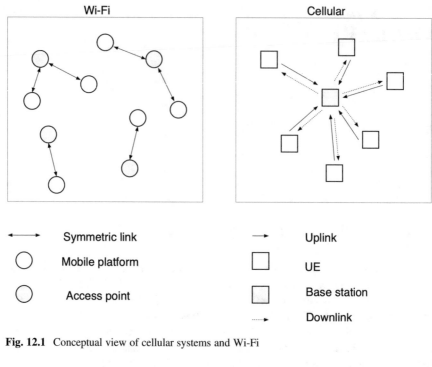

Fig. 12.1 Conceptual view of cellular systems and Wi-Fi

Fig. 12.2 Cellular concept

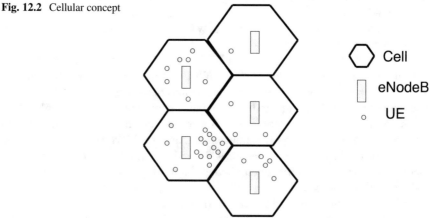

geographical area served by the cellular network is divided into cells which may or may not be of equal sizes depending on the standard.

Resources available to users in a cell are frequency and time. Both frequency and time can be reused once in a distant cell because the distance is large enough to rule out interference. Specifically, a user in a cell cannot use a band or time slot occupied by another user in the same cell but can use them if they are being used by a user in a different (or at least sufficiently distant) cell.

Each cell has at least one base station called an extended Node B, or eNodeB in LTE. It also has users with mobile terminals called UEs (user equipments). Cellular service providers design the cells carefully and place eNodeBs at strategic locations. As users move from cell to cell, the base stations will hand over the UEs between each other. The base stations are high-powered and nonmobile. They are planned a priori, and the network architecture is not ad hoc.

Thus, in cellular systems, it makes more sense for communication to be asymmetric. The base station has a full image of what the entire cell looks like. It can divide resources accordingly. This impacts every decision about the network; it affects modulation and coding, multiple access, and the frame structure. But ultimately, it makes it possible and desirable to have asymmetric links. So, it is important to define what we mean by uplink and downlink. An uplink is when the UE is transmitting and the eNodeB is receiving. The downlink is the opposite.

12.2   Are Wi-Fi and Cellular Really That Different? Yes!

Table 12.1 contrasts Wi-Fi and LTE. Because we have discovered a lot of these aspects for Wi-Fi in detail, this contrast will help expose LTE. We will investigate most of these topics for LTE for the rest of the chapter.

In Wi-Fi, communication takes place on a per packet basis. In Sect. 2.5, we saw that variable-length packets are sent from the transmitter to the receiver. A "correct" transmission is when an entire packet is correct, and a bad transmission is where a single bit is wrong in the packet. Time between packets is random, and so are their lengths. The packet structure is a result of the ad hoc nature of Wi-Fi. Because there is no central authority, and no way to know the network architecture a priori, we cannot force a rigid data frame on everyone.

In LTE, there is a central authority to which all UEs connect, namely, eNodeB. Thus, in LTE we do not have to use packets. Instead, we use frames of data. The frames are centrally planned by the eNodeB so that it is predictable who will be sending what when. We will investigate the frame structure and resource allocation in detail in Sect. 12.3.

Table 12.1 Comparison of Wi-Fi and LTE

Wi-Fi	LTE
Packet-based	Frame-based
Entire channel per user	Subcarriers assigned by eNodeB per user
CSMA-CA	OFDMA/SC-FDM
Header and pilots for estimation	Pilots only
Simple ad hoc network	Centrally planned network
Inefficient	Efficient

Because there is no central authority to assign resources, once a user gets the channel in Wi-Fi, the user can and has to use all the channel. This means that all the subcarriers in the channel are occupied by a single user for the duration that said user is transmitting. In LTE, the eNodeB can assign different subcarriers in the same channel to different users. This is particularly true in downlink communication where it is combined with OFDM to perform multiple access.

In Wi-Fi, the ad hoc nature requires a stochastic multiple access scheme. There is no central body that can assign bands or time slots to different users. Thus, users will randomly try to access the channel and the best we can do is to reduce the (inevitable) probability of collision. This leads to CSMA-CA (Sect. 1.4).

In LTE, the eNodeB has complete image of the state of the network. It can assign resources to different users in a combination of time-division multiple access and frequency-division multiple access. The FDMA structure is particularly interesting because it uses orthogonal carriers for frequency multiplexing (Sect. 12.5).

The packet-based nature of Wi-Fi lends itself to a physical layer header (Sects. 2.5 and 11.1). The header has preambles which can be used to perform synchronization and estimation (Chap. 11). In LTE, there is a frame rather than a packet, and thus there is no header. Instead, pilot subcarriers are used for both synchronization and channel estimation. Pilots are also part of the Wi-Fi packet structure, but there they are used in refining or updating offset estimates.

Efficiency in LTE is much higher than Wi-Fi. Because of the nature of packet-based communication, if a single bit error occurs in Wi-Fi, the entire packet is discarded; this is not the case in LTE. In Wi-Fi multiple access involves a stochastic process where collisions inevitably occur, requiring regular retransmissions. In LTE multiple access uses a combination of FDM and TDM and collisions can be eliminated. Finally, in Wi-Fi there is a lengthy physical layer header that represents an overhead over the MAC MPDU. In LTE, we use only pilots and there is no header.

In Sect. 7.14, we discussed how the data rate obtained from the MCS in 802.11n is not exactly the throughput observed by the user. The reasons were the overhead of the header, packets dropped due to bit errors, and collisions/retransmissions. Thus, in LTE the actual throughput observed by the user will be much closer to the theoretical rate than in Wi-Fi. This is what we mean by saying LTE is more efficient.

The opposite side of the coin is that LTE network architecture is complicated. There are asymmetric links, bulky and expensive eNodeBs, and the need to centrally plan everything. Wi-Fi networks are much simpler to understand, every platform is the same, and all links are identical. This difference is clearest when we need to expand the network by adding more terminals. When you install a new access point for Wi-Fi, that is a network expansion. You take no special steps to do it, and anyone can set up an AP through a simple user interface. Expanding a cellular network is significantly more complicated. It requires installing complex equipment, significant setup, possible adjustment to surrounding cells, and special permits.

12.3 Downlink Frame, PRB, and Bandwidth Allocation

The LTE downlink can be challenging to understand because the eNodeB uses both time-domain and frequency-domain multiplexing to assign resources to users. To understand the frame structure, we will consider it in time domain, then in frequency domain, and then in both in tandem.

Figure 12.3 shows the LTE frame. This is the structure of downlink data in time domain. A frame is 10 ms long. Each frame is split into ten sub-frames, and each is 1 ms long. Each sub-frame is divided into two slots, and each slot is 0.5 ms. Each slot is divided into several OFDM symbols. This could be six or seven symbols because the standard allows two values of CP. With a short CP, there are seven OFDM symbols per slot, two slots per sub-frame, and ten sub-frames per frame. With a long CP, we can only fit six OFDM symbols in a slot.

Figure 12.4 shows the downlink with time on the horizontal axis and frequency on the vertical axis. The unit of frequency is subcarriers, and the unit of time is OFDM symbols. There are seven OFDM symbols on the horizontal axis forming one slot, which indicates the figure shows a case where short CP is used. The vertical axis shows assignment of subcarriers to different users. The total number of subcarriers available in the channel (NBW) will be discussed below, but the question now is how this channel can be divided by the eNodeB among multiple UEs.

The eNodeB can assign resources with a granularity of physical resource block or PRB. As shown in Fig. 12.4, a PRB is 12 subcarriers in frequency domain and one slot in time domain. Different UEs can be given different numbers of PRBs based on their requirements, but our smallest unit must always be a PRB. Note that we are performing both TDM and FDM here. TDM comes from the fact that different slots are assigned to different users. FDM comes from the fact that different subcarriers in

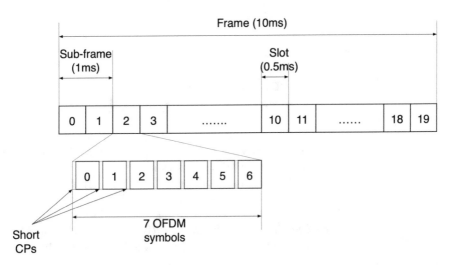

Fig. 12.3 Frame structure (time domain) in LTE downlink

Fig. 12.4 Downlink PRB structure in both frequency and time domains

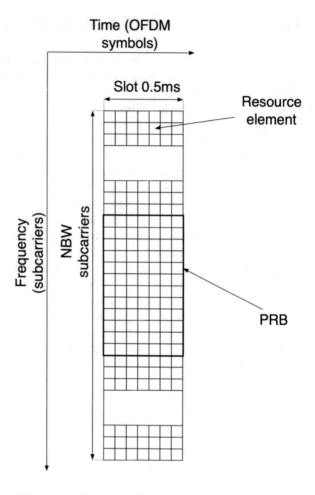

Table 12.2 PRB resource calculation for different bandwidths of LTE

Bandwidth NBW (MHz)	1.25	2.5	5	10	15	20
Subcarrier BW (kHz)	15					
PRB BW (kHz)	180					
Available PRBs	6	12	25	50	75	100

the same channel are assigned to different users. The assignment of PRBs can differ from frame to frame as the situation in the network evolves.

NBW in Fig. 12.4 is the total channel bandwidth. Table 12.2 lists the different values of channel bandwidth that the LTE standard permits. The table also lists the total number of PRBs available for the different channel bandwidths, ranging from 6 PRBs for a channel bandwidth of 1.25 MHz to 100 PRBs for a channel bandwidth of 20 MHz. This makes it possible for different eNodeBs to be adapted to different cell conditions.

The only constant in Table 12.2 is the subcarrier bandwidth. This is the bandwidth of a single subchannel, and it is specifically 15 kHz. Because the PRB consists of 12 subcarriers, the PRB bandwidth is also constant at $15*12 = 180$ kHz. Thus, the downlink channel contains PRBs that range from 6 in 1.25 MHz to 100 in 20 MHz. Notice that $\frac{1.25 \times 10^3}{180} \neq 6$ and also $\frac{20 \times 10^3}{180} \neq 100$. This is because some of the channel bandwidth is used for filter roll-off and DC coupling and is thus not available for assignment to users.

It is important to understand how TDM and FDM work in tandem in LTE. Assume, for example, a setup where there are 1200 subcarriers. A certain UE is assigned four PRBs as follows:

- In time slot 0, it is assigned 24 subcarriers.
- In time slot 1, it is assigned 24 subcarriers.

In a figure where time and frequency are shown on the x- and y-axes as in Fig. 12.4, this would lead to the four PRBs forming a square. Based on local regulation, the PRBs assigned to a certain user may or may not have to be contiguous in either or both time and frequency. For now, we will assume that both are, meaning that subcarriers 0 through 23 are assigned to the user in slots 0 and 1.

This UE will get the entire downlink frame of 1200 subcarriers seven times in slot 0. It will perform FFT on each 1200 sample time-domain sample to obtain 1200 sample frequency-domain frames. Out of each of the seven 1200 sample frequency-domain frames, it will pick samples 0 through 23 and discard the rest. It will then repeat this for slot 1, where it will pick only its own subcarriers for each of the seven OFDM symbols. In total, the UE will procure $12*4*7$ frequency-domain samples. This can be generalized into $12 * PRBs * OFDM\_symbols/slot$.

12.4   Pilots and MIMO Decoding in Cellular Systems

Figures 12.5 and 12.6 show one sub-frame over 12 subcarriers, in other words two adjacent PRBs in time. Certain subcarriers are carved out in certain OFDM symbols and marked as R for reference signals. Reference is what LTE calls pilots. These pilot subcarriers are used to send known (reference) data for frequency offset calculation and channel estimation.

Figure 12.5 is the single antenna case. In the first OFDM symbol of the slot, subcarriers 0 and 6 are references. At the fifth OFDM symbol in the slot, subcarriers 3 and 9 are references. These are the only known data that the receiver gets from the transmitter. If we assume relatively low mobility so that the channel does not change much within a slot, we have references for four out of 12 subcarriers. We must figure out the remaining eight channel estimates in the PRB by interpolating these four estimates.

Fig. 12.5 Reference signals in SISO LTE

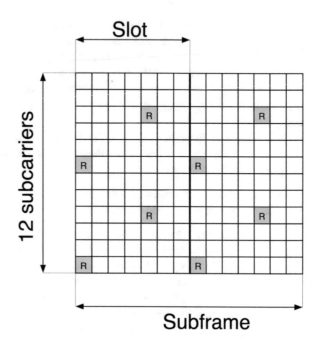

In cellular systems, mobility is often much higher than in Wi-Fi; thus, you cannot assume that the channel will remain constant for more than a single slot. Fortunately, according to Fig. 12.5, references are repeated at most seven OFDM symbols, or one slot, apart.

Figure 12.6 shows an extension of this reference scheme for multiple antennas. This figure refers to 2 × 2 MIMO. The first antenna will transmit a reference on subcarriers 0 and 6 in the first OFDM symbol and on subcarriers 3 and 9 in the fifth OFDM symbol. Meanwhile, the second antenna transmits reference signals on subcarriers 3 and 9 in the first OFDM symbol and subcarriers 0 and 6 in the fifth OFDM symbol. The first antenna will transmit a null on subcarriers 3 and 9 in the first OFDM symbol and 0 and 6 in the fifth OFDM symbol. Meanwhile, the second antenna transmits a null on subcarriers 0 and 6 in the first OFDM symbol and 3 and 9 in the fifth. Thus, the two antennas are never transmitting a reference simultaneously on the same subcarrier. When one is transmitting a reference, the other is transmitting a null. This allows channel coefficients to be readily available in frequency domain along the same lines of the use of HT-LTF in 802.11n (Sect. 11.5).

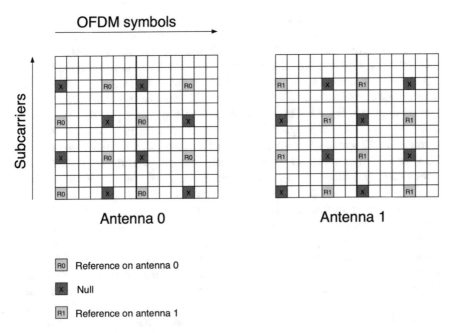

Fig. 12.6 Reference signals in 2 × 2 LTE.

12.5   OFDMA and SC-FDMA: Using the Asymmetric Links to Our Advantage

In Sect. 12.3, we saw how the eNodeB allows multiple access within the cell. TDM is performed by assigning different slots in the frame to different users. FDM is performed by assigning different *subcarriers* to different UEs. This is already hinting at the nature of frequency division. In other words, how are the subcarrier frequencies chosen, and how does a UE extract its assigned subchannels?

Consider the 20 MHz mode from Table 12.2. It has 100 PRBs available. Each PRB has 12 subcarriers. There are thus 1200 subcarriers in the entire channel. This is similar to the example we saw in Sect. 12.4. A UE could be assigned anything from 1 to 100 PRBs in a time slot, meaning it could be assigned from 12 up to 1200 subcarriers in multiples of 12.

Refer to Sect. 7.3 for a thorough discussion of the advantages of OFDM. Because eNodeB is free to choose subcarrier frequency allocation, it is advantageous to choose them so that the subcarrier frequencies are orthogonal. This gives us all the advantages of OFDM over the whole 1200 subcarriers. But we are doing more than modulation; we are doing frequency-division multiple access because no single user will (normally) be assigned all 1200 subcarriers. This downlink scheme is called OFDMA for orthogonal frequency-division multiple access. OFDMA reduces to conventional OFDM between eNodeB and the UE if all 100 PRBs are assigned to a single UE.

At each UE, FFT is performed over an entire OFDM symbol (e.g., 1200 samples) to demodulate the OFDMA subcarriers. The UE then extracts the subcarriers corresponding to its assigned PRBs and discards the rest. Refer to Figs. 12.3 and 12.4 and Sect. 12.4 to ensure that no confusion arises about samples in the time and the frequency domains. When the 1200 samples in frequency domain are transformed into time domain by eNodeB, all 1200 samples are contained in a single OFDM symbol in Fig. 12.3. In the downlink, the UE will only pick those frequency-domain samples it was assigned by the eNodeB after the FFT. This number is in multiples of 12 and is generally less than 1200. Thus, LTE multiple access is both FDM and TDM in a way. Users are assigned different subcarriers in different slots. They will only demodulate in slots where they were assigned subcarriers, and after the FFT, they will only pick subcarriers that belong to them.

Modulation in the uplink is a bit more problematic. If we try to use OFDMA, we run into a couple of problems. Frequency offset is a disastrous problem for OFDM (Chap. 7). Without very careful frequency offset estimation and correction, insurmountable ICI will destroy BER. In the downlink, frequency offsets can be handled as in Wi-Fi (Chap. 11). Assume eNodeB has a transmitting carrier ω_{cBt}. Assume also that there are N UEs and that the receiving carrier for UE i is ω_{cir}. This means that at each UE, there will be a different frequency offset $\Delta\omega_i = \omega_{cBt} - \omega_{cir}$. This is not a problem because each UE will perform an independent frequency offset estimation and correction operation. At each UE, $\Delta\omega_i$ impacts all NBW subcarriers identically. In other words, there are N copies of Sect. 11.6, each happening at a different UE.

In the uplink, frequency offset is a completely different story. Each UE will transmit all NBW subcarriers, with only a subset of active subcarriers. Every UE will have a transmitting carrier frequency of ω_{cit}. The receiving carrier at the eNodeB is ω_{cBr}. eNodeB will receive a composite signal formed of the superposition of all the UE transmitted signals. These signals *should* not interfere in frequency domain because they are frequency multiplexed. However, each signal carries a different transmitting carrier ω_{cit}, and they are all downcoverted by the same receiving carrier, ω_{cBr}. This means that the received baseband frame has N different and simultaneous frequency offsets added on top of each other. This destroys the orthogonal frequency-division and introduces a form of ICI that can never be resolved.

Another major issue is PAPR and PA biasing in the radio section of the UEs. In Sect. 7.3, we showed that high PAPR is a major problem for any OFDM and OFDMA system. So why does it stop us from using OFDM only in the cellular uplink?

- In Wi-Fi, OFDM is used in all links. PAPR is an issue, but mobile platforms in a Wi-Fi network are relatively close to the access points. This means that their transmitters are not pushing excessive power out anyway and we can find a middle ground where we use OFDM while handling PAPR.
- In cellular downlink, PAPR is very high, but the transmitter in this case is eNodeB. The base station is stationary and high-powered. It has no problem pushing a lot of power into PA bias and PAPR is not an impediment.

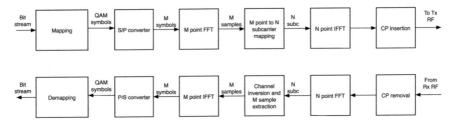

Fig. 12.7 SC-FDMA transmitter and receiver

- In the cellular uplink, UEs are transmitters, and their PAs are turned on. A cell can cover a significant geographical area relative to area covered by a Wi-Fi AP. This means that UEs can be significantly far from eNodeB. This is a situation where a battery-powered mobile platform must perform a high-powered high path loss communication. PA bias will immediately drain all UE batteries if we use OFDM.

The answer to this in LTE uplink is SC-FDMA. This stands for single carrier frequency-division multiple access. The transmitter and receiver for SC-FDMA are shown in Fig. 12.7. They look nearly identical to OFDM systems, with the difference being an initial M-point FFT in the transmitter and a final M-point IFFT in the receiver. The value of M is always less than or equal to N. In LTE M is 12 or multiples of 12 representing the subcarriers assigned to the UE. N are the total subcarriers available in the uplink channel (NBW), which can range from 72 to 1200 depending on total channel bandwidth.

Let us consider extreme cases for N and M:

- If $M = 1$ and $N = $ NBW, the transmitter FFT and the receiver IFFT are redundant and can be removed. The block diagram in Fig. 12.7 reduces to traditional M point OFDM.
- If $N = M = $ NBW, then the transmitter FFT and IFFT cancel out. The receiver IFFT and FFT also cancel out. The block diagram in Fig. 12.7 reduces to a single carrier system.

So, if $M < N$ but M is not equal to 1, we are somewhere in the middle between OFDM and single carrier. This is exactly what SC-FDMA is: a middle ground between the two.

Let us look at this another way. Assume $M = 12$ and $N = 128$. The initial 12-point FFT in the transmitter does not perform a time-domain to frequency-domain transform. It takes the 12 frequency-domain samples of data and combines them into 12 output samples still in frequency domain. From the DFT equation in Sect. 7.5, every sample at the output of the transmitter FFT in Fig. 12.7 contains data from every sample at the input. These 12 "mixed-up" samples at the output of the transmitter FFT are then mapped into a symbol 128 samples long (the M point to N subcarrier mapping block in Fig. 12.7). The remaining 116 samples are null. The standard allows the 12 active samples to be grouped together in a certain sub-band of the channel or interleaved through the whole channel with nulls in between.

The 128 samples at the input of the N-point IFFT in Fig. 12.7 are in frequency domain, and when they pass through the IFFT, they are transformed to time domain. However, this IFFT does not perform OFDM. In OFDM, the frequency-domain inputs are separate QAM symbols, with each subcarrier containing a single QAM symbol. The frequency-domain symbols at the input of IFFT in the transmitter in Fig. 12.7 each contain data from **all** the 12 QAM symbols. This is due to the FFT precoding.

The time-domain samples of orthogonally multiplexed QAM symbols in OFDM add up randomly, leading to occasional peaks and generally low averages, which is why OFDM PAPR is high. In SC-OFDMA, the time-domain samples are pre-correlated by the M-point FFT precoding. This pre-correlation allows the highs and lows of the frequency-domain samples to be smoothed out, and thus high peaks are avoided, and the time-domain signal has less spread than the corresponding OFDM signal. Similarly, the single carrier nature of SC-OFDMA leads to relative insensitivity to frequency offsets and ICI.

12.6 LTE to 5G: Things That Change and Things That Do Not

Most cellular systems have migrated or are in the process of migrating from 4G LTE to 5G. As in the evolution of Wi-Fi (Sect. 2.6), we move from a cellular standard to the next to achieve one thing: more useful throughput for users. This can be achieved in a variety of ways including increasing raw throughput, increasing the reliability of transmission (reducing BER), or giving users more reliable access to the channel (better MAC results).

Evolution of cellular standards is slower than Wi-Fi and their adoption is even slower. This is because cellular system evolution requires massive and expensive upgrades to base stations as well as a significant legal and standardization infrastructure. But at their core, cellular standards do the same things that Wi-Fi does to improve user experience.

The 5G physical layer is incredibly similar to 4G LTE. The same core technologies are used, the PRB and frame structure are nearly identical, and even the concept and placement of reference signals is unchanged. This is by design. When setting a new standard, especially in a system as cumbersome to upgrade as cellular, it is useful to keep the new system as close to the old system as possible.

This makes sense in a few ways. First, if it is not broken, do not fix it. None of the aspects of 4G that were left unchanged in 5G were responsible for throttling LTE throughput. Second, keeping the changes limited helps alleviate the cost of equipment upgrade and aids in adoption, especially in systems with modular subsystems. Finally, keeping core parts of the standard the same makes backward compatibility easier to support.

But 5G differs from 4G in fundamental ways. 4G used carrier frequencies that all lied below 6 GHz in the ISM band. 5G supports carriers up to 60 GHz spanning the SHF and EHF bands. This is perhaps the most fundamental evolution. 5G preserves support for old bands used by 4G-LTE. This helps with adoption in areas where communication in EHF is hard due to legal or technical reasons. It is also necessary because longer range communication in large cells is only possible in SHF.

But it is in situations where communication in EHF is possible that 5G shines. Center frequencies in the 60 GHz band means that there is plenty of bandwidth, which means throughput can be increased to a degree not imaginable in ISM. The high center frequency, however, leads to worse propagation characteristics (Sect. 4.5). This is especially true at the extreme high end of allowable band around 60 GHz. This means that 5G cells at such extremely high frequencies must be much smaller. Providing coverage to the same total geographical area requires many more cells and thus many more base stations than in LTE.

But the small cells combined with the short-range propagation of the 60 GHz band could be an opportunity rather than a challenge. Antenna size is directly proportional to wavelength, and thus antennas in the 60 GHz range are extremely small. This allows us to include many more antennas on the same wireless platform compared to LTE. The plentiful antenna arrays allow for beamforming. Because cells are small, the base station can be lower powered. The low power of the base station combined with the directionality of beamforming reduces interference between cells and allows channels and time slots to be reused more often.

As we saw in Sect. 12.3, LTE allowed scalable channel bandwidth. The maximum channel in LTE was 20 MHz. In 5G the widest channel possible is 100 MHz in bands below 6GHz. This means that even in a comparable band, 5G uses five times the bandwidth of LTE, providing proportionally more maximum throughput.

It is above 6GHz though that the difference becomes astounding, with bandwidth in 5G potentially going up to 400 MHz. This is the main advantage of going to ultrahigh carrier frequency: there is more bandwidth available. Near the beginning of this book, we said that the most direct way to provide more throughput is to increase bandwidth, and this will always be true no matter how much technology improves.

There are other physical layer departures in 5G, but none are as fundamental as band and bandwidth. But as an example, subcarrier spacing in 4G is constant. In fact, it is the only constant in Table 12.2. In 5G, subcarrier spacing can be picked from a variety of possibilities. In 4G, every UE must perform FFT on all the downlink subcarriers, even though each was assigned only a subset of the NBW subcarriers. 5G allows UEs to perform OFDM on a subset of subcarriers in the channel. This is necessary because 5G bandwidth is significantly larger than in 4G, and with some subcarrier spacing settings, the number of subcarriers can be enormous, requiring FFTs that could destroy the latency requirements of 5G.

12.7   MU-MIMO: The Future

Multi-user MIMO is not part of the LTE standard but is part of 5G and is guaranteed to be part of any future Wi-Fi or cellular standard. MU-MIMO is contrasted with traditional or single user MIMO which was covered in detail in Chap. 9. In SU-MIMO there was only a single transmitter and a single receiver, each having multiple antennas and using spatial multiplexing or diversity to improve performance. In MU-MIMO, there are more than two terminals. Some of the terminals may contain single antennas and some may contain multiple antennas. The multiple receiving stations and/or the multiple transmitting stations act as cooperative MIMO units.

This is a paradigm shift that can have a huge impact on performance. MU-MIMO allows a much more comprehensive view of the channel to be used to improve communication for everyone. In some ways, MU-MIMO is more like a multiple access method than a diversity technique. MU-MIMO thus shines when it is used in a cross-layer manner to achieve requirements in both the PHY and link layer by improving throughput, reducing BER, addressing interference, and improving access to the channel.

MU-MIMO can be categorized into three situations based on the number of stations transmitting or receiving. Because MU-MIMO is envisioned to mainly impact cellular systems, we will use the terms base station and UE to describe the network. The three cases shown in Fig. 12.8 will differ substantially in how they operate and in their complexity. The source of the difference is the kind of CSI needed and who should know it.

The first situation is when there is a single transmitter (base station) and multiple receivers (UEs). This is called MIMO broadcast or MIMO BC. There are multiple antennas on the base station and potentially single or multiple antennas on each of the UEs. But together, the multiple UEs form a virtual multiple antenna receiver, even if each carries only a single antenna.

The channel matrix in MIMO-BC has as many columns as the base station has antennas and as many rows as the sum of UE antennas. The base station can use this matrix to make informed decisions about how to transmit. The simplest way it can do this is by beamforming to focus beams on different users while reducing their mutual interference on each other. In a more general sense, the base station can pass data through a precoding stage which manipulates data before transmission to improve communication.

The main challenge in MIMO-BC is that the base station needs to know CSIT. This is Channel State Information at the Transmitter. That is to say, the base station needs to know CSI at itself. The CSI we have talked about throughout this book is CSI at the receiver or CSIR. This is where the receivers estimate the channel that they observe from known data the transmitter sends in a header or pilots (Chap. 11 and Sect. 12.4). Nor is this the kind of transmitter CSI we discussed in Sect. 9.1 where CSI was fed back to the transmitter, because this is still CSIR. In this case, we need the transmitter to know the channel at itself so that it can decide how to best

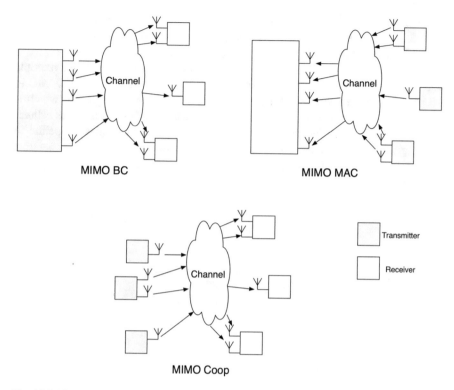

Fig. 12.8 The three cases of MU-MIMO

preprocess data. This can only be achieved if the UEs send the base station known data that the transmitter then uses to evaluate CSIT. This requires a special uplink communication for CSIT estimation before downlink communication can happen. Channel estimation in MIMO-BC uplink is going to be very challenging because the signal comes from multiple UEs.

The opposite of MIMO-BC is shown in the upper right corner of Fig. 12.8. There are multiple transmitters, potentially with single or multiple antennas, and a single receiver with multiple antennas. This reflects the cellular uplink with the transmitters being UEs and the receiver the base station. This is called MIMO multiple access channel or MIMO-MAC. The naming makes sense because the multiple UEs are using spatial diversity to introduce multiple access into the channel.

The channel matrix in MIMO-MAC will have as many columns as the sum of antennas in the UEs and as many rows as the antennas in the base station. In its simplest form, MIMO-MAC will not require UEs to perform precoding. The base station can invert the matrix and detach each UE channel from the others. This matrix can be huge, and the computational load of the inversion can be substantial, but fortunately the base station is a high-power immobile platform that can afford to do this. Figuring out the channel matrix in MIMO-MAC requires knowledge of the channel from the UEs to the base stations. This would be CSIR at the base station. This does not require a special transmission as in MIMO-BC. Pilots within the downlink of MIMO-MAC can be used to obtain channel estimates.

The third case in Fig. 12.8 is a more general case where there are multiple transmitters and multiple receivers. This is variably called true MU-MIMO, MU to MU-MIMO, cooperative MIMO, or just MU-MIMO. This is a situation which is suitable for ad hoc networks and is thus applicable to Wi-Fi just as much as cellular systems. The idea behind cooperative MIMO is that the terminals cooperate to exchange CSI and then use this to remove each other's interference. Cooperative MIMO is currently an emerging technology and research is active on how to practically use it.

Index

Printed in the United States
by Baker & Taylor Publisher Services